Register No

THE NATIONAL SCHEDULE OF RATES

INFORMATION FOR PURCHASERS

The National Schedules are part of a complete option for undertaking term maintenance contracts. Purchasers rest in the knowledge that assistance is available from consultants who are all able, from experience in the field of measured term contracting, to assist users in the setting up and operation of this type of contract.

The National Schedules are fully resourced, updated and incorporate new items annually. They are providing an increasing number of owners and contractors with an efficient, consistent and accountable system for the commissioning of building repairs and maintenance works.

The cost of undertaking repairs and maintenance work can be estimated in advance of orders being placed. A further benefit to building owners and occupiers lies in the fact that standards of materials and workmanship can be firmly controlled and is likely to be of higher quality than the work valued on a daywork basis.

Apart from the printed version, the major National Schedules are available in a data format with integrated software. Samples of which can be downloaded from our website www.nsrm.co.uk

As a purchaser, you will receive the benefit of being able to share the experiences of other purchasers in operating measured term contracts by registering in the space provided below.

Those wishing to receive further detailed advice and assistance should contact the Administrator on: 01296 339966, or by fax 01296 338514 or email: nsr@nsrmanagement.co.uk.

It will help us to help you if you complete and return the information below:

REGISTRATION OF PURCHASERS - 08/09   No: [_____]   [from back of book]

Name: _____

Organisation: _____   Job title: _____

Address: _____
_____
_____

Telephone No: _____   Fax No: _____

email address: _____

How long have you been using NSR? _____

Named contractor using NSR on your behalf _____

NSR Management
22-28 Cambridge Street
Aylesbury
HP20 1RS

Affix
stamp
here

The Administrator
NSR Management
Pembroke Court
22-28 Cambridge Street
Aylesbury
HP20 1RS

NATIONAL SCHEDULE of Rates 2008-2009

ACCESS AUDIT SCHEDULE

Part 1
Introduction
Guidance to Users
Preliminaries/General Conditions
Materials and Workmanship Preambles
Measurement and Pricing Preambles
Compiler's Notes

Part 2
Schedule of Descriptions and Prices

NSR Management

ISBN 978-1-904739-41-8

Published and Compiled Annually by:

NSR Management Ltd

General, Publications, Data Licence and Software available direct from:

NSR Management Ltd
Pembroke Court
22-28 Cambridge Street
Aylesbury HP20 1RS

Tel: 01296 339966
Fax: 01296 338514
www.nsrm.co.uk

Whilst all reasonable care has been taken in the compilation of the Access Audit Schedule the Co-authors and NSR Management will not be under any legal liability of any kind in respect of any mis-statement error or omission contained herein or for the reliance any person company or authority may place thereon.

Important

© All rights reserved. No part of any of the National Schedule of Rates may be reproduced stored in a retrieval system or transmitted in any form or by means electronic mechanical photocopying recording or otherwise without the prior consent of the Copyright owners.

This book is sold subject to the condition that it shall not, by way of trade or otherwise, be lent, resold, hired out or otherwise circulated without the publisher's prior written consent in any form of cover or binding other than that in which it is published and without a similar condition including this condition, being imposed upon the subsequent purchaser.

PART 1

Save money and complete your financial management jigsaw

ABOUT US

For over 12 years NSR Management have been promoting and developing the National Schedule of Rates on behalf of the Co-authors, the Society of Construction and Quantity Surveyors and the Construction Confederation.

The National Schedule of Rates have become synonymous with the effective management of Measured Term Contracts, thus enabling organisations to be more efficient and save money.

THE NATIONAL SCHEDULE OF RATES

The National Schedule of Rates are in use throughout the whole of the UK, helping hundreds of organisations in both the public and private sector manage and maintain their diverse property needs.

They have also become an invaluable tool for benchmarking and ensuring that planned maintenance agreements are operated as cost effectively as possible.

NSR Management offer a whole range of National Schedule of Rates in book, data and internet subscription format.

The Schedules comprise, in total, approximately 16,000 items of work, covering Housing Maintenance, Building, Mechanical, Electrical, Access Audit, Roadworks and Painting and Decorating

FINANCIAL BENEFITS FOR YOU AND YOUR ORGANISATION

- **Keeping Costs Down** — More competitive tenders are obtainable from contractors as they become more familiar with the schedules and the level of pricing within them.
- **Information Up Front** — By instructing contractors to use our schedules you will have the benefit of 'up front' cost estimation.
- **Piece of Mind** — The schedules are in use throughout the whole of the UK. and are updated annually to ensure the information is a useful as possible.
- **Good Benchmarking** — The use of our schedules is extensive, hence they are an ideal benchmarking **tool**.

AND FINALLY

NSR Management offer training in the use of the National Schedule of Rates and our 'One Stop Shop' consultancy service is available to assist with contract advice, benchmarking, QS services, condition surveys, contract management and DDA surveys.

Our after sales service is second to none, well that's what our customers tell us, so technical support is on hand for all of our schedules, or any of their associated computer applications.

After all *'Our Business is Your Property'*.

Please contact us for more information

NSR Management Limited
Pembroke Court
22-28 Cambridge Street
Aylesbury, Bucks HP20 1RS

tel: 01296 339966
fax: 01296 338514

nsr@nsrmanagement.co.uk

www.nsrm.co.uk

GENERAL INDEX

PART 1

SECTION A	Introduction	1
SECTION B	Contract Applications	3
SECTION C	General Guidance to Users	7
SECTION D	The Advisory Panel	15
SECTION E	Tender Forms and Contract Documents	17
SECTION F	Preliminaries/General Conditions	21
SECTION G	Material and Workmanship Preambles	41
SECTION H	Compiler's Notes	241

Allow us to help you with the Property Maintenance Jigsaw

- Contract Advice
- QS Services
- Benchmark Services
- Condition Surveys
- DDA Surveys
- Management of Contracts

One STOP Shop

QS Services, Benchmarking, Condition Surveys

For over 12 years, NSR Management have been promoting and developing the National Schedule of Rates on behalf of the Co-authors, the Society of Construction and Quantity Surveyors and the Construction Confederation. During this time the portfolio of National Schedule of Rates have become synonymous with the management of Measured Term Contracts, and from 2006 NSR Management will compile the National Schedules.

Together with the appropriate contract conditions, these National Schedules allow customers to issue a series of works orders, confident that the charges for the work will be based on a pre-determined and agreed basis of measurement and pricing. Over the years the National Schedules have saved time and money for hundreds of organisations just like yours.

We like to think we listen carefully to our customers needs, and in response to their requests NSR Management now offer a comprehensive **One STOP Shop** consultancy service for those organisations involved in term contracting. Our team have over 70 years experience in the construction industry and we thought it time to let our customers have the benefit of this expertise.

Just look at some of the ways in which our comprehensive consultancy service can help you:

- Comprehensive advice on all aspects of Term Contracting
- Efficient preparation of Tender Documents to suit clients needs
- Sound Technical Advice
- Thorough Benchmarking and value for money assessments
- A wide range of other QS Services
- Smooth management of Measured Term Contracts
- Effective Stock Condition and DDA Surveys

NSR Management can offer flexible cost effective solutions to all of your property management needs. Combine this with the range of National Schedule of Rates, including our new web based schedules, and it is not hard to see why we lead the way...

Please contact us for more information

NSR Management

NSR Management Limited
Pembroke Court
22-28 Cambridge Street
Aylesbury, Bucks HP20 1RS

tel: 01296 339966
fax: 01296 338514

nsr@nsrmanagement.co.uk

www.nsrm.co.uk

'Our business is your property'

SECTION A

INTRODUCTION

The National Schedule

The National Schedule was first introduced by the Co-Authors - the Society of Chief Quantity Surveyors and the Construction Confederation in 1982 for use on maintenance and minor building work to meet the competition requirements of the Local Government Planning and Land Act 1980.

It is now in its twenty sixth year and this year's edition is updated to include prices of materials, labour and plant current at spring/early summer 2008.

The users' attention is drawn to the materials prices used in this document. Wherever possible, the costs of materials are based on standard published price lists current at April/May 2008. Where these are not available, indicative quotations have been obtained for current trade prices. No discounts have been allowed against the materials costs and users are advised to check with their local suppliers to ascertain material costs and make due allowance in their tender percentage adjustment for such materials against those contained in the National Schedule.

The labour rates in this schedule include adjustments to the annual awards announced by the Construction Industry Joint Council and by the Joint Industrial Boards (JIB) for Plumbing & Mechanical Engineering Services and for the Electrical Contracting Industry. Detailed calculations of all allowances added to the basic rates of wages for craftsmen and labourers are given in the Compiler's Notes.

Whereas there is a requirement that all parties to the contract are required to comply with all relevant legislation current at the time of execution of the Works, users' attention is particularly drawn to the 'Landfill tax', these regulations came into force on 1 October 1996, and were revised with effect from 1 April 1999, increasing the cost for disposal of active waste, the costs of which are to be included in the tendered percentage adjustment to schedule rates. (see page 1/9 for details of other percentage adjustment costs.)

The rates contained in this schedule do not include for any costs associated with 'The Construction (Design and Management) Regulations 1994'. Although it is envisaged that the majority of works undertaken will not require compliance with the CDM regulations, we would recommend that Clients appoint a CDM Co-ordinator and ensure that the tendered percentage adjustment includes for complying with the CDM Regulations and all as detailed in the approved Code of Practice – managing construction for Health and Safety, Construction (Design and Management) Regulations 1994.

This will ensure that a framework for complying with the CDM Regulations is in place and that the Contractor will be recompensed for his costs.

The rates contained in this schedule do not include for any costs associated with working with asbestos. When carrying out work on asbestos based materials the contractor shall comply with the Control of Asbestos Regulations 2002 and appoint a licensed Specialist Subcontractor. Where minor work with asbestos is encountered and which can be carried out by a contractor who does not normally require an HSE license as defined in the Approved Code of Practice and Guidance published by the Health and Safety Executive, the contractor may carry out the work as requested by the Contract Administrator at agreed rates or on a daywork basis (see page 1/30).

As an increasing number of property managers in both public and private sectors are forced to review their costs of building maintenance and repair they are considering the benefits to be gained from using a measured term contract, a comprehensive schedule of rates and a computerised management system such as is available through the National Schedule.

The value of the National Schedule to any public or private property owner or estate manager is best understood by considering how it may be used and the advantages it has over alternative methods of commissioning and paying for maintenance and minor works.

The use of the pre-priced and widely accepted National Schedule together with appropriate Contract Conditions allows the Client to issue a series of works orders, confident in the knowledge that the charges for the work when carried out will be made on a predetermined and agreed basis of measurement and pricing. Without the National Schedule the client either has to produce his own schedule, or invite a series of competitive quotations for the works before placing orders, or issue orders and subsequently be charged on a lump sum or daywork basis without prior knowledge of the end cost. In the latter case there may be no clearly defined contract conditions covering such matters as insurance, payment, or standards of workmanship and materials.

The published National Schedule data is produced and stored using computer software and is available under a licence to subscribers for use on their computer systems. For further details contact the Administrator on 01296 339966.

The Form of Tender is available from NSR Management Ltd. The JCT Measured Term Contract, together with Practice Note and Guide is available from NSR Management, RIBA, RICS etc. Standard Method of Measurement exception clauses are incorporated in the Preambles where appropriate.

The use of the National Schedule and the JCT Form of Measured Term Contract has been explained at conferences across the country organised by the Society of Construction and Quantity Surveyors (formally the Society of Chief Quantity Surveyors), Local Government Associations and other organisations.

Other seminars covering the use of the National Schedule and the importance of Energy Efficient Design of buildings have taken place on a number of occasions in the South of England, promoted by Southern Electricity.

The Access Audit Schedule

The *Access Audit Schedule* is a new *composite* schedule along the same lines as the *National housing Maintenance Schedule* and contains approximately 900 items and uses the *National Schedule* for Building Works as the source document with specific additions. These composite items have been carefully selected and compiled for use in work commissioned to comply with the Disability Discrimination Act.

At present, the measurement of a multitude of items is generally required to achieve a total work cost. By the careful selection and compilation of items, the measurement of one or two composite items could achieve approximately the same ends.

SECTION B

CONTRACT APPLICATIONS

The *Access Audit Schedule* is produced as an annual publication comprising of single items and composite type items. It will thus be kept up-to-date annually and in line with the *National Schedule* prices. Unlike the *National Housing Maintenance Schedule* the information is included as one easy to read description followed by a full description showing the item rates used from the *National Schedule* to make up the composite

Building owners and managers are increasingly setting up contracts based upon schedules of rates. *Daywork* and *cost-plus* are no longer considered to be viable and cost-effective methods for procuring contracts for this type of work. They are becoming increasingly concerned about obtaining value for money and realise that this can best be achieved if the cost of work required can be calculated, if only approximately, in advance of it being carried out. This was recognised by the Joint Contracts Tribunal for the building industry by publishing the Standard Form of Measured Term Contract in 1989 (subsequently superseded by both the 1998 edition and the new 2006 edition).

There are a number of options to consider in setting up a measured term contract. For example, the grouping of buildings under a particular contract may be made with reference to their location within the total area administered by an organisation or they may be grouped by building type. The type of work to be carried out can be all that which can be reasonably undertaken by a general contractor, whether it be programmed or un-programmed work.

Whichever course of action is adopted it will be necessary to ensure that the size and scope of the packaged work is commercially attractive to those contractors who are invited to tender, since they will be required to commit the required resources in terms of operatives, direct supervision and contract management, should they be successful in winning the contract.

By tailoring their package to suit the requirements of the industry the building owners and managers will obtain competitive prices from contractors of repute and should also obtain a high quality of work coupled with satisfactory performance and response times.

The building owner or manager should expect to receive the following **benefits**:-

a) An **ability to forecast** more accurately the cost of individual works orders and whole programmes of work in advance of a commitment to undertake such works or programmes.

b) The setting up of a **single point of contact** within the whole area administered by the manager, or each defined division of the area, rather than a number of different contractors' representatives within a range or disparate companies.

c) **Saving** in staff time in their not having to prepare specifications for individual contracts.

d) **Reduction of lead-in time** for individual contracts.

e) An opportunity to **plan programmes** of work with the appointed contractor, who can arrange current and future programmes.

f) The facility to monitor costs and use for benchmarking purposes.

g) The facility for monitoring an established system of **quality control** with the appointed Contractor.

h) The **swift resolution of difficulties** in the light of experience of similar difficulties on previous orders.

i) An opportunity to set up a **methodical routine** system for analysing costs of works orders and making payments, particularly by the adoption of computerised systems.

j) An opportunity to invite contractors to **negotiate tenders** for other works and services using the accepted tender rates as a basis.

Disability Discrimination Act and Special needs Adaptations

The *Access Audit Schedule* (which also incorporates Special Needs Adaptations) will assist you in meeting the requirements of the Disability Discrimination Act. As of October 2004 the DDA act requires service providers to take reasonable steps, to make premises more accessible.

The *Access Audit Schedule* will assist you in meeting these requirements and it will make it easier for you to cost out the implications of meeting your legal, and to the same extent, moral obligations to your customers.

It will be of invaluable use to organisations who wish to keep an eye on budgets as well as meeting the legal obligations of the DDA legislation.

Key features

The new *Access Audit Schedule* has been formulated following extensive research and assistance from organisations in the commercial sector.

The schedule is extensive containing over 900 items, over two thirds of which are composite items, thus making it easy to use.

The schedule covers an extensive range of possible works that may be required both internally and externally. Everything from internal finishes, doors, kitchens, access aids, alarms and communications, car parks, lighting and electrical and much more besides. However user are asked to provide feedback on the content of the schedules so any modifications/improvements can be made.

The layout is simple, and wherever possible item descriptions are expressed in plain English.

The 'single item concept' mirrors that of the *National Housing Maintenance Schedule*, obviating the need, in most cases, to look up several unitary items for one job.

Benefits for you

Helping You Stay Legal
The schedule will be an invaluable tool in helping you comply with current legislation.

Information Up Front
By instructing contractors to use this schedule you will have the benefit of 'up front' cost estimation.

Keeping Costs Down
More competitive tenders are obtainable from contractors as they become more familiar with this schedule and the level of pricing within it.

Peace of Mind
This schedule joins the portfolio of National Schedule of Rates that are in use throughout the whole of the UK. It will be updated annually to ensure the information is a useful as possible.

Benchmarking
The use of the new schedule will be extensive, hence it will become an ideal benchmarking tool.

© AAS 2008 - 2009

Contract Arrangement Options

It is essential for the contractor to understand fully the obligations which are set out in the contract documents. If the contract is to run smoothly it is important that co-operation between those involved in its administration exists from the outset. Attention needs to be given to communications between the contract administrator and the contractor, with back-up procedures in place in the event of the contract administrator or the contractor's representative not being available, for any reason.

If, in the case of programmed works, it is intended that the contractor produces a programme, this should be made available no later than the pre-contract meeting. It should be sufficiently detailed for the contract administrator to be aware of the time that each element of the contract is to be commenced and completed. The programme may be in the form of a bar chart of the contract works. It is essential that a record of the actual timing of the events is noted against the programme and that the contract administrator is kept informed if any significant deviation from the time scale envisaged in the programme as presented to the pre-contract meeting.

With regard to access to occupied dwellings, it is recommended that a procedure for gaining access is set up under the contract by specific clauses in the preliminaries and that the responsibility for obtaining access remains with the contractor. The procedure under which access will be applied for will need to be agreed with the housing director or manager, whose co-operation will be required to gain access if the contractor experiences difficulties in gaining access.

It is recommended that a premises log book, giving relevant information concerning the site, particularly with regard to incoming services, together with details of mechanical and electrical installations, be kept on site and be available at all times for the use of the contract administrator, the contractor's representative, consultants and specialists. All events concerning the structure and its fittings should be recorded in the premises book, together with details of maintenance instructions.

Standard Forms of Contract

The Joint Contracts Tribunal's *Standard Form of Measured Term Contract 2006* represents current working practice and a useful initial exercise for new users is to define the strategy which is appropriate for their organisation and then working through the Appendix to the Form of Contract.

Measured term contracts are often based upon other forms of contract, such as the JCT Intermediate Form. As such forms are devised for use in connection with building works contained within one site, they are not suitable for use in connection with measured term contracts.

A detailed description of JCT Measured Term Contract is not included in the Schedule but three features of this contract are worth mentioning.

The first is the inclusion of the **"no fault *Break Clause*"**, Clause 7, which allows either the Client or the Contractor to end the contract after its operation for six months having given three months notice of their intention to do so. The successful operation of a measured term contract is dependent on a **conciliatory approach** by the parties to any difficulties that may arise. The inclusion of a break clause removes uncertainty for contractors and provides a sounder basis for more competitive tendering than would otherwise have been the case. Similarly, it is of benefit to building owners and managers if there is a dramatic change in their building requirements or the availability of funds.

The second feature is there is no **liquidated and ascertained damages** clause due to the difficulty, if not the impossibility, of setting a genuine pre-estimate of losses which might be sustained by the building owner or manager as a result of default on the part of the contractor. In the case of unsatisfactory performance by a contractor a number of **sanctions** are open to the Building Owner or Manager. These are set out below in ascending order of severity:-

a) Pressure applied at progress meetings.

b) Direct intervention by the Employer.

c) Cancellation of works orders – not necessarily those that are the subject of delay.

d) Appointment of another contractor to undertake the completion of the works orders after giving the notice required under the contract conditions. The additional costs will be offset against monies due to the defaulting contractor.

e) Operation of the *break clause*, if such provision is retained within the conditions of the contract, and the subsequent determination of the contractor's employment.

The third feature is that there are no clauses dealing with Named or Nominated Suppliers or Sub-contractors. Much of the work ordered may be carried out by Sub-contractors who will be treated as Domestic Sub-contractors and their work will be paid for as if it had been carried out by the Contractor's own work force regardless of how they were actually paid. However Specialist Sub-contractors work, for which there are no items in the Schedule, will be paid for on terms that have been agreed by the Contractor and the Client

SECTION C

GENERAL GUIDANCE TO USERS

Descriptions

The *Access Audit Schedule* contains two sets of descriptions, the 'General' descriptions and 'Full' descriptions.

The 'General' description is followed by the 'Full' descriptions, which, as the name implies, contains the full textual description together with the National Schedule codes and the quantity of the items used in the compilation of the composite item e.g.-

Y2030
110 Form reinforced concrete ramp to abut existing wall, n.e. 200mm high; 1800mm wide at gradient of 1:15; include for 1800mm long landings at top and bottom; including 100mm high kerbs to open side; (Handrails and surface finishes measured elsewhere)

To reduce levels 0.25 m max depth (D2030/240 - 2.51 m3), Excavated material off site (D2052/005 - 2.51 m3), Filling to excavations ne 250 mm av thick (D2070/035 - 1.25 m3), Compacting filling, blinding with sand (D2082/080 - 12.54 m2), Slabs ne 150 mm thick (E1040/355 - 2.63 m3), Sides of ground beams and edges of beds ne 250 mm high (E2030/090 - 8. m), Square mesh ref A142 2.22 kg/m2 (E3030/210 - 12.54 m2), Trowelling to falls (E4110/072 - 12.54 m2)

nr   437.64   557.88   188.35   1183.87

These codes and quantities enable all parties to examine and satisfy themselves as to the content of each description.

Single items

In addition to the composite items, the *Access Audit Schedule* contains *single* items. Apart from items that are singular by their nature, items taken from the build ups to the composite items are also included. This enables the user to create their own composite item where the schedule items does not match their requirements, without the need to refer to the National Schedule as described below.

Adding items

There may be items that do not cover the envisaged work to be undertaken. In these cases further items may be added by the user with reference to the *National Schedule* and/ or, possibly, an in-house Schedule. These should follow the same descriptive and coded format.

For example it could be that the ramp description (Y2030/110) is not appropriate as the configuration and construction of the ramp is slightly different. In this case the existing composite description can be copied and items added or excluded and assigned a new code. The new item code will follow that in the schedule to indicate that it is a different item.

Where new items, such as that illustrated above are introduced, they should be flagged as they will not be automatically updated in future years if the computerised application is used to update the prices included in the Schedule.

Where work items are required that are not similar to any item in the *Access Audit Schedule* so that it cannot be used as a basis for new items. These can be assembled by the user to form a local schedule which might also include the type of items referred to in the previous paragraph. Such a schedule will require manual updating with reference to the main schedules.

The procedure in compiling a new item is as follows:-

Write Description - Keep this as brief as possible whilst still conveying the intention of the item to minimise the possibility of disputes on site.

Create Code - The coding structure will need to be different to that of the Access Audit Schedule so that *local* schedules can be easily identified for manual updating using recognisable prefix or suffix. Gaps should be left between codes to allow for the insertion of further items in their appropriate position.

Determine Quantities - These can be calculated either by scanning previous job tickets or by site measurement of similar works items to establish average quantities.

Determine Rates - The required work items identified from the job tickets or site measurement are selected from the relevant Schedule and adjusted for the required quantities to complete the description and calculate the rate.

Amending composite rates

If it is clear at the outset that new composite type items will be required to cover the work envisaged it is recommended that these be compiled before tenders are invited and included in the *local* schedule before tenders are invited.

Amending composite rates on a job-by-job basis is not acceptable without the agreement of both parties to the contract who should satisfy themselves as to the scope and content of the items prior to their inclusion in the *local* schedule.

The concept of *swings and roundabouts* should be accepted by both parties to the contract as the time spent on adjusting schedule rates may be longer than that likely to be taken in measuring items in the traditional manner.

Term Contracts

Schedules of rates have been widely used for term contracts for maintenance and repair work for many years. The National Schedule is suitable for use with measured term contracts which include both reactive or day-to-day maintenance as well as pre-planned repair and improvement works. If the tender documents indicate that programmed cyclical maintenance is to be included it is probable that the tendered percentages will be lower as a result and it will also save the client the necessity of pre-pricing and inviting a number of individual tenders and contracts. Such works should not subsequently be taken out of the measured term contract to achieve more competitive bids for remaining reactive maintenance and repair work.

Where an employer wishes to engage a contractor to carry out ad hoc maintenance and repair work and also requires competitive quotations for the work, experience has shown that term contracts based on a priced schedule of rates may be suitable. However the following factors should be noted:-

(a) Within certain broad parameters, such as overall value, location and time, the term contract is for an unspecified amount of work at an unspecified location at an unspecified time.

(b) This type of contract is intended to cope with large numbers of small orders placed over the whole term of the contract and this allows the statistical "swings and roundabouts" to operate to reduce the overall pricing risk to reasonable proportions.

(c) In arriving at his percentage on or off the schedule of rates the contractor will have to price what he thinks will be a representative selection of items of work to be done. The tender invitation should include a clear description of the types of work to be done and the pricing restricted to those sections of the schedule covering the intended work. For example it would be wrong to invite tenders for work which is described as predominantly electrical work and issue substantial orders for fire protection. Such requirements for unrepresentative work could seriously affect the pricing calculations and could not be held to come within the normal "swings and roundabouts" of the system.

(d) Both contractors and employers need experience with a schedule system before prices can be stabilised. Users are therefore reminded that they should look upon the introduction of a schedule of rates as a long term project and ensure that they have adequate technical support services in their organisation.

(e) Contract conditions are specifically designed for this type of work.

Single Projects of Refurbishment or Modernisation Work

Where a single substantial package of work can be identified it might be advantageous to tender separately on the basis of the schedule. In these circumstances more specific information may be given about scope, timing and location of the work and the contractor will price with greater confidence. For example, the kind of project envisaged for this approach might be an estate of houses or a school or college which are to be modernised and redecorated. The description of the work would be backed up by drawings, specifications, schedules and project preliminaries.

The contract conditions for this type of project could be carried out under the JCT 2005 Forms with approximate quantities where the quantities are priced at National Schedule rates to which tenderers will apply a percentage adjustment.

Where it is not possible to measure in sufficient detail and the Measured Term Contract is used, it is nevertheless recommended that some pre-measurement or assessment is undertaken to evaluate the scope of the contract to attract the best tenders.

Single New Work Projects

As explained above the more specific the description of the work, the less the risk in pricing for both contractor and employer. It is possible to fully describe new work through drawings, specifications and bills of quantities, which may be priced relative to the National Schedule if desired.

Selection of Contractor

The Form of Tender for use with Measured Term Contracts let on the National Schedule provides for tenderers to insert a percentage adjustment to the rates listed in the National Schedule for work to be measured and valued in accordance with the contract and for other percentage adjustments to the prime cost of labour, materials, plant and subcontractors for work carried out or valued under the contract as daywork.

It is recommended that the client informs the tenderers of the value of dayworks that will be assessed in the selection process as different combinations may give different results.

In the table, contractor A is the lowest price by virtue of the more competitive percentages, after allowing for the assessed values of measured work and dayworks, as is seen from the calculation of anticipated final account costs.

TABLE

Contractor	PERCENTAGES IN TENDER			
	Percentage Adjustment to schedule rates	DAYWORK		
		Material	Plant	Subcontract
A	+10	15	10	10
B	+11	10	5	5
C	+12	5	5	5

ASSESSMENT

If, for example, the client estimates that in every £100,000 of the final account £80,000 will be valued at schedule rates or pro rata, and that there will be; £5,000 labour; £4,000 materials; £3,000 plant and £8,000 sub-contract values in dayworks, the estimated final account values of the three tenders will be:-

Net Value of Work			Final Account Values		
		£	Contr. A £	Contr. B £	Contr. C £
Schedule		80,000	88,000	88,800	89,600
Dayworks -	Labour	5,000	5,000	5,000	5,000
	Materials	4,000	4,600	4,400	4,200
	Plant	3,000	3,300	3,150	3,150
	Sub-Contr	8,000	8,800	8,400	8,400
Total		100,000	109,700	109,750	110,350

This Assessment assumes a contract where little overtime will be instructed by the Contract Administrator. Where a contractor is expected to carry out a significant part of the contract outside normal working hours an allowance in respect of overheads and profit on non profit overtime rates [Percentage adjustment of Clause 5.7 of the JCT Measured Term Contract] should be incorporated into the tender assessment.

Another factor recommended to clients who order a regular amount of this work is to maintain a separate select list of contractors for schedules. In this way the contractor may establish a workforce with knowledge and experience of schedules, have surveyors and managers who understand the system and the client's requirements, and thus provide the most competitive price.

Certainly, contractors are now available who specialise in National Schedule work and these should be sought. The other advantage in having a specialist list is that it gives the inevitable "swings and roundabouts" a chance to work.

It is significant that tenders are now stabilising as experience of methods and pricing levels is obtained. Encouragement and stimulation given to 'schedule keen' contractors will provide an increasing pool of knowledge which will benefit all parties and is certain to reduce costs and prices in the long term.

Tendered Percentage Adjustment

The tendered percentage is deemed to include but not limited to:-

Contractor's overheads and profit
Travelling time where applicable
Permit to work
Setting out and checking for dimensions
Off-loading, manhandling and transport and delivery of materials and plant
Square cutting and waste where different from allowance included in rates
All necessary security precautions
Protection, drying and cleaning – see item in Section F on page 1/29
Compliance with Regulations
Safety Health and Welfare regulations and policies
CDM Regulations
Landfill tax
Control of noise
Any site establishment costs necessary including temporary telephones
Rubbish removal including tipping
Management and staff

Disbursements arising from the employment of work people
Service and facilities
Hand tools and mechanical plant where not included in Part 2
Temporary works
Scaffolding –   temporary staging up to 1.5 metres and ladders for access to buildings of not more than 2 storeys in height
Client training
Adjustment to standard specification requirements notified to tenderers and/or variations for local conditions
Records and drawings
Identification and labelling of services
Builders work information (rates are provided for chases and holes etc)
Work in conjunction with ceiling installations
Painting and priming of services
General bonding and earthing
Inspection testing and commissioning of the works (see also references to testing on pages 1/35 and 1/97
Complying with preliminaries particulars, conditions of contract and preambles
Trade discount adjustments

Specialist Goods and Work not specified within the Schedule

There will be some items of work which are not suitable for inclusion in a standard schedule of rates because the circumstances on which they are required and consequently the cost will vary considerably e.g. asbestos removal etc. There are other non-standard items which are required only rarely. Sometimes it will be better for a client to have separate contracts for such works, however where these works are included as non-standard items on orders given to the Term Contractor the following procedures are to be adopted.

The Contractor is entitled to a 5% cash discount on the nett cost of specialist goods and a 2.5% cash discount on the nett cost of specialist work. Appropriate adjustments to achieve these percentages will be made when compiling the final account.

The nett cost of specialist goods and work will NOT be subject to any tendered percentage (omission or addition) but the Contractor is entitled to a percentage addition for overheads, profit and general attendance on the cost of specialist goods and work **after** the adjustment to provide cash discount described above.

Specialist attendance shall be measured separately in accordance with the provisions of the Standard Method of Measurement.

The nett cost is defined as the cost charged to the contractor after deduction of all trade or other discounts except for cash discounts referred to above.

NOTE: Where items containing Prime Cost sums have an identical fix only item then the materials used for the fix only item will NOT be treated as specialist goods but adjusted for waste factors, cash discounts and the tender percentage (omission or addition) as for Prime Cost items.

Prime Cost Items specified within the Schedule

All Prime Cost Items will be adjusted in the final account. The Contractor is entitled to a 5% cash discount on the nett cost of goods supplied under instruction against a Prime Cost Item specified within the Schedule. Appropriate adjustment to achieve this percentage will be made when compiling the final account.

The effect of the adjustment of the cost of the Prime Cost Item (after the adjustment to provide cash discount described above) will be subject to any tendered percentage (omission or addition).

The nett cost is defined as the cost charged to the contractor after deduction of all trade or other discounts except for cash discounts referred to above.

No waste factors have been allowed relative to prime cost items.

Overtime Payments

Overtime payments are to be paid in accordance with Clause 5.7 of the Standard Form of Measured Term Contract which refers to Working Rule 4, published by Construction Industry Joint Council, which states when overtime payments should commence.

The rates for overtime should be paid as follows:-

Mondays - Fridays

First four hours	-	Time and a half
After first four hours until normal starting time next morning	-	Double time

Saturdays

First four hours	-	Time and a half
After first four hours	-	Double time

Sundays

All hours	-	Double time

Notwithstanding the requirements of the Standard Form an easy administrative way of dealing with out of hours work is to adopt the measures set out in the following example.

Work carried out being valued using Schedule Rates:-

Example 1:

	£	£
Y 2910/036		27.38
Add tender percentage adjustment say 10%		2.74
		30.12
Add non productive overtime (time and a half) 50% of labour £7.78 x 50%	3.89	
Add tender percentage for overtime adjustment say 15%	0.58 +	4.47
	£	34.59

Work carried out being valued using Daywork Rates:-

Example 2

	£	£
3 hours @ £17.28		48.87
Add non productive overtime (time and a half) 50 % x 3 hours x £ 17.28 =	25.92	
Add tender percentage adjustment, say 60%	15.55 +	41.47
	£	93.31

Procedures

Utilising the documents requires the client to think carefully about procedures and staffing, especially in the context of term contracts. This can only be done at local level but is preferably defined in the tender documentation. No national system can be devised to satisfy all requirements, and the National Schedule is not a panacea for this headache. The matters of progress of job tickets, priorities, emergency work, valuations, dayworks, supervision and access are all specific to the individual client. Clauses resolving these items may be inserted in the preliminaries or within any relevant explanatory procedure document. Suggested headings for preliminaries clauses are given elsewhere (see Section F).

Appendix A of Section F contains a list of supplementary clauses, which a client may wish to include in contract documentation dependent on particular circumstances. Advice on the incorporation of these clauses and other more particular advice on the contents of Conditions of Contracts and Preliminaries is available from NSR Management on a consultancy basis.

User Feedback

In order to continuously update and improve the Schedules, feedback is required from users. Please return the User Response Form issued with this documentation to the Administrator, NSR Management, Pembroke Court, 22-28 Cambridge Street, Aylesbury, HP20 1RS.

SECTION D

THE ADVISORY PANEL

The panel has been established as an independent body providing a forum for discussion and feedback to the Co-Authors.

The Panel's role is seen to be one which provides users with an independent committee capable of directing the needs of the user. The Co-Authors are committed to providing the Panel with the facilities to act as such.

Members of the Advisory Panel represent the following bodies concerned with the execution of construction and maintenance work:-

1. British Constructional Steelwork Association Ltd
2. British Gas
3. Construction Confederation
4. Chartered Institute of Public Finance and Accountancy
5. Chief Building Surveyors Society
6. Society of Chief Architects
7. Society of Construction and Quantity Surveyors
8. Society of Electrical and Mechanical Engineers in Local Government

SECTION E

TENDER FORMS AND CONTRACT DOCUMENTS

The Form of Tender and the Contract Conditions are available as separate publications. The Forms of Tender are available from NSR Management Ltd and the JCT Conditions of Contract are published by RIBA Publications and available through their usual outlets and NSR Management Ltd.

It is recognised that some alterations may be made in applying these forms to particular contractual requirements, but as printed they provide a model format suitable for most requirements.

In view of the variety of possible combinations of tender the following must be considered as a guide. It is possible to tabulate the requirements of a tender thus:-

(a) **Which sections of the National Schedule will apply?**

(b) **Are dayworks to be appended?**

(c) **How is scaffolding to be paid for?**

(d) **How is travelling to be paid for?**

(e) **How is plant to be paid for?**

(f) **How are the CDM regulations to be paid for?**

(g) **What is the definition of the area of operation of the contract?**

(h) **How is emergency work to be paid for?**

(i) **What is duration of the contract?**

(j) **Is the contract fixed price or fluctuating?**

(k) **How is the work to be measured?**

(l) **What type of work is to be executed?**

(m) **What preliminaries and preambles apply?**

(n) **What comprises the Contract Documents for tendering?**

NOTES

(a) **Which sections will apply?** - Rather than have a tender which has a different percentage against each section it is advised that either a single percentage is applied to the National Schedule or it is split in trades or groups so that a single adjusting rate will be applicable to each of those groups.

(b) **Dayworks** - Daywork should be calculated in accordance with JCT MTC clause 5.4.

(c) **Scaffolding** - Further details on this may be found in the Preliminaries in Section F; rates for scaffolding are included in Part 2 (Schedule of Descriptions and Prices).

The rates contained within this schedule, together with the Percentage adjustment include for working within high rise buildings and for working from scaffolding up to a height of 10 metres above ground level.

Where external operations are to be carried out from scaffolding or other methods of access above 10 metres the tender documents should give specific information on how such orders are to be executed and, where not deemed included in the Percentage adjustment, allow for an additional percentage adjustment to the rates depending on the height eg:

    10 - 15 metres 2.5%
    15 - 20 metres 3%
    20 - 25 metres 3.5%
    25 - 30 metres 4%

(d) **Travelling** - The cost of this item should be included in the tender percentage adjustment. Where the properties to be maintained are spread over a wide area this could be a significant overhead cost and relevant information should be given to tenderers. Some contracts provide for adjustments to the tendered percentage to allow for different distances travelled to jobs.

(e) **Plant** - The cost of plant such as mechanical excavators, mixers and small plant etc. is included where appropriate and identified in the break down of the rates shown in Part 2. If the client intends to make some plant available to the contractor it should be clearly identified in the preliminaries giving details of the terms under which it may be used.

(f) **CDM Regulations** - The Client should appoint a CDM Co-ordinator and ensure the Principal Contractor prepares a construction phase plan. The cost to the Contractor in complying with the CDM Regulations should then be included in the tender percentage.

(g) **Definition of area** - The tenderer must be advised of the location and conditions affecting his work. For example, an 'area' could be an estate, a building or a geographical area.

(h) **Emergency Work** - During normal working hours emergency work may be paid for under the provisions of clause 5.8 of the JCT Measured Term Contract. Such work outside normal working hours may be paid at daywork rates and may involve a minimum call-out charge. The tender documents should give specific information on how such orders will be executed and paid for and what levels of priority will be used. Priorities may be listed as footnote (7) to Contract Appendix item 6.

Further requirements for the provision of emergency services may be given as follows:

Emergency Services

The contractor must provide and maintain an out-of-hours emergency service to avoid danger to the health and safety of residents and the public or serious damage to buildings and other structures. The service must be provided 24 hours each day including at weekends, and during holiday periods.

The names, addresses and telephone numbers of suitable competent persons who may be contacted outside working hours shall be provided by the contractor. The telephone must be manned and not an answering machine or answering service or mobile service.

(i) **Duration** - The exact dates for commencement and completion of the contract must be defined.

(j) **Fixed price or fluctuating** - Alternative methods may be adopted.

For fixed price work it is only necessary to identify the rates which will apply by reference to a particular edition of the schedule.

For fluctuating contracts, which are expected to be for periods over one year duration, it is recommended that a fixed date be specified when the rates will be varied, i.e. all work executed after 1st August will be priced at that schedule rate until 31st July the following year.

It may be more convenient to apply a change date to the date the work is measured but this must be agreed by the user.

(k) **How is the work to be measured?** - The work may be measured by the Contract Administrator; alternatively the contract provides for the Contractor to carry out the appropriate measurement for subsequent verification of the account by the Contract Administrator. Where the client intends that the contractor shall measure the works in whole or in part this is to be stated in item 9 of the Appendix to the contract.

(l) **What type of work?** - The National Schedule does not define what information must be given at tender stage about the type of work involved and the range expected, eg. "replace guttering and downpipes" or "all repairs to joinery, brickwork and subsequently complete redecoration". Further information may be given as follows:-

"The scope of work comprised in any order or in the whole contract cannot be pre-determined and no undertaking is given regarding continuity or overall value of the work and the contractor must allow in his tender for all intermittent or abnormal workloads.

As an indication of the anticipated scope of the work the following information is given as a guide:

No. of properties..............
No. of orders per week...............
Average value of order...............

Average classification of work in main trade groups:-

Bricklaying and labouring	....%
Joinery	....%
Plumbing	....%

Work in other trades will also be required."

(m) **Preliminaries and Preambles** - Model Preliminaries and Standard Preambles are included as part of the Schedule. It is important that the Contract Administrator should define the conditions and the character of work as carefully as possible at tender stage and the circumstances in which it is executed.

(n) **The Contract Documents for Tendering** - Contractors prepared to tender will be expected to subscribe to the National Schedule of Rates. Letters of invitation from the Client will then be required which will incorporate the tender documentation set out below.

Recommended Tender Documentation from the Client:

This would include:

(i) **Letter of invitation**

(ii) **Tender Form**

This must note the Contract Conditions if different from those included in the schedule of rates.

(iii) **Schedule of Works**

The Model Preliminaries allow for Description of Works, however, as much detail as possible of the actual work to be done should be provided and this is best achieved by way of an accurate Schedule of Works. The Schedule of Works may be the job ticket(s) issued during the contract, a prepared Schedule of the Works actually required, or an overall definition of the work to be done when and if circumstances arise. As much information as possible at tender stage will assist in obtaining a relevant and competitive tender.

(iv) **Preliminaries & Preambles**

The relation of all documents to one another is most important and any adaptation of Model Preliminaries and Standard Preambles should be done with care to ensure that they are compatible. The appendix to the contract must be completed by the client.

(v) **Drawings**

Any drawings that are available will assist the tenderer.

(vi) **Other information**

All attached correspondence and information should relate to the main documents and indicate the exact content of the Tender and Contract Documentation. The date and time for receipt of Tenders and a plain addressed envelope for their return to the client should be provided.

Further information on the procurement and administration of term contracts is contained in a Code of Procedure for Building Maintenance works published jointly by the Society of Construction and Quantity Surveyors and NSR Management.

SECTION F

PRELIMINARIES/GENERAL CONDITIONS

The following information is given to provide general and particular information relating to the proposed contract which could affect the tender and to assist in the completion of the Contract Particulars in the JCT Measured Term Contract Conditions pages 5 - 10.

Appendix A contains a list of supplementary clauses which may be applicable on a particular contract dependent on circumstances.

CONTENTS

	Page
PROJECT PARTICULARS	1/24
DRAWINGS	1/24
THE SITE/EXISTING BUILDINGS	1/24
DESCRIPTION OF THE WORK	1/24
THE CONTRACT/SUB CONTRACT	1/25
EMPLOYERS REQUIREMENTS	1/25
: Tendering/Sub letting/Supply	1/25
: Provision, content and use of documents	1/26
: Management of the Works	1/26
: Quality Standards/Control	1/26
: Security/Safety/Protection	1/27
: Specific limitations on method/sequence/timing	1/32
: Facilities/Temporary work/Services	1/33
: Operation/Maintenance of the finished works	1/35
CONTRACTOR'S GENERAL COST ITEMS	1/35
WORK/MATERIALS BY THE EMPLOYER	1/36
NOMINATED SUB CONTRACTORS AND SUPPLIERS	1/36
WORK BY STATUTORY AUTHORITIES	1/37
APPENDIX A LIST OF SUPPLEMENTARY CLAUSES	1/38
APPENDIX B GENERAL ADVICE ON TUPE	1/40

© AAS 2008- 2009

PROJECT PARTICULARS

Employer : Name and address of Employer to be given

Location : The site(s) is/are situated *

Access : Provision for access will be arranged by the Contract Administrator in accordance with clause 3.4 of the Contract Conditions, access to the site(s) is by .... *

The Contractor is deemed to have visited the site(s) and ascertained the means of access and any limitations thereto: no claims for additional costs caused by access difficulties arising from the execution of the works and which were identified at the date of tender will be allowed.

The Contractor is to agree access to the Site(s) with the Contract Administrator before commencement of the works.

DRAWINGS

The following drawings will form part of this contract:

Drawing No	Description
....	....
....	....

Further contract drawings may be issued during the term of the contract.

THE SITE/EXISTING BUILDINGS

Site boundaries: Where works are to be carried out on open sites the boundaries and other details must be provided with the order.

Existing buildings: Work to existing buildings should be clearly defined by room or dwelling number or drawing as necessary.

Existing mains or services: Contractor's attention should be drawn to services in the vicinity of proposed works and any limitations regarding their use or protection should be given.

DESCRIPTION OF THE WORK

General Description: The works covered by this contract and these schedules of rates comprise ....

....

....*

---

* Contract Administrator to complete/delete as applicable

© AAS 2008 - 2009

THE CONTRACT/SUB-CONTRACT

Conditions of Contract: The Form of Contract will be the JCT Measured Term Contract 2006 Edition with amendments unless otherwise stated. Note - This contract makes no provision for Nominated Suppliers or Sub-Contractors.

(Where an alternative form of contract is selected a schedule of the clause headings should be given).

Special conditions or amendments to standard conditions should be stated.

All insertions to the contract appendix are to be made including employers insurance responsibility.

Performance Guarantee Bond ++

Note ++ This may not be required for measured term contracts where completion of works ordered occurs regularly throughout the term of the contract. Similarly a retention percentage of stage payments is not needed for the same reason.

EMPLOYERS REQUIREMENTS

Tendering/Sub-letting/Supply:

Details to be given of any restrictions or requirements e.g.

MATERIALS to be supplied by client.

TENDERING to be in accordance with the principles of the Code of Procedure for Single Stage Selective Tendering 1996 (Alternative 1 or 2).

AMENDMENTS: No amendments to be made to the schedules or other contract documents by tenderers. Tenders containing amendments or qualifications may be rejected.

COSTS TO BE INCLUDED: The contractor is to allow for costs of fulfilling all liabilities and obligations referred to in the preliminaries, preambles and other tender documents as part of his tendered percentages.

FIXED PRICE: All works carried out under the contract shall be valued in accordance with the published edition of the schedule of rates named in the tender and contract subject to the tendered percentage adjustment *

or

FLUCTUATING PRICE: All works carried out under the contract shall be valued in accordance with the edition of the schedule of rates current at the date the work is carried out subject to the tendered percentage adjustment. *

(Note this does not apply to dayworks)

TRANSFER OF UNDERTAKINGS (PROTECTION OF EMPLOYMENT) REGULATIONS 1981 AND THE ACQUIRED RIGHTS DIRECTIVE 1977: The requirements of these regulations may or may not* apply to this contract and professional advice should be sought. General advice is contained in Appendix B.

---

* Contract Administrator to complete/delete as applicable

Provision, content and use of documents

All tender documents may be inspected by appointment during normal office hours at the employer's offices situated at *

....

....

Tel....................................

Tenderers are recommended to obtain their own copy of the National Schedule which includes the Conditions of Contract, Preliminaries, Preambles, the schedule of descriptions and prices.

SCHEDULE COMPLIES WITH SMM7: the Schedule of Rates has been prepared in accordance with the Standard Method of Measurement of Building Works, 7th Edition (amendment 3), subject to amplifications inherent in any of the descriptions therein. Items or methods of measurement contrary to the Standard Method of Measurement of Building Works are stated in the Preambles relating to the Schedule of Rates.

PREAMBLES QUALIFY DESCRIPTIONS: items in the Preambles relating to the Schedules of Rates are deemed to qualify and to be part of every description to which they refer.

RECIPROCAL USE OF RATES; rates may be used reciprocally between sections of the Schedules of Rates in the settlement of accounts.

Management of the Works

PERSON IN CHARGE: During the carrying out of the works, the Contractor is to keep on the works a competent person in charge who shall be empowered to receive and act upon any instructions given by the contract administrator or his representative.

Quality Standards/Control

QUALITY CONTROL: the contractor is to establish and maintain procedures to ensure that the works including the work of all subcontractors, comply with specified requirements. Maintain full records, keep copies for inspection by the Contract Administrator and submit copies on request. The records must include:-

a) The nature and dates of inspections, tests and approval
b) The nature and extent of any nonconforming work found
c) Details of any corrective action

SETTING OUT: the Contractor is to take the dimensions from existing premises and check with dimensions given on the drawings. Allow for setting out the works and providing all instruments and attendance required for checking by the Contract Administrator.

MATERIALS, LABOUR AND PLANT: provide all materials, labour and plant and all carriage, freightage, implements, tools and whatever else may be required for the proper and efficient execution and completion of the works. All materials are to be new unless otherwise specified.

OFF-LOADING AND MANHANDLING: the Contractor shall allow for all off-loading and manhandling into position including placing into and removing from temporary site storage prior to final positioning, all general materials, plant and items of equipment.

---

* Contract Administrator to complete/delete as applicable

SAMPLES AND STANDARDS OF MATERIALS: the Contractor shall allow for obtaining samples of materials as required by the Contract Administrator. Such samples to be approved by the Contract Administrator before use or application in the works. All material subsequently used in the works is to be of equal quality in all respects to the approved sample.

MANUFACTURER'S RECOMMENDATIONS: means the manufacturer's recommendations or instructions, printed or in writing current at the date of tender.

COMPLIANCE WITH REGULATIONS: the Contractor shall allow in his tendered percentage adjustment to the schedule of rates for and ensure that the works and components thereof comply with all recommendations, requirements and current editions of the following:-

a) British Standard Specifications

b) British Standard Codes of Practice

c) The Factories Act

d) The Health and Safety at Work Act

e) The Gas Act and Regulations

f) Electricity at Work Act and Regulations

g) The Regulations for Mechanical Installations issued by the Heating, Ventilation and Air Conditioning Association

h) The Regulations for Electrical Installations issued by the Institution of Electrical Engineers

i) The Electrical Supply Regulations

j) The Rules and Regulations of the Local Electricity, Gas and Water Authorities

k) Fire Precaution Act and Regulations and the requirements of the Local Fire Authorities

l) Portable Appliance Regulations

m) Regulations relating to the control of Asbestos and Legionnaires Disease

n) Regulations relating to Waste Management

PROPRIETARY NAMES: the phrase 'or other approved' is to be deemed included whenever products are specified by proprietary name.

INCOME AND CORPORATION TAXES ACT 1988 or any statutory amendment or modification thereof: the attention of the Contractor is drawn to this Act which replaces the Finance (NO.2) ACT 1975. The provisions of this Act are explained in the Board of Inland Revenue pamphlets.

The Contractor is also reminded that it is his duty and responsibility to satisfy himself that all Sub-Contractors are approved by the Contract Administrator and hold an appropriate Sub-Contractor's Certificate from the Inland Revenue.

DIMENSIONS: dimensions stated or figured dimensions on the drawings are to be adhered to. Any discrepancy between the drawings is to be brought to the notice of the Contract Administrator for clarification and instruction.

GENERALLY: all standards referred to within these documents shall be held to be the latest edition published at the date of tender. A reference to any Act of Parliament, Regulation, Code of Procedure or the like shall include a reference to any amendment or re-enactment of same.

CO-ORDINATION OF ENGINEERING SERVICES: The site organisation staff must include at least one person with appropriate knowledge and experience of mechanical and electrical engineering services to ensure compatibility between engineering services, one with another and each in relation to the works generally.

REINSTATEMENT: the Contractor shall make good and re-instate in working order any existing security system switched off, damaged or otherwise rendered inoperable by the works.

Security/Safety/Protection

SECURITY PRECAUTIONS: the Contractor is to allow for any security precautions that may become necessary in relation to the adjoining properties during the course of the works and is to allow for adequate measures to prevent access from scaffolding or similar means.

The Contractor shall issue ID badges to all employees who are engaged on the Works and these badges are to be displayed at all times. Contractor employees shall wear suitable clothing to identify them as employees of the Contractor.

SAFEGUARDING THE WORKS, MATERIALS AND PLANT AGAINST DAMAGE AND THEFT: the Contractor shall provide for all necessary watching and lighting and care of the whole works from weather, theft or other damage. All materials on site shall be protected from damage or loss.

STABILITY: the contractor shall accept responsibility for the stability and structural integrity of the works and support as necessary. Prevent overloading.

TRESPASS AND NUISANCE: all reasonable means shall be used to avoid inconveniencing adjoining owners and occupiers. No workman employed on the works shall be allowed to trespass upon adjoining properties. If the execution of the Works requires that workmen must enter upon adjoining property, the necessary permission shall be first obtained by the Contractor.

The Contractor shall not obstruct any public way or otherwise permit to be done anything which may amount to a nuisance or annoyance, and shall not interfere with any right of way or light to adjoining property.

TRAFFIC AND POLICE REGULATIONS: all traffic and police regulations particularly relating to unloading and loading of vehicles must be complied with and all permits properly obtained in due time for the works.

CONTROL OF NOISE: ensure that all measures are taken to control noise levels in accordance with the Noise at Work Regulations 1989, the control of Pollution Act 1974, the control of Noise (Code of Practice for Construction Sites) Order 1975 and BS 5228 including complying with DOE advisory leaflet 72 – noise control on building sites. Compressors, percussion tools and vehicles shall be fitted with effective silencers of a type recommended by the manufacturers of the compressors, tools or vehicles. The use of pneumatic drills and other noisy appliances must have the Contract Administrator's consent. The use of radio or other audio equipment will not be permitted.

POLLUTION: take all reasonable precautions to prevent pollution of the works and the general environment. If pollution occurs inform the appropriate Authorities and the Contract Administrator without delay and provide them with all relevant information.

NUISANCE: take all necessary precautions to prevent nuisance from smoke, dust, rubbish, vermin and other causes.

SAFETY HEALTH AND WELFARE: allow for complying with all Safety, Health and Welfare Regulations appertaining to all workpeople on site including those employed by sub-contractors and professional advisers, including but not limited to the following and to include any future amendments or re-enactments thereto:-

    i)   The Construction (Design and Management) Regulations 1994

    ii)   The Construction Regulations 1961, 1966 and 1996

iii) The Factories Act 1961

iv) The Offices, Shops and Railway Premises Act 1963

v) Work Equipment Regulations 1998

vi) Management of Health and Safety at Work Regulations 1999

vii) The Health and Safety at Work, etc Act 1974

viii) Manual Handling Operations Regulations 1992

ix) Workplace (Health Safety and Welfare) Regulations 1992

x) Personal Protective Equipment at Work Regulations 1992

xi) Display Screen Equipment Regulations 1992

xii) Special Waste Regulations 1996

xiii) Control of Asbestos at Work Regulations 2002

xiv) Construction (Head protection) Regulations 1989

xv) Construction (Health and Safety and Welfare) Regulations 1996

xvi) Control of Substances Hazardous to Health Regulations 2002

xvii) Environmental Protection Act 1990

xviii) Reporting of Injuries, Diseases and Dangerous Occurrences Regulations 1995

xix) Waste Management Licensing Regulations 1994

xx) Gas Safety (Installation and Use) Regulations 1994

xxi) Fire Precautions (Workplace) Regulations 1997

xxii) Lifting Operations & Lifting Equipment Regulations 1998

xxiii) The provision & Use of Work Equipment Regulations 1998

xxiv) The Secure Tenancies (Right to Repair) Regulations 1985

xxv) The Equal Pay Act 1970

xxvi) The Employment Protection (Consolidation) Act 1994

xxvii) The Enterprise Act 2002

xxviii) The Sex Discrimination Act 1975 and Disability and Discrimination Act

xxix) The Race Relations Act 1976 and the Race Relations Amendment Act 2000

xxx) Health and Safety (Safety, Signs and Signals) and (Consultation with Employees) Regulations 1996

xxxi) Management of Health and Safety at work and Fire Precautions Regulations 2003

xxxii) Working at height regulations 2005

A copy of the Contractor's Health and Safety Policy/Health and Safety Plan as applicable shall be produced for inspection by the Contract Administrator/CDM Co-ordinator.

The Contractor shall be responsible for ascertaining whether execution of any order for work complies with their requirements under the CDM Regulations, and must notify the Contract Administrator/ CDM Co-ordinator of same and obtain approval prior to commencement of site activities

The Contractor will be responsible where applicable, for providing all information required by the Contract Administrator/ CDM Co-ordinator to update the Health and Safety File.

In the event of the Contractor ascertaining that execution of an order will or may involve interference with any hazardous substance or installation, then the Contractor shall forthwith notify the same to the Contract Administrator and in so doing shall notify him in writing of any precautions proposed to be taken in consequence of the hazard which may affect the use of the premises or the comfort or freedom of movement of any person likely to be in or near the premises during execution of the order.

The Contractor shall likewise notify in writing the occupant of the premises, or the person in charge of the occupants or users of the premises on which works are in progress or about to be carried out, all restrictions guidance or other precautions which are desirable or necessary for the safety of all persons occupying or using the premises in consequence of the works. The Contractor shall provide all barriers and warning notices required for that purpose and shall make effective arrangements for the occupant or person in charge to consult and communicate with the Contractor, throughout the duration of the works, on the effects and nature of such precautions.

ASBESTOS SAMPLES: in the case of asbestos, the Contractor shall comply with the Control of Asbestos at Work Regulations 2002 and arrange for any necessary sampling and analysis before undertaking any work effecting the suspect material, and shall give the necessary notice to the Health and Safety Executive.

WORKING WITH ASBESTOS: when carrying out work on asbestos based materials, the Contractor shall comply with the Control of Asbestos at Work Regulations 2002 and appoint a licensed specialist sub-contractor. This requirement will be strictly enforced.

Any work with asbestos cement (e.g. cleaning, painting repair or removal); any work with materials of bitumen, plastic, resins or rubber which contain asbestos, the thermal and acoustic properties of which are incidental to its main purpose and minor work with asbestos insulation, asbestos coating and asbestos insulating board which because of its limited extent and duration does not require a license (e.g. drilling holes, repairing minor damage, painting, removal of a single panel etc) shall be carried out in accordance with the Approved Code of Practise and Guidance entitled Work with Asbestos which does not normally require a license (Fourth edition) published by the Health and Safety Executive. The rates for such work shall be agreed with the Contract Administrator.

EMPLOYER'S SAFETY POLICIES: without prejudice to the Contractor's general obligations to ensure compliance with all statutory requirements relating to health and safety, the Contractor shall in particular observe and comply with:

(a) any specific condition, warning or direction given by the Contract Administrator on any matter relating to health and safety;

(b) the relevant provisions of any Employer's Safety Policy applicable to operations of the type in question when undertaken by Employer's employees, being a Safety Policy of which a copy has been given to the Contractor at or before the start of the work:

and

(c) any method statement agreed with the Contractor before the work begins identifying the safety precautions to be taken.

FIRE PRECAUTIONS: the contractor shall take all necessary precautions to prevent personal injury, death and damage to the works or other property from fire. Comply with Joint Code of Practice 'Fire Precaution on Construction Sites' published by the Construction Confederation and the Fire Protection Agency. Smoking will not be permitted on the works.

CANCELLATION ON DEFAULT: in the event of default by the Contractor in the proper observance of any necessary health and safety requirements, cancellation of the written order by the Contract Administrator shall not result in the Employer being obliged to reimburse either any costs incurred by the Contractor or the value of any abortive work except to such extent (if any) as those costs or that abortive work were incurred or performed without contravention of the health and safety requirement in question.

MAINTENANCE OF PUBLIC ROADS: the Contractor shall make good all damage to public roads, kerbs and footpaths, lawns etc, occasioned by exceptional traffic, delivery of materials and building operations generally to the reasonable satisfaction of the Contract Administrator and the local authority.

EXISTING MAINS AND SERVICES: the Contractor shall maintain during the progress of the works, the existing drainage system, water, gas, sewers, electric and other services and is to make arrangements for their continuance and take all necessary steps to protect and prevent damage to them. Should any mains, services ducts or lines be found to be in the way of new works, or require any attention, the Contractor is to seek instructions from the Contract Administrator.

Where it is necessary to interrupt any mains or services for the purpose of making either temporary or permanent connections or disconnections, prior written permission shall be obtained from the Contract Administrator and where appropriate from the local authority or public undertaking and the duration of any interruption kept to a minimum.

PROTECTION, DRYING AND CLEANING: the Contractor shall undertake the following, the cost of which shall be included in the tendered percentage adjustment to the prices in the schedule of rates:-

1. protect all work and materials on site, including that of Sub-Contractors, during frosty or inclement weather.

2. protect all parts of existing buildings which are to remain using polythene/dust sheets and make good any damage caused. Items C1073 to C1090 are only to be charged where they are specifically instructed by the Contract Administrator.

3. prevent damage to existing furniture, fittings and equipment left in the property. Cover and protect as necessary.

4. protect the adjoining properties by screens, hoardings or any other means to prevent damage or nuisance caused by the Works.

5. dry out the Works as necessary to facilitate the progress and satisfactory completion of the Works. The permanent heating installation may be used for drying out the works and controlling temperature and humidity levels, but:

    a) the Employer does not undertake it will be available
    b) the Contractor must take responsibility for operation maintenance and remedial work, and arrange supervision by and indemnification of the appropriate subcontractors and pay costs arising.

6. protect and preserve all trees and shrubs except those to be removed.

7. treat or replace any trees or shrubs damaged or removed without approval.

8. clean the Works thoroughly removing all splashes, deposits, rubbish and surplus materials.

DAMAGE: the Contractor shall exercise great care at all times to prevent damage to the building structure, fittings, furniture, equipment, finishes or the like and shall make good any damage caused by him at his own expense. In this connection all carpets, desks and other furniture or equipment including telephones in the vicinity of the work shall be covered by the Contractor with protective dust sheets or the like prior to any work commencing. Where it is necessary to use any naked flame or welding equipment in executing the work and where combustible materials are in use, adequate protection shall be given to other adjacent materials and personnel. Suitable fire extinguishers shall be provided and readily available at the position where such work is proceeding. The Contractor shall maintain the designated escape routes and exit doors within any building clear of all materials and plant at all times. The Contractor shall consult the Premises
Manager at any of the locations in respect of precautions which should be taken for the safety of other occupants prior to the commencement and whilst work is in progress.

PERMIT TO WORK: before commencing any portion of the work the Contractor must establish the need for and if necessary obtain a Permit to Work.

The Contractor's tendered percentage adjustment to the Schedules of Rates shall be deemed to include allowances for time spent in obtaining permits for each portion of the work and claims made by the Contractor shall not be entertained by the Employer in respect of time lost in connection with the issue of permits. However the nett cost of fees and charges by local authorities is reimbursible to the Contractor by the Employer.

Permits to Work will be required for, but not limited to:-

        a) Excavation Work
        b) Hot Work/ Welding
        c) Confined space entry
        d) Cutting through or disconnecting existing services
        e) Access to roofs

Specific limitations on method/sequence/timing

    SITE VISITS: before tendering the Contractor shall visit the site(s) and ascertain:-

1. Local conditions.

2. Means of access to the site(s).

3. The confines of the site(s).

4. Restrictions in respect of loading and unloading vehicles.

5. Factors affecting the order of execution of the work and the time required for the execution of the works.

6. The supply of and general conditions affecting labour, materials and plant required for the execution of the work.

    POSSESSION OF THE SITE: no restriction. *

OR

    POSSESSION OF THE SITE: possession of the site by the Contractor will be restricted as follows:-

    WORKING SPACE : working space is limited to .... *

    WORKING SPACE : take reasonable precautions to prevent workmen, including those employed by Sub-Contractors from trespassing on adjoining owner's property and any part of the premises which are not affected by the Works.

---

* Contract Administrator to complete/delete as applicable

WORKING HOURS: working hours are limited to the normal working hours defined in National Working Rule 3 for the Building Industry as appropriate to the area in which the Works are located. Overtime shall not be worked by operatives on the site without the prior express permission in writing from the Contract Administrator.

OCCUPIED PREMISES: where work is done in occupied premises the Contractor shall take all reasonable care to avoid damaging the property or contents and shall make good all damage which arises from his work.

PROGRAMME OF WORKS: the Contractor shall prepare and submit for the approval of the Contract Administrator, a programme covering all aspects of the works.

SEQUENCE OF WORK OR OTHER RESTRICTION: .... [Note: any restriction on the work (eg. sequence or timing) must be given]. *

USE OF SITE: the site is not to be used for any purpose other than the execution of the contract.

REINSTATE SITE: confine to as small an area as practicable any operations which may affect the surface of the site and reinstate the site after the works are completed.

APPROVAL TO SITING: obtain the approval of the Contract Administrator to the siting of permanent soil disposal and notify him of the proposed siting of materials, of temporary spoil and rubbish deposits and of temporary buildings and the like.

MATERIALS OBTAINED FROM SITE: sand, gravel, vegetable soil, turf and other materials including those arising from demolition or alteration works obtained from the site shall remain the property of the Employer until removed from the site in accordance with the Contract. Excavations are not to be made or enlarged for the purpose of obtaining such materials.

TIPPING: no allowance for tipping charges and Landfill tax charges in connection with materials obtained from site including those arising from demolition or alteration works has been included in the schedule of rates and costs should be included in the Contractors tendered percentage.

OVERTIME: where overtime is ordered in writing by the Contract Administrator, the Contractor will be paid the net additional cost, subject to the addition of 10% for overhead costs provided the contractor's returns are certified each week in relation thereto by the Contract Administrator. ⊕

OVERTIME, NIGHT WORK AND INCENTIVES: all costs of overtime or night work at the discretion of the Contractor must be borne by the Contractor and no claims for additional payment in this respect will be allowed.

DAYWORKS: no work will be allowed as daywork unless previously authorised by the Contract Administrator and confirmed in writing. All vouchers specifying the time daily spent upon the work (and, if required by the Contract Administrator, the workmen's names) and the materials used properly priced and extended, shall be signed by the Contract Administrator.

Where daywork is authorised, the Contract Administrator shall be notified of the commencement and completion of the work, and the items of plant and workpeople concerned are to be solely engaged thereon and not employed upon any other work during progress of the daywork.

BUILDING OPERATIONS IN WINTER: the Contractor must be conversant with the measures and operations described in the booklet 'Winter Building' published on behalf of the DOE and obtainable from HMSO for ensuring continuity of work and productivity during inclement weather. The operations and measures described in the booklet shall be taken wherever practicable and having regard to nature, scope and programme of the Works.

⊕ This percentage will only apply if no other percentage is inserted in appendix item 12 of the JCT MTC Conditions or is otherwise included in the Contract.

* Contract Administrator to complete/delete as applicable

Facilities/Temporary work/Services

NOTICES AND FEES TO LOCAL AUTHORITIES AND PUBLIC UNDERTAKING: such fees, charges, rates and taxes paid by the Contractor shall be reimbursed nett to him by the Employer. (See also WORKS BY PUBLIC BODIES and PERMIT TO WORK)

WORKS BY PUBLIC BODIES: the .... * (name of local authority or public undertaking) will carry out the following work which will be covered by provisional sums as follows:

Description of Work	Provisional Sum
....	....
....	.....................*

The Contractor is to allow for general attendance and overheads.

The .... (name of local authority or public undertaking) will be carrying out, in accordance with their statutory obligations, the following work which is not part of the Contract Works: *

Description of Work .... *

[NB: this description should include the scope and timing of the work and its effect on the Contractor's operations]

OFFICES FOR PERSON-IN-CHARGE AND FACILITIES FOR EMPLOYER: the Contractor is to provide, erect and maintain suitably equipped offices for the Person-in-Charge and other necessary staff including Clerk of Works and provide heating and lighting and attendance throughout the duration of the Contract and remove and clear away upon completion and make good all work disturbed.

SITING OF TEMPORARY BUILDINGS ETC: all offices, messrooms, storage sheds, sanitary accommodation and temporary buildings shall be sited to the approval of the Contract Administrator. All areas so used must be made good on completion.

SANITARY ACCOMMODATION: the Contractor is to provide adequate suitable and proper sanitary accommodation which must be water-borne to the satisfaction of the Authority's Health Department and washing facilities for workmen to the standard required by the current Working Rule Agreement and to the satisfaction of the Health Department of the local authority and keep same in a clean and sanitary condition and remove and make good upon completion.

WORKING PLATFORMS: are to be provided to enable the work to be safely and effectively carried out. In all cases the Contractor shall comply with the requirements of the Construction (Health and Safety and Welfare) Regulations 1996 together with other relevant regulations and requirements, consulting the Contract Administrator where differing provisions of scaffolding are possible. Agreement by the Contract Administrator to a particular method of scaffolding shall not relieve the Contractor of his responsibility for fully complying with Health and Safety at Work provisions. The Contractor's percentage addition to the prices in the Schedule of Rates is to include for:-

1. temporary staging to provide a working platform up to a height of 1.5 metres.

2. ladders for access to buildings of not more that two storeys in height.

* Contract Administrator to complete/delete as applicable

SCAFFOLDING: scaffolding or towers or mobile towers to provide working platforms greater than 1.5 metres in height will be dealt with as follows:-

1. by the use of the rates for scaffolding contained in Part 2 (Schedule of Descriptions and Prices).

2. by agreement between the Contract Administrator and Contractor.

3. by daywork.

4. by quotations from not less than three specialist firms tendering in competition.

SCAFFOLDING must be constructed in accordance with the requirements of the Health and Safety at Work Act 1974, the Management of Health and Safety at Work Regulations 1992 etc and subsequent amendments or re-enactments thereto and comply with:-

1. BS 5973: 1981 "Access and Working Scaffolds and Special Scaffold Structures in Steel".
2. BS 5974: 1982 Code of Practice for Temporary Installed Scaffold and Access Equipment.

SHORING, SCREENS, FENCING AND HOARDINGS: will be dealt with as follows:-

1. by the use of the rates for such non-mechanical plant contained in Part 2 (Schedule of Descriptions and Prices).
2. by agreement between Contract Administrator and Contractor.
3. by daywork.
4. by quotations from not less than three specialist firms tendering in competition.

NOTICE BOARD: upon written application the Contractor may display and maintain in an approved position a notice board stating his name and that of authorised Sub-Contractors.

PROVISION OF SKIPS: application must be made to the appropriate local authority department for the siting of any skips required for the collection and removal of contractors' waste and rubbish. No allowance for charges in connection with the use of skips has been included in the schedule of rates and costs should be included in the Contractor's tendered percentage. The cost of removal of rubbish on site not arising from the contract works such as fly tipping or spoil from occupants or other contractors is reimbursable, provided adequate security measures are taken to prevent it from re-occuring.

The Contractor is to ensure that his tender provides for removing rubbish from the site both as it accumulates from time to time and at completion of the works.

Ensure that non-hazardous material is disposed of at a tip approved by a Waste Regulation Authority. Remove all surplus hazardous materials and their containers regularly for disposal off site in a safe and competent manner as approved by a waste regulation authority and in accordance with relevant regulations. Retain waste transfer documentation.

TEMPORARY TELEPHONE: the Contractor is to allow for providing temporary telephone facilities to the site and defray all charges in connection therewith, including the costs of all calls made by his own employees and those of any Sub-Contractors. No provision need be made for telephones for the Employer's representatives.

WATER FOR THE WORKS: the Contractor shall provide all water required for use in the works, by him or by his Sub-Contractors, together with any temporary plumbing, standpipes, storage tanks and the like, and remove on completion. He shall pay all fees and charges in connection therewith and make good all work disturbed.

LIGHTING AND POWER FOR THE WORKS: the Contractor shall provide all artificial lighting and power (electricity and/or gas) for the works, including that required by Sub-Contractors, together with any temporary wiring, switchboards, distribution boards, poles, brackets, etc. and remove same on completion, and pay all fees and charges in connection therewith and make good all work disturbed.

*Note: Where miscellaneous or improvement works to services installations is carried out under a subcontract the services contractor should identify the extent and nature of temporary lighting and power which will be provided by the general contractor at no charge.

METER READINGS: where charges for service supplies need to be apportioned ensure that meter readings are taken by relevant authority at possession and/or completion as appropriate. Ensure that copies of readings are supplied to interested parties.

Operation/Maintenance of the finished works

Where appropriate the Contractor will provide the Contract Administrator with a free copy of the manufacturers' maintenance/operation manuals for installed equipment.

Submit a copy of each test certificate to the Contract Administrator as soon as practical and keep copies of all certificates.

CONTRACTOR'S GENERAL COST ITEMS

The contractor is to allow in his tendered percentage for the cost of the following:

MANAGEMENT AND STAFF including all overheads, offices, equipment, insurances, travel and expenses, supervision, programming, quantity surveying support staff and the like.

SITE ACCOMMODATION including erection, dismantling, hire charges, maintenance, services, charges, insurances etc. for offices, stores, canteens, compounds, sanitary facilities and the like.

SERVICES AND FACILITIES where not provided by the Employer at no charge to the Contractor, including power, lighting, fuels, water, telephone, security and the like.

MECHANICAL PLANT where not shown as included in Part 2 Schedule of Descriptions and Prices or provided by the Employer at no charge to the Contractor, including cranes, hoists, transport and other mechanical plant.

TEMPORARY WORKS including temporary access roads, hardstandings, hoardings, fans, fences and the like. (The Contract Administrator should define the exact requirements at tender stage).

TRAVELLING AND TRANSPORT OF LABOUR, PLANT AND MATERIALS within the boundaries specified as the 'Contract Area' to the properties listed within the Appendix item 1a including all expenses and vehicle costs.

CLIENT TRAINING: explain and demonstrate to the Employer's premises staff the purpose, function and operation of the installations including all items and procedures listed in the Building Manual.

SPECIFICATION REQUIREMENTS: any adjustments required to the schedule to comply with the specification, drawings or site requirements.

RECORDS AND DRAWINGS of the work as required by the Contract Administrator

GENERAL BONDING AND EARTHING as required by the IEE Regulations for any electrical work

INSPECTION TESTING AND COMMISSIONING of the works. Where testing, balancing and commissioning exceeds the scope of works, the Contract Administrator is to notify the Contractor of his requirements, and an hourly charge or lump sum agreed with the Contractor, if required.

WORK/MATERIALS BY THE EMPLOYER

Plant or materials may be provided by the Employer under the provision of contract clause 2.3.

NOMINATED SUB CONTRACTORS AND SUPPLIERS

Although the JCT Measured Term Contract does not provide for them the Employer may wish to nominate specialist sub-contractors and/or suppliers to carry out works and or supply goods which are not included in the schedule of descriptions and prices. The Contractor will be entitled to 2½% cash discount on the nett cost of the specialist work and 5% on the nett cost of specialist goods after deduction of trade and other discounts, and will also be allowed a percentage addition for profit, attendance and all overheads on Sub-Contractor's accounts after the adjustment to provide the cash discounts described above. [*]

Attendance includes general attendance and unloading, storing and placing goods in position for fixing, returning crates and packing etc. (Prices are included for fixing only items which may have been ordered from a nominated supplier or supplied by the Employer). Separate provision shall be made for special attendance in accordance with the Standard Method of Measurement.

WORK BY STATUTORY AUTHORITIES

Where the Contract Administrator orders the Contractor to instruct Statutory Authorities to carry out works under the contract, the Contractor is entitled to recover the full cost of any fees and charges payable in consequence thereof and will be allowed 10% on the net charge to be added for profit, attendances and all overheads.

---

[*] This percentage does not apply to Sub Contractors' work ordered under the daywork provision in condition 11 of JCT MTC conditions unless percentage adjustment for this item has not been included.

Appendix A

List of supplementary clauses which may be applicable on a particular contract.

1. Abortive Calls

2. Administration: Call out procedures
   Reports
   Evaluation
   Complaints procedure, penalties, damages, right of appeal

3. Certificate of Non-collusion

4. Contracting Associations (NICEIC, CORGI and HVCA)

5. Provision of a Contract Bond

6. Communications

7. Data Protection Act

8. Emergency services

9. Employer's obligations and restrictions

10. Good practice and examples

11. Identification

12. Inspections

13. Tenderers proposed method of working

14. Materials, goods and workmanship

15. Notice to occupiers

16. Obligations for statutory tax deduction scheme

17. Annual performance review

18. Potential hazards

19. Protection of furniture

20. Protection of gardens

21. Rehabilitation of Offenders' Act 1974

22. Definition of repairs : Day to day responsive repairs
    Package maintenance works
    Void property
    Out of hours emergency work

23. Response times

24. Security arrangements

25. Safety of children

Appendix A (Continued)

26. Spot audits
27. Smoking policy
28. Tests
29. Work equipment
30. Use of chlorofluorocarbons

Appendix B

1.0 Transfer of Undertakings and Protection of Employment Regulations

1.1 The present contract is being performed by an outside contractor. The Employer has no view as to whether or not the European Acquired Rights Directive No 77/187 and/or the Transfer of Undertakings (Protection of Employment) Regulations 1981 ("TUPE") applies to this Contract. It is up to each tenderer to form his own view on this. The Employer proposes to ask the present contractor whether he is of the opinion that TUPE might apply to this Contract and if so to provide a list of posts and details relating to it which he anticipates might transfer should TUPE apply. The Employer does not accept any responsibility for whether this information is made available, or if it is correct or not. The Tenderer may be required to complete a confidentiality agreement in respect of this information before it is made available to him.

1.2 The Tenderer must indicate whether his tender is based on TUPE applying or not applying. If the Tenderer has indicated that his tender is based on TUPE applying, he will be taken to have accepted that he will accept a transfer of any staff that should transfer to his contract. If the Tenderer's tender is successful, he must take any issues that may arise about transferring staff (including any questions as to who might transfer to his contract) directly with the present contractor. The Employer will not be willing to become involved in this.

1.3 If TUPE applies the Tenderer should take into account the following requirements in respect of transferring staff,

   1. The need to consult with recognised trade unions.
   2. The need to maintain existing rates of pay and conditions of employment.
   3. The need to provide pension arrangements broadly comparable to those provided at present to transferring employees. (Optional requirement)
   4. That liability will transfer to the successful contractor for any claims by transferring employees for redundancy, unfair dismissal or arising out of their previous employment even before transfer.

   It will be the Tenderer's responsibility, not the Employer's, to do this.

1.4 The Tenderer is expected in pricing his tender to make his own allowances for and accept the risk of fluctuations in his staffing availability or requirements. The Employer will not accept any tender in which the Tenderer's pricing varies according either to the number, identity or pension status of the staff he requires to perform the Contract or to any changes in the Tenderer's wage rates except so far as they may be reflected directly or indirectly in any method provided in the Contract Conditions for an annual review of the Tenderer's prices.

1.5 If the Tenderer has indicated that his tender is based on TUPE not applying, he must submit with his tender a written statement explaining why he believes TUPE will not apply if his tender is successful. Although this is primarily a matter between the outgoing and incoming Contractor, the Employer is concerned that the transition between contracts should be as seamless as possible. In the interests of continuity if the Employer does not agree with the Tenderer's contention that TUPE does not apply it reserves the right either to reject the tender or to evaluate it according to the Employers own assessment of the financial and other implications based on the Employers' own views as to the applicability to TUPE.

1.6 If the Employer accepts a bid on the basis that TUPE does not apply, it will require the successful tenderer in writing.

   1. To accept the risk that TUPE and/or the Directive might apply.
   2. To accept that if either is held to apply it will indemnify the Council against any cost that may fall on the Council in respect of any claims under TUPE or the Directive.
   3. To agree that if TUPE or the Directive are held to apply, the tenderer will not seek to rescind, repudite, terminate or amend the Contract.

Appendix B (Continued)

2.0　　TUPE and the expiry of this contract

2.1　　The Employer cannot and does not propose to commit itself as to,

1. What will be its Service requirements after this contract has expired.
2. What arrangements it may propose to make to procure the Service, or
3. What the legislative regime will be at that time either as to procurement of services or transfer of staff.

2.2　　It therefore will not enter into any commitment as to what might happen to the successful tenderer's staff at the expiry of the Contract.

SECTION G

PREAMBLES/SPECIFICATION

GENERALLY

Unless otherwise stated or contradicted Preambles are to apply reciprocally between Work Groups.

Unless otherwise stated or contradicted the rates contained in Part 2 (Schedule of Descriptions and Prices) are to apply reciprocally between Work Groups/Sections. For example, the prices for doors in Windows/Doors/Stairs are to apply in Demolition/Alteration/Renovation.

MATERIALS

Where and to the extent that materials are not fully specified they are to be, in order of priority, suitable for the purposes of the Works stated in or reasonably to be inferred from the contract, in accordance with good building practice, and complying strictly with **current** British Standards or in general with British Standards.

Proprietary brand names have been specified in certain circumstances. Equivalent (other equal or approved) may be used if approved, in writing, by the Contract Administrator.

Proprietary materials are to be handled and stored strictly in accordance with manufacturer's instructions and recommendations. Such materials are to be obtained direct from the manufacturers or through their accredited distributors.

Where the specification states that material is to be to the approval of the CA, then the contractor should submit details of the materials that are proposed to the CA for approval before ordering. Such materials should be in the price range of the schedules to which they relate

WORKMANSHIP

All workmanship is to conform to British Board of Agrement Specification.

Where and to the extent that workmanship is not fully specified it is to be, in order of priority, suitable for the purposes of the Works stated in or reasonably to be inferred from the contract, in accordance with good building practice, and complying with **current** British Standard Codes of Practice.

Proprietary materials are to be used and proprietary processes are to be carried out strictly in accordance with the manufacturer's instructions and recommendations.

Workmanship is to be of a high standard throughout particularly with regard to the accuracy of dimensions, lines, planes, levels and the quality of surface textures. The Contractor is to do everything necessary to ensure that the standard of finish demanded by the contract is achieved.

Work liable to damage by frost is not to be carried out at temperatures less than 5 degrees Celsius unless precautions are taken against low temperatures. Submit details of such precautions to the Contract Administrator.

A: CONTRACTORS GENERAL COST ITEMS

A44 Temporary Works

NOTE: The scaffolding rates in this section allow for basic forms of scaffolding. Where more complex forms of scaffolding are required, sub-contract rates should be used.

Scaffolding will only be paid for in accordance with this section when the height of the working platform exceeds 1.50m above ground level. Independent and putlog scaffolds are to be measured the perimeter of the structure to be scaffolded x the height of the scaffold to comply with the current appropriate regulations.

The rates in this section allow in each item for the cost of scaffolding from ground level to the working platform including boards and double guard rails at one level only. The rates for chimney scaffolds allow for free standing stacks commencing at ground level. Where scaffolding is required to chimneys above roof level it is recommended that rates for aluminium access units A4435 and A4490 are used in conjunction with rates for any other appropriate method of access from ground level to these items.

Hire costs for independent and putlog scaffolding are for 1 weeks hire, However it is recognised that a minimum hire periods of 4 weeks is applicable. Where this is the case users should apply the rate by the minimum hire period. This would also apply in other cases where minimum hire periods apply.

100   CONTRARY TO SMM
  -   Scaffolding has been measured in m2 or nr as applicable (SMM A44.1.3.*.*)

C: DEMOLITION/ALTERATION/RENOVATION

C20 DEMOLITION

Where appropriate, items in the preambles in other sections shall apply equally to this section

GENERAL REQUIREMENTS

100 CONTRARY TO SMM
- Demolition items have been measured in m3, m2, m or nr as applicable (SMM C20.*.*.*).
- In order to maintain the existing coding structure, C20 in the schedules is numbered C10

115 DEFINITIONS
- "Demolishing". Includes removing incidental items including windows, doors, skirtings and plaster and clearing the site.

110 DESK STUDY/ SURVEY
- Scope: Before starting demolition work, examine all available information, carry out a survey of the structures, site and surrounding area.
- Report and method statements: Submit describing:
- Form, condition and details of the structures.
- Form, location and removal methods of any flammable, toxic or hazardous materials.
- Type and location of adjoining or surrounding premises which may be adversely affected by noise, vibration, dust or removal of structure.
- Identification and location of services above and below ground, including those required for the Contractor's own use. Arrangements for disconnection and removal of services.
- Type and location of any features of historical, archeological or geological importance.
- Sequence and method of demolition including details of any specific pre-weakening.
- Arrangements for protection of personnel and the public including exclusion of unauthorized persons.
- Arrangements for control of site transport and traffic.

120 RESPONSIBILITY
- With regard to the actual items of demolition, forming and blocking up of openings and similar works of structural alteration, assess the full implication of such works and allow for any further work to such items as may be judged necessary and not contained within the descriptions.

130 GROUNDWORKS
- Old foundations, slabs and the like: Break out where and to the extent stated.
- Contaminated earth: Remove and disinfect as required by Local Authority.

140 BENCH MARKS
- Unrecorded bench marks and other survey information: Give notice when found.
- Do not remove or destroy.

SERVICES AFFECTED BY DEMOLITION

210 SERVICES REGULATIONS
- Work carried out to or which affects new or existing services: Carry out in accordance with the Byelaws or Regulations of the relevant Statutory Authority.

220 LOCATION OF SERVICES
- Services affected by the Works: Locate and mark positions.
- Mains services: Arrange with the appropriate authorities for location and marking of positions.

230 DISCONNECTION
- General: Arrange with the appropriate authorities for disconnection of services and removal of fittings and equipment prior to starting demolition.
- Service connections for appliances which are to be removed or relocated are to be cut back and stopped off out of sight wherever possible.

240 DISCONNECTION OF DRAINS
- General: Locate disconnect and seal disused drain connections.
- Sealing: Within the site and permanent.

250 DRAINS IN USE
- General: Protect drains, manholes, inspection chambers, gullies, vent pipes and fittings still in use and ensure that they are kept free of debris.
- Damage: Make good any damage arising from demolition work. Leave clean and in working order at completion.

260 BYPASS CONNECTIONS
- General: Provide as necessary to maintain continuity of services to occupied areas of the same and adjoining properties.
- Notice: Give a minimum 72 hours notice to occupiers if shutdown is necessary during changeover.

270 SERVICES WHICH ARE TO REMAIN
- Damage: Give notice and notify service authority or owner of any damage arising from the execution of the works.
- Repairs: To the satisfaction of the CA and service authority or owner.

DEMOLITION WORK

310 WORKMANSHIP
- Standard: Demolish structures in accordance with BS 6187.
- Operatives: Appropriately skilled and experienced for the type of work and hold or be training to obtain relevant CITB Certificates of Competence.
- Site staff responsible for supervision and control of work: Experienced in the assessment of risks involved and methods of demolition to be used.
- Explosives may not be used without prior approval of the Contract Administrator and all relevant authorities.

320 GAS OR VAPOUR RISKS
- Precautions: Adequate to prevent fire or explosion caused by gas or vapour.

330 DUST CONTROL
- Method: Reduce by periodically spraying demolition works and skips with water.

340 HEALTH HAZARDS
- Precautions: Protect site operatives and general public from hazards associated with vibration, dangerous fumes and dust arising during the course of the Works.

345 BURNING ON SITE
- Burning on site of materials arising from the work will not be permitted without prior approval.

350 ADJOINING PROPERTY
- Temporary support and protection: Provide. Maintain and alter as necessary as work progresses.
- Damage: Minimise. Promptly repair.
- Leave no unnecessary or unstable projections.
- Make good to ensure safety, stability, weather protection and security.
- Support to foundations: Do not disturb.
- Defects: Report any exposed or becoming apparent.

360 STRUCTURES TO BE RETAINED
- Parts which are to be kept in place: Protect.
- Extent of work: Cut away and strip out with care to reduce the amount of making good to a minimum.

365 REMOVAL OF CHIMNEY STACKS
- Allow for all necessary protection to the property and adjoining premises and curtilages, the careful demolition of the stack and the removal of all debris. Seal flues with two layers of slate bedded in cement mortar 1:3. Form the timber infilling to the opening of the same minimum dimension and section as the adjacent existing rafters, trimmers, battens, boarding and the like. Fix all rafters and battens at same centres and gauge as the existing. Lay roofing felt to provide a minimum of 150 mm lap to the existing and ensure that the roof tiles or slates match the existing in colour and dimension and are fixed, with non-ferrous screws, to the same gauge. Allow for the provision of special tiles or slates as appropriate. Pre-treat all timbers with preservative.

370 PARTLY DEMOLISHED STRUCTURES
- General: Leave in a stable condition, with adequate temporary support at each stage to prevent risk of uncontrolled collapse. Keep safe outside working hours.
- Temporary works: Prevent debris from overloading.
- Unauthorised persons: Prevent access.

372 STABILITY OF THE WORKS
- Do not permit anything to be done which may injure the stability of the works or the existing buildings; no cutting through floors or walls or under foundations will be allowed, other than that required by the drawings or schedules without the sanction of the Contract Administrator. Any such works other than those required by the drawings or schedules, sanctioned by the Contract Administrator, shall be made good to his entire satisfaction and without extra charge.

375 TEMPORARY PROTECTION: items are included for adequate temporary waterproof protection where work involves alterations to external walls, doors or windows or roofs

380 DANGEROUS OPENINGS
- General: Illuminate and protect. Keep safe outside working hours.

391 ASBESTOS BASED MATERIALS
- Discovery: Give notice immediately of any suspected asbestos based materials discovered during demolition work. Avoid disturbing such materials.
- Methods for safe removal. Submit details.

410 UNFORESEEN HAZARDS
- Unrecorded voids, tanks, chemicals, etc. discovered during demolition: Give notice.
- Methods for safe removal, filling, etc: Submit details.

420 OPEN BASEMENTS, ETC
- Temporary support: Leave adequate buttress walls or provide temporary support to basement retaining walls up to ground level.
- Safety: Make the remaining sections of any retaining and buttress walls safe and secure.
- Water movement: Make holes in basement floors to allow water drainage or penetration (depending on water table). Provide a hole for every 10 m^2, not less than 600 mm in diameter.

440 SITE CONDITION AT COMPLETION
- Debris: Clear away and leave the site in a tidy condition.

MATERIALS ARISING

510 CONTRACTOR'S PROPERTY
- Components and materials arising from the demolition work: Property of the Contractor except where otherwise provided.
- Remove from site as work proceeds.

520 RECYCLED MATERIALS
- Materials arising from demolition work: May be recycled or reused elsewhere in the project, subject to compliance with the appropriate specification.
- Evidence of compliance: Submit full details and supporting documentation.

D: GROUNDWORK

D20 EXCAVATING AND FILLING

Where appropriate, items in the preambles in other sections shall apply equally to this section

GENERALLY

100 DEFINITIONS
- "Bottoms of excavations" are to relate to all excavations other than excavation to reduce levels. "Ground" is to relate to virgin ground and to excavations after reducing levels.

110 CONTRARY TO SMM
- Notwithstanding clause D20.2.4 pits have not been numbered.
- Notwithstanding clause D20.8.1 a separate item has not been included, the cost is deemed to be included in the rates for excavation

120 METHODS OF MEASUREMENT
- Where topsoil is specified to be preserved, the existing ground level as SMM clause 2.*.*.1 for subsequent excavation, is deemed to be the ground level after topsoil has been excavated.
- Where reduced level dig is measured, existing ground level as SMM clause 2.*.*.1 for subsequent excavation, is deemed to be the ground level after reduced level dig.
- Unless otherwise stated the starting level is existing ground level or, if levels are to be reduced, reduced level.
- Notwithstanding the provisions of SMM Clause D7 (Earthwork support), the Contractor is to provide for everything to uphold the sides of excavation by whatever means necessary including any interlocking driven steel sheet piling deemed necessary by the Contractor (interlocking driven steel piling that is the subject of a specific design requirement will be measured in accordance with the rules of SMM Work Section D32).

130 PRICES TO INCLUDE
- Hand and/or mechanical excavation and disposal in whatever types of ground and fillings, drain pipes, roots and other obstructions are encountered but otherwise excluding brickwork, concrete and rock and the like as defined by SMM. (For example excavation in hardcore and other fillings is not measured separately).
- Earthwork support left in at the Contractor's volition.
- Temporary retention of fillings.
- Compacting filling of any material excluding top soil in layers not exceeding 150 thick unless stated otherwise.
- Lightly compacting, digging over and leaving top soil filling for planting free of weeds, large stones and debris.
- Bulking of excavated material.

WORKMANSHIP

140 EXCAVATION
- Is to be done by hand or machine at the discretion of the Contractor. The Contractor is to confirm his method of operation with the Contract Administrator prior to the execution of the work.

CLEARANCE/ EXCAVATING

170 REMOVING TREES, SHRUBS AND HEDGES
- Identification: Clearly mark trees to be removed.
- Safety: Comply with HSE/ Arboriculture and Forestry Advisory Group Safety Guides.
- Felling: Cut down and grub up roots of shrubs and smaller trees. Fell larger trees as close to the ground as possible.
- Work near retained trees: Take down trees carefully in small sections to avoid damage to adjacent trees that are to be retained, where tree canopies overlap and in confined spaces generally.
- Stumps: Obtain approval before removing stumps by winching and do not use other trees as supports or anchors.

225 HANDLING TOPSOIL
- Aggressive weeds:
- Species: Included in the Weeds Act, section 2 or the Wildlife and Countryside Act, Schedule 9, part II.
- Give notice: Obtain instructions before moving topsoil.
- Earthmoving equipment: Select and use to minimize disturbance, trafficking and compaction.
- Contamination: Do not mix topsoil with:
  - Subsoil, stone, hardcore, rubbish or material from demolition work.
  - Oil, fuel, cement or other substances harmful to plant growth.
  - Other grades of topsoil.
- Multiple handling: Keep to a minimum. Use topsoil immediately after stripping.
- Wet conditions: Handle topsoil in the driest condition possible. Do not handle during or after heavy rainfall or when it is wetter than the plastic limit as defined by BS 3882, Annex N2.
- Vegetable soil: remove and set aside or remove from site as specified. No top soil shall be buried or used for filling.

240 EXCAVATED FORMATIONS
- Are to be inspected by the Contract Administrator before new work is laid on them.
- Give the Contract Administrator not less than 24 hours notice of when excavations will be ready for inspection.
- Place concrete or other fill as soon as possible after inspection to prevent deterioration due to water, frost or other causes.

280 TRENCH FILL FOUNDATIONS
- Excavation: Form trench down to formation in one operation.
- Safety: Prepare formation from ground level.
- Inspection of formations: Give notice 24hrs before commencing excavation.
- Shoring: Where inspection of formation is required, provide localised shoring to suit ground conditions.
- Concrete fill: Place concrete immediately after inspection and no more than four hours after exposing the formation.

290 FOUNDATIONS IN MADE UP GROUND
- Depth: Excavate down to a natural formation of undisturbed subsoil.
- Discrepancy: Give notice if this is greater or less than depth given.

300 EARTHWORK SUPPORT
- Erect, alter as necessary support to sides of excavations and remove on completion unless otherwise instructed. Take measurements of any supports which are to be left in position and inform the Contract Administrator (NOTE: Earthwork support does not include interlocking driven steel sheet piling).
- Remedial works necessitated by the absence or failure of earthwork support are to be carried out at the Contractor's expense.

310 UNSTABLE GROUND
- Generally: Ensure that the excavation remains stable at all times.
- Give notice: Without delay if any newly excavated faces are too unstable to allow earthwork support to be inserted.
- Take action: If instability is likely to affect adjacent structures or roadways, take appropriate emergency action.

320 RECORDED FEATURES
- Recorded foundations, beds, drains, manholes, etc: Break out and seal drain ends.
- Contaminated earth: Remove and disinfect as required by local authority.

330 UNRECORDED FEATURES
- Give notice: If unrecorded foundations, beds, voids, basements, filling, tanks, pipes, cables, drains, manholes, watercourses, ditches, etc. not shown on the drawings are encountered.

350 EXISTING WATERCOURSES
- Diverted watercourses which are to be filled: Before filling, remove vegetable growths and soft deposits.

450 WATER
- Generally: Keep all excavations free from water until:
  - Formations are covered.
  - Below ground construction is completed.
  - Basement structures and retaining walls are able to resist leakage, water pressure and flotation.
- Drainage: Form surfaces of excavations and fill to provide adequate falls.
- Removal of water: Provide temporary drains, sumps and pumping as necessary. Do not pollute watercourses with silt laden water and do not disturb material in and around excavations

454 GROUND WATER LEVEL/ RUNNING WATER
- Give notice: If it is considered that the excavations are below the water table.
- Springs/ Running water: Give notice immediately if encountered.

457 PUMPING
- General: Do not disturb excavated faces or stability of adjacent ground or structures.
- Pumped water: Discharge without flooding the site, or adjoining property.
- Sumps: Construct clear of excavations. Fill on completion.

FILLING

500 PROPOSED FILL MATERIALS
- Details: Submit full details of proposed fill materials to demonstrate compliance with specification, including:
  - Type and source of imported fill.
  - Proposals for processing and reuse of material excavated on site.
  - Test reports as required elsewhere.

510 HAZARDOUS, AGGRESSIVE OR UNSTABLE MATERIALS
- General: Do not use fill materials which would, either in themselves or in combination with other materials or ground water, give rise to a health hazard, damage to building structures or instability in the filling, including material that is:
  - Frozen or containing ice.
  - Organic.
  - Contaminated or noxious.
  - Susceptible to spontaneous combustion.
  - Likely to erode or decay and cause voids.
  - With excessive moisture content, slurry, mud or from marshes or bogs.
  - Clay of liquid limit exceeding 80 and/ or plasticity index exceeding 55.
  - Unacceptable, class U2 as defined in the Highways Agency 'Specification for highway works', clause 601.

520 FROST SUSCEPTIBILITY
- General: Except as allowed below, fill must be non frost-susceptible as defined in Highways Agency 'Specification for Highway Works', clause 801.17.
- Test reports: If the following fill materials are proposed, submit a laboratory report confirming they are non frost- susceptible:
  - Fine grained soil with a plasticity index less than 20%.
  - Coarse grained soil or crushed granite with more than 10% retained on a 0.063 mm sieve.
  - Crushed chalk.
  - Crushed limestone fill with average saturation moisture content in excess of 3%.
  - Burnt colliery shale.
- Frost-susceptible fill: May only be used within the external walls of buildings below spaces that will be heated. Protect from frost during construction.

530 PLACING FILL
- Excavations and areas to be filled: Free from loose soil, rubbish and standing water.
- Freezing conditions: Do not place fill on frozen surfaces. Remove material affected by frost. Replace and recompact if not damaged after thawing.
- Adjacent structures, membranes and buried services:
    - Do not overload, destabilise or damage.
    - Submit proposals for temporary support necessary to ensure stability during filling.
    - Allow 14 days (minimum) before backfilling against in situ concrete structures.
- Layers: Place so that only one type of material occurs in each layer.
- Earthmoving equipment: Vary route to avoid rutting.

610 COMPACTED FILLING FOR LANDSCAPE AREAS
- Fill: Materials, capable of compaction by light earthmoving plant.
- Filling: Layers not more than 200 mm thick. Lightly compact each layer to produce a stable soil structure.
- Extend each layer over the whole width to be filled before spreading the next layer.
- Compact and consolidate final layer to required levels.

617 HIGHWAYS AGENCY TYPE 1 GRANULAR FILLING
- Fill: To Highways Agency 'Specification for highway works', clause 803:
    - Crushed rock (other than argillaceous rock).
    - Crushed concrete.
    - Recycled aggregates.
    - Crushed non-expansive slag to clause 801.2.
    - Well-burned non-plastic colliery shale.
- Filling: To Highways Agency 'Specification for highway works', clauses 801.3 and 802.

618 HIGHWAYS AGENCY TYPE 2 GRANULAR FILLING
- Fill: To Highways Agency 'Specification for highway works', clause 804:
    - Crushed rock (other than argillaceous rock).
    - Crushed concrete.
    - Crushed non-expansive slag to clause 801.2.
    - Well-burned non-plastic colliery shale.
    - Natural gravel.
    - Natural sand.
- Filling: To Highways Agency 'Specification for highway works', clauses 801.3 and 802.

700 BACKFILLING AROUND FOUNDATIONS
- Under oversite concrete and pavings: Hardcore as clause 710.
- Under grassed or soil areas: Material excavated from the trench, laid and compacted in 300 mm maximum layers.

710 HARDCORE FILLING
- Fill: Granular material, free from excessive dust, well graded, all pieces less than 75 mm in any direction, minimum 10% fines value of 50 kN when tested in a soaked condition to BS 812-111, and in any one layer only one of the following:
    - Crushed rock (other than argillaceous rock) or quarry waste with not more binding material than is required to help hold the stone together.
    - Crushed concrete, crushed brick or tile, free from plaster, timber and metal.
    - Crushed non-expansive slag.
    - Gravel or hoggin with not more clay content than is required to bind the material together, and with no large lumps of clay.
    - Well-burned non-plastic colliery shale.
    - Natural gravel.
    - Natural sand.
- Filling: Spread and level in 150 mm maximum layers. Thoroughly compact each layer.

© AAS 2008 - 2009

715 **VENTING HARDCORE LAYER**
- Fill: Clean granular material, well graded, passing a 75 mm BS sieve but retained on a 20 mm BS sieve. In each layer only one of the following:
  - Crushed hard rock.
  - Crushed concrete, crushed brick or tile, free from plaster, timber and metal.
  - Gravel.
- Filling: Spread and level in 150 mm maximum layers. Thoroughly compact each layer whilst maintaining enough voids to allow efficient venting.

720 **GRANULAR MATERIAL**
- approved well graded natural sands, gravels, rock or slag fines or mixtures thereof. The grading is to be within the following limits:-

BS Sieve Size	Percentage by Weight Passing
75 mm	100
37.5 mm	85 - 100
9.5 mm	45 - 100
4.75 mm	25 - 85
600	8 - 45
75	0 - 10

- Granular filling is to be laid and compacted only at a moisture content not more than 1% above or 2% below the optimum moisture content for maximum compaction or other moisture content as may by directed.

730 **BLINDING**
- Surfaces to receive sheet overlays or concrete:
  Blind with:
  - Concrete where shown on drawings; or
  - Sand, fine gravel, or other approved fine material applied to fill interstices. Moisten as necessary before final rolling to provide a flat, closed, smooth surface.
- Sand for blinding: To BS 882 table 4, type C or M, or to BS EN 12620, grade 0/4 or 0/2 (MP).
- Permissible deviations on surface level: +0 -25 mm.

E: INSITU CONCRETE/LARGE PRECAST CONCRETE

E10 MIXING/ CASTING/ CURING IN SITU CONCRETE

NOTE: The rates in this section allow for ready mixed concrete delivered to a point not exceeding 50m horizontally to placing point and all hoisting by non mechanical means.

GENERALLY

100 PRICES TO INCLUDE
- Monolithic finishes to include for maintaining all arrises and formations and for any necessary treatment of the exposed surface following removal of the formwork to obtain the specified finish.
- Reinforced concrete is to include working around all types of reinforcement, ties and the like.

CONCRETE

101 SPECIFICATION
- Concrete generally: To BS EN 206-1 and BS 8500-2.

132 DESIGNED CONCRETE
- Concrete mixes to be C15, C20 or C25 as specified

135 CONSISTENCY
- to be such that the concrete can be readily worked into the corners and angles of the forms and around the reinforcement without permitting the materials to segregate or free water to collect at the surface.

MATERIALS, BATCHING AND MIXING

215 READY-MIXED CONCRETE
- Production plant: Currently certified by a body accredited by UKAS to BS EN 45011 for product conformity certification of ready-mixed concrete.
- Source of ready-mixed concrete: Obtain from one source if possible. Otherwise, submit proposals.
  - Name and address of depot: Submit before any concrete is delivered.
  - Delivery notes: Retain for inspection.
- Declarations of nonconformity from concrete producer: Notify immediately.

221 INFORMATION ABOUT PROPOSED CONCRETES
- Details listed in BS 8500-1 clause 5.2: Submit when requested.

305 DRYING SHRINKAGE
- Drying shrinkage of concrete (maximum): 0.075%.
- Test method: To BS EN 1367-4.

310 CEMENT
- Ordinary Portland cement to BS 12. DO NOT USE High Alumina Cement or Extra Rapid Hardening cement. Deliver in sealed bags or other containers or purpose-made bulk delivery vehicles. Store in dry conditions and use in order of delivery.

315 AGGREGATES
- TO BS 12620

320 MARINE (SEA DREDGED) AGGREGATES
- To BS 1260 and to the following requirements:

1. Sodium chloride content not to exceed either
    a. for fine aggregate 0.15% of the dry weight of the aggregate and for coarse aggregate 0.05% of the dry weight of the aggregate

   OR

    b. 0.5% of the weight of cement in the mix

2. Shell content shall not exceed :

Nominal Size aggregate	Shell content as calcium carbonate of (CaCO3) % by weight of dry aggregate
40 mm	5
20 mm	10
10 mm	20
Fine	45

- Hollow shell shall not exceed 50% of the total shell content in 20 and 40 mm nominal size aggregates

325 AGGREGATES FOR EXPOSED VISUAL CONCRETE
- Limitations on contaminants: Free from absorbent particles which may cause 'popouts', and other particles such as coal and iron sulfide which may be unsightly or cause unacceptable staining.
- Colour: Consistent.
- Supply: From a single source and maintained throughout the contract.
- Samples: Submit on request.

330 MATERIALS FOR EXPOSED VISUAL CONCRETE
- Alterations to sources, types and proportions: Submit proposals.

415 ADMIXTURES
- Do not use unless approved by Contract Administrator.
- Calcium chloride and admixtures containing calcium chloride: Do not use.

420 WATER
- To be clean and uncontaminated. Obtain approval for other than mains supply.

490 PROPERTIES OF FRESH CONCRETE
- Adjustments to suit construction process: Determine with concrete producer. Maintain conformity to the specification.

PLACING/ COMPACTING/ CURING AND PROTECTION

620 TEMPERATURE OF WATER RESISTANT CONCRETE
- Objective: Limit maximum temperature of concrete to minimize cracking during placing, compaction and curing. Take account of:
  - High temperatures and steep temperature gradients: Prevent build-up during first 24 hours after casting. Prevent coincidence of maximum heat gain from cement hydration with high air temperature and/ or solar gain.
  - Rapid changes in temperature: Prevent during the first seven days after casting.
- Proposals for meeting objective: Submit.

630 PREMATURE WATER LOSS
- Requirement: Prevent water loss from concrete laid on absorbent substrates.
  - Underlay: Select from:
    Polyethylene sheet: 250 micrometres thick.
    Building paper: To BS 1521, grade B1F.
- Installation: Lap edges 150 mm.

640 CONSTRUCTION JOINTS
- Locations of construction joints: Submit proposals where not shown on drawings.
- Preparation of joint surfaces: Select from:
  - Brushing and spraying: Remove surface laitance and expose aggregate finish while concrete is still green.
  - Other methods: Submit proposals.
- Condition of joint surfaces immediately before placing fresh concrete: Clean and damp.

650 SURFACES TO RECEIVE CONCRETE
- Cleanliness of surfaces immediately before placing concrete: Clean with no debris, tying wire clippings, fastenings or free water.

660 INSPECTION OF SURFACES
- Notice: Give notice to allow inspections of reinforcement and surfaces before each pour of concrete.
- Period of notice: Obtain instructions.

670 TRANSPORTING
- General: Avoid contamination, segregation, loss of ingredients, excessive evaporation and loss of workability. Protect from heavy rain.
- Entrained air: Anticipate effects of transport and placing methods in order to achieve specified air content.

680 PLACING
- Records: Maintain for time, date and location of all pours.
- Timing: Place as soon as practicable after mixing and while sufficiently plastic for full compaction.
- Temperature limitations for concrete: 30°C (maximum) and 5°C (minimum). Do not place against frozen or frost covered surfaces.
- Continuity of pours: Place in final position in one continuous operation up to construction joints. Avoid formation of cold joints.
- Discharging concrete: Prevent uneven dispersal, segregation or loss of ingredients or any adverse effect on the formwork or formed finishes.
- Thickness of layers: To suit methods of compaction and achieve efficient amalgamation during compaction.
- Poker vibrators: Do not use to make concrete flow horizontally into position, except where necessary to achieve full compaction under void formers and cast-in accessories and at vertical joints.

690 COMPACTING
- General: Fully compact concrete to full depth to remove entrapped air. Continue until air bubbles cease to appear on the top surface.
  - Areas for particular attention: Around reinforcement, under void formers, cast-in accessories, into corners of formwork and at joints.
- Consecutive batches of concrete: Amalgamate without damaging adjacent partly hardened concrete.
- Methods of compaction: To suit consistence class and use of concrete.

720 VIBRATORS
- General: Maintain sufficient numbers and types of vibrator to suit pouring rate, consistency and location of concrete.
- External vibrators: Obtain approval for use.

730 PLASTIC SETTLEMENT
- Settlement cracking: Inspect fresh concrete closely and continuously wherever cracking is likely to occur, including the top of deep sections and at significant changes in the depth of concrete sections.
  - Timing: During the first few hours after placing and whilst concrete is still capable of being fluidized by the vibrator.
- Removal of cracks: Revibrate concrete.

810 CURING GENERALLY
- Evaporation from surfaces of concrete: Prevent, including from perimeters and abutments, throughout curing period.
  - Surfaces covered by formwork: Retain formwork in position and, where necessary to satisfy curing period, cover surfaces immediately after striking.
  - Top surfaces: Cover immediately after placing and compacting. If covering is removed for finishing operations, replace it immediately afterwards.
- Surface temperature: Maintain above 5°C throughout the specified curing period or four days, whichever is longer.
- Records: Maintain details of location and timing of casting of individual batches, removal of formwork and removal of coverings. Keep records on site, available for inspection.

811 COVERINGS FOR CURING
- Sheet coverings: Suitable impervious material.
- Curing compounds: Selection criteria:
  - Curing efficiency: Not less than 75% or for surfaces exposed to abrasion 90%.
  - Colouring: Fugitive dye.
  - Application to concrete exposed in the finished work: Readily removable without disfiguring the surface.
  - Application to concrete to receive bonded construction/ finish: No impediment to subsequent bonding.
- Interim covering to top surfaces of concrete: Until surfaces are in a suitable state to receive coverings in direct contact, cover with impervious sheeting held clear of the surface and sealed against draughts at perimeters and junctions.

812 PREVENTING EARLY AGE THERMAL CRACKING
- Deep lifts or large volume pours: Submit proposals for curing to prevent early age thermal cracking, taking account of:
  - Temperature differentials across sections.
  - Coefficient of thermal expansion of the concrete.
  - Strain capacity of the concrete mix (aggregate dependent).
  - Restraint.

813 CURING COMPOUNDS:
- To be approved by the Contract Administrator.

815 **ADDITIONAL CURING REQUIREMENT - WATER CURING**
- Commencement of water curing: As soon as practicable after placing and compacting concrete.
  - Surfaces covered by formwork: Expose to water curing as soon as practicable.
  - Top surfaces: Cover immediately with impermeable sheeting to prevent evaporation before commencement of water curing.
- Water curing: Wet surfaces continuously throughout curing period.
  - Select methods from:
    Mist spray.
    Wet hessian covered with impermeable sheeting.

820 **CURING PERIODS**
- General: Curing periods are in days (minimum).
  - Definition of 't': The average number of degrees Celsius air temperature during the curing period.
- Curing periods for concrete surfaces which, in the finished building, will be exposed to the elements; concrete wearing surface floors and pavements; water resistant concrete:

	Concrete made using CEM1; SRPC (BS 4027); IIA	Concrete made using IIB; IIIA; IIIB; IVB
Drying winds or dry, sunny weather	$\frac{140}{t+10}$	$\frac{180}{t+10}$
Intermediate conditions	$\frac{100}{t+10}$	$\frac{140}{t+10}$
Damp weather, protected from sun and wind	$\frac{100}{t+10}$	$\frac{100}{t+10}$

- Curing periods for other structural concrete surfaces (cements/ combinations as above):

Drying winds or dry, sunny weather	$\frac{80}{t+10}$	$\frac{140}{t+10}$
Intermediate conditions	$\frac{60}{t+10}$	$\frac{80}{t+10}$
Damp weather, protected from sun and wind	No special requirements	No special requirements

- Curing periods for concretes using admixtures or other types of cements/ combinations: Submit proposals.

840 **PROTECTION**
- Prevent damage to concrete, including:
  - Surfaces generally: From rain, indentation and other physical damage.
  - Surfaces to exposed visual concrete: From dirt, staining, rust marks and other disfiguration.
  - Immature concrete: From thermal shock, physical shock, overloading, movement and vibration.
  - In cold weather: From entrapment and freezing expansion of water in pockets, etc.

E20 FORMWORK FOR IN SITU CONCRETE

GENERALLY/ PREPARATION

100 METHODS OF MEASUREMENT
- Notwithstanding SMM clause E20.M14 recesses, nibs and rebates are measured extra over on linear items of formwork.
- The heights to the soffits required to be given under SMM E20.8.*.* and E20.10.*.* have been measured from the general floor level or ground level immediately below the soffit. The Contractor is to allow for any strutting he elects to provide below this level (e.g. from the top of foundations).
- Formwork to sides of foundations, sides of ground beams and edges of beds, edges of suspended slabs, sides of upstands, steps in to surfaces, steps in soffits and machines bases and plinths deemed to be plain vertical unless stated otherwise.
- Formwork to soffits of slabs, soffits of landings, soffits of coffered or troughed slabs and top formwork deemed to be horizontal unless stated otherwise.

102 CONTRARY TO SMM
- The number of soffits to landings has not been stated (SMM E20.9.*.*).
- The number of members has not been stated (SMM E20.13.*.* - E20.16.*.*).
- Beams attached to slabs or walls have been grouped together (SMM E20.13-14.1.* and E20.13-14.2.*).
- The number of members has not been stated (SMM E20 17-19.*).
- The number of stairflights has not been stated (SMM E25.*.*).

104 PRICES ALSO TO INCLUDE
- For plain or dovetailed shaped mortices.
- Formwork is to include battens, struts and other supports, reverse-cut strings, bolting, wedging, easing, striking and removing and coating with mould oil.
- Making good concrete after removal of formwork.

110 LOADINGS
- Requirement: Design and construct formwork to withstand the worst combination of the following:
    - Total weight of formwork, reinforcement and concrete.
    - Construction loads including dynamic effects of placing, compacting and construction traffic.
    - Wind and snow loads.

132 PROPPING
- General: Prevent deflection and damage to the structure. Carry down props to bearings strong enough to provide adequate support throughout concreting operations.
- Method statement: Submit proposals for prop bearings and sequence of propping/ repropping and backpropping.

140 TEMPORARY SUPPORTS TO PROFILED STEEL SHEETS
- Location: Continuous along the centre of each span.
- Removal of temporary supports: Obtain instructions.

170 WORK BELOW GROUND
- Casting vertical faces of footings, bases and slabs against faces of excavation: Obtain consent from CA.
- Casting walls against faces of excavation: Use formwork on both sides.
- Protect soil faces to the Contractor Administrator's satisfaction to prevent contaminating the concrete.
- All protective material to be withdrawn as concreting proceeds unless the Contractor chooses to leave it in with Contract Administrator's agreement.

CONSTRUCTION

310 ACCURACY
- General requirement for formwork: Accurately and robustly constructed to produce finished concrete to the required dimensions.
- Formed surfaces: Free from twist and bow (other than any required cambers).
- Intersections, lines and angles: Square, plumb and true.

320 JOINTS IN FORMS
- Agree form and location with CA
- Requirements including joints in form linings and between forms and completed work:
  - Prevent loss of grout, using seals where necessary.
  - Prevent formation of steps. Secure formwork tight against adjacent concrete.
  - clean face of formwork immediately before assembly of forms. Keep clean until concrete is placed.

330 INSERTS, HOLES AND CHASES
- Positions and details: Submit proposals.
  - Positioning relative to reinforcement: Give notice of any conflicts well in advance of placing concrete.
- Method of forming: Fix inserts or box out as required. Do not cut hardened concrete without approval.

340 KICKERS
- Method statement: Submit proposals including means of achieving quality of concrete consistent with that specified for the column or wall.
  - Kicker height (minimum): 100 mm.

350 FORM TIES
- Metal associated with form ties/ devices: Prohibited within cover to reinforcement.

351 PROHIBITION OF FORM TIES
- Do not use wire form ties

361 FORM TIES FOR WATER RESISTANT CONCRETE
- General: Maintain water resistance of construction.
- Tie type and sealing system: Submit proposals.

380 VOID FORMERS
- Manufacturer: Contractors choice .

470 RELEASE AGENTS
- General: Achieve a clean release of forms without disfiguring the concrete surface.
- Product types: Compatible with formwork materials, specified formed finishes and subsequent applied finishes. Use the same product throughout the entire area of any one finish.
- Protection: Prevent contact with reinforcement, hardened concrete, other materials not part of the form face, and permanent forms.

480 SURFACE RETARDERS
- Use: Obtain approval.
- Reinforcement: Prevent contact with retarder.

STRIKING

510 STRIKING FORMWORK
- Timing: Prevent any disturbance, damage or overloading of the permanent structure.

521 MINIMUM PERIODS
- Determination of minimum periods for retaining formwork in position shown in the following table:

	Average Air Temperature (degrees Celsius)		
	16°C or more	7°C	3°C or less
Vertical formwork to columns, walls or large beams	9 Hours	12 Hours	Increase period in 7°C column by 24 hours for each day on which the average temperature is 3°C or less
Soffit formwork to slabs	4 Days	7 Days	
Props to slabs	11 Days	14 Days	
Soffit formwork to beams	8 Days	14 Days	
Props to beams	15 Days	21 Days	

Interpolation between 16°C and 7°C and between 7°C and 3°C is permitted.

FORMED FINISHES

615 FINISH TO RECEIVE ASPHALT TANKING
- Finish: Even and suitable to receive asphalt.
- Permissible deviation of surfaces:
  - Sudden irregularities (maximum): 3 mm.
  - Gradual irregularities when measured from underside of a 1 m straightedge, placed anywhere on surface (maximum): 3 mm.
- Surface blemishes:
  - Permitted: Blowholes less than 10 mm in diameter.
  - Not permitted: Voids, honeycombing, segregation and other large defects.
- Projecting fins: Remove.
- Formwork tie holes: Filled with mortar.

620 PLAIN SMOOTH FINISH
- Finish: Even with panels arranged in a regular pattern as a feature of the surface.
- Permissible deviation of surfaces:
  - Sudden irregularities (maximum): 5 mm.
  - Gradual irregularities when measured from the underside of a 1 m straightedge, placed anywhere on surface (maximum): 5 mm.
- Variations in colour:
  - Permitted: Those caused by impermeable form linings.
  - Not permitted: Discoloration caused by contamination or grout leakage.
- Surface blemishes:
  - Permitted: Blowholes less than 10 mm in diameter and at an agreed frequency.
  - Not permitted: Voids, honeycombing, segregation and other large defects.
- Formwork tie holes: In a regular pattern and filled with matching mortar.

F: MASONRY

F10 BRICK/ BLOCK WALLING

GENERALLY

100 DEFINITIONS
- "Composite work". Walls etc built of more than one type of brick or block in the thickness of the wall.

105 METHODS OF MEASUREMENT
- Notwithstanding SMM D1, nominal thickness is to be related to co-ordinating sizes for clay and calcium silicate bricks and work size for concrete bricks and blocks.
- Unless otherwise stated bonding is deemed to be stretcher bond for blockwork and 102mm brick walls and English bond for walls over 102mm.
- Unless otherwise stated pointing is deemed to be with a struck or flush joint as the work proceeds.

107 PRICES ALSO TO INCLUDE
- For any approved method of bonding to suit the particular circumstances.
- where a mix of mortar is not given in the description, the mix is to be primarily that used in the work generally. Alternatively, cement mortar is to be 1:3 and cement lime mortar is to be 1:1:6.
- Substitution of hollow blocks for solid blocks where switch and outlet drops occur.
- Building paper protection at junction of cast insitu concrete and brick facework.
- Reinforced brickwork is to include all additional labours implied by the use of reinforcement.
- Fabric reinforcement is to include for all additional labours and cuttings.
- Holes and openings not exceeding 225 mm wide for building in air bricks, ventilating gratings, soot doors and the like and for pipes and the like are to include Welsh arches or slate lintels.
- Bricks described as "from the demolitions" are to include extra over the descriptions of "demolish" or "take down" for handling, cleaning, stacking, covering, protecting and the like, and movement about the site.

TYPES OF WALLING

110 CLAY FACING BRICKWORK
- Bricks: TO BS 3921 of the type strength and quality as specified in schedules.
  - Mortar: As section Z21.
  - Standard: as specified in schedules.
  - Mix: as specified in schedules.
  - Bond: English unless otherwise stated.
- Joints: struck or flush as work proceeds unless otherwise stated.

250 CONCRETE BLOCKWORK
- Blocks: To BS 6073-1.
  - Type: as specified in schedules.
  - Average compressive strength (minimum): as specified in schedules.
  - Work sizes (length x height x thickness): as specified in schedules.
- Mortar: As section Z21.
  - Standard: as specified in schedules.
  - Mix: as specified in schedules.
- Bond: stretcher unless otherwise stated.
- Joints: struck or flush as work proceeds unless otherwise stated.

310 CLAY COMMON BRICKWORK
- Bricks: Flettons To BS 3921. approved, hard, burnt, local brick or equivalent.
- Mortar: As section Z21.
  - Standard: as specified in schedules.
  - Mix: as specified in schedules.
- Bond: as specified in schedules..

350 CONCRETE COMMON BLOCKWORK
- Blocks: To BS 6073-1.
  - Type: as specified in schedules.
- Mortar: As section Z21.
  - Standard: as specified in schedules.
  - Mix: as specified in schedules.
- Bond: as specified in schedules.

380 ENGINEERING BRICKWORK
- Engineering bricks: To BS 3921.
  - Class A or B as specified in schedules.
- Mortar: As section Z21.
  - Standard: as specified in schedules.
  - Mix: as specified in schedules.
- Bond: as specified in schedules.
- Joints: Flush.

WORKMANSHIP GENERALLY

400 WORK
- comply with the general recommendations of BS 5628: Part 2.

430 CONDITIONING OF CLAY AND CALCIUM SILICATE BRICKS
- Bricks delivered warm from manufacturing process: Do not use until cold.
- Absorbent bricks in warm weather: Wet to reduce suction. Do not soak.

440 CONDITIONING OF CONCRETE BRICKS/ BLOCKS
- Autoclaved concrete bricks/ blocks delivered warm from manufacturing process: Do not use.
- Age of non autoclaved concrete bricks/ blocks: Do not use until at least four weeks old.
- Avoidance of suction in concrete bricks/ blocks: Do not wet.
- Use of water retaining mortar admixture: Submit details.

460 MORTAR GROUPS
- Mix proportions: For a specified group select a mix design from the following:

Group	1	2	3	4
PC*:lime:sand with or without air entraining additive	1:0-0.25:3	1:0.5: 4-4.5	1:1:5-6	1:2:8-9
Masonry cement:sand containing PC* and lime in approx ratio 1:1, and an air entraining additive	-	1:3	1:3.5-4	1:4.5
Masonry cement:sand containing PC* and inorganic materials other than lime and air entraining additive	-	1:2.5-3.5	1:4-5	1:5.5-6.5
PC*:sand and air entraining additive	1:3	1:3-4	1:5-6	1:7-8

PC* = Portland cement

- Batching: Mix proportions by volume.
- Mortar type: Continuous throughout any one type of masonry work.

500 LAYING GENERALLY
- Mortar joints: Fill vertical joints. Lay bricks, solid and cellular blocks on a full bed.
- Bond where not specified: Half lap stretcher.
- Vertical joints in facework: Even widths. Plumb at every fifth cross joint.

520 ACCURACY
- Courses: Level and true to line.
- Faces, angles and features: Plumb.
- Permissible deviations:
  - Position in plan of any point in relation to the specified building reference line and/ or point at the same level ± 10 mm.
    - Straightness in any 5 m length ± 5 mm.
    - Verticality up to 3 m height ± 10 mm.
    - Verticality up to 7 m height ± 14 mm.
    - Overall thickness of walls ± 10 mm.
    - Level of bed joints up to 5 m (brick masonry) ± 11 mm.
    - Level of bed joints up to 5 m (block masonry) ± 13 mm.

535 HEIGHT OF LIFTS IN WALLING USING CEMENT GAUGED OR HYDRAULIC LIME MORTAR
- Quoins and advance work: Rack back.
- Lift height (maximum): 1.2 m above any other part of work at any time.
- Daily lift height (maximum): 1.5 m for any one leaf.

545 LEVELLING OF SEPARATE LEAVES USING CEMENT GAUGED OR HYDRAULIC LIME MORTAR
- Locations for equal levelling of cavity wall leaves: As follows:
  - Every course containing vertical twist type ties or other rigid ties.
  - Every third tie course for double triangle/ butterfly ties.
  - Courses in which lintels are to be bedded.

560 COURSING BRICKWORK
- Gauge: Four brick courses including bed joints to 300 mm.

561 COURSING BRICKWORK WITH EXISTING
- Gauge: Line up with existing brick courses.

580 LAYING FROGGED BRICKS
- Single frogged bricks: Frog uppermost.
- Double frogged bricks: Larger frog uppermost.
- Frog cavity: Fill with mortar.

585 LAYING CELLULAR BRICKS
- Orientation: Cavities downward.

595 LINTELS
- Bearing: Ensure full length masonry units occur immediately under lintel ends.

610 SUPPORT OF EXISTING WORK
- Joint above inserted lintel or masonry: Fully consolidated with semidry mortar to support existing structure.

615 BRICKWORK TO RECEIVE ASPHALT DPC
- Substrate: Mortar bed finished flush, smooth and level.

620 BLOCK BONDING NEW WALLS TO EXISTING
- Pocket requirements: Formed as follows:
  - Width: Full thickness of new wall.
  - Depth (minimum): 100 mm.
  - Vertical spacing:
    Brick to brick: 4 courses high at 8 course centres.
    Block to block: Every other course.
- Pocket joints: Fully filled with mortar.

635 JOINTING
- Profile: Consistent in appearance.

645 ACCESSIBLE JOINTS NOT EXPOSED TO VIEW
- Jointing: Struck flush as work proceeds.

665 POINTING
- Joint preparation: Remove debris. Dampen surface.
- Mortar: As section Z21.
  - Standard: as specified in schedules.
  - Mix: as specified in schedules.
- Profile: as specified in schedules.

671 FIRE STOPPING
- Avoidance of fire and smoke penetration: Fit tightly between cavity barriers and masonry. Leave no gaps.

690 ADVERSE WEATHER
- General: Do not use frozen materials or lay on frozen surfaces.
- Air temperature requirements: Do not lay bricks/ blocks:
  - In cement gauged mortars when at or below 3°C and falling or unless it is at least 1°C and rising.
  - In hydraulic lime:sand mortars when at or below 5°C and falling or below 3°C and rising.
  - In thin joint mortar glue when outside the limits set by the mortar manufacturer.
- Temperature of walling during curing: Above freezing until hardened.
- Newly erected walling: Protect at all times from:
  - Rain and snow.
  - Drying out too rapidly in hot conditions and in drying winds.

ADDITIONAL REQUIREMENTS FOR FACEWORK

710 THE TERM FACEWORK
- Definition: Applicable in this specification to brick/ block walling finished fair.
  - Painted facework: The only requirement to be waived is that relating to colour.

730 BRICK/ CONCRETE BLOCK SAMPLES
- General: Before placing orders with suppliers submit for approval of appearance labelled samples.
- Selection of samples: Representative of the range in variation of appearance.

750 COLOUR CONSISTENCY OF MASONRY UNITS
- Colour range: Submit proposals of methods taken to ensure that units are of consistent and even appearance within deliveries.
- Conformity: Check each delivery for consistency of appearance with previous deliveries and with approved reference panels; do not use if variation is excessive.
- Finished work: Free from patches, horizontal stripes and racking back marks.

760 APPEARANCE
- Brick/ block selection: Do not use units with damaged faces or arrises.
- Cut masonry units: Where cut faces or edges are exposed, cut with table masonry saw.
- Quality control: Lay masonry units to match relevant reference panels.
  - Setting out: To produce satisfactory junctions and joints with built-in elements and components.
  - Coursing: Evenly spaced using gauge rods.
- Lifts: Complete in one operation.
- Methods of protecting facework: Submit proposals.

780 GROUND LEVEL
- Commencement of facework: Not less than 150 mm below finished level of adjoining ground or external works level.

790 PUTLOG SCAFFOLDING
- Use: Not permitted in facework.

800 TOOTHED BOND
- New and existing facework in same plane: Bond together at every course to achieve continuity.

830 CLEANLINESS
- Facework: Keep clean.
- Mortar on facework: Allow to dry before removing with stiff bristled brush.
- Removal of marks and stains: Rubbing not permitted.

F30 ACCESSORIES/ SUNDRY ITEMS FOR BRICK/ BLOCK/ STONE WALLING

GENERALLY

100 PRICES ALSO TO INCLUDE
- Forming cavities is to include forming weep holes at base of cavity when directed and for keeping cavity clear of mortar droppings.
- Building in ends of wall ties is to include straightening ties as necessary.
- **Extra ties at openings and angles in hollow walls.**

CAVITIES

110 CONCRETE FILL TO BASE OF CAVITY
- Concrete generally: To BS EN 206-1 and BS 8500-2.
  - Designated concrete: C20.
    Workability: High.
- Extent: Maintain 75 mm between top of fill and external ground level and a minimum of 225 mm between top of fill and ground level dpc.
- Placement: Compact to eliminate voids.

120 CLEANLINESS
- Cavity base and faces, ties, insulation and exposed dpcs: Free from mortar and debris.

130 PEREND JOINT WEEP HOLES
- Form: Open perpend joint.
- Locations: Through outer leaf immediately above base of cavity, at cavity trays, stepped dpcs and external openings. 75 mm above top of cavity fill at base of cavity.
- Provision: At not greater than 1000 mm centres and not less than two over each opening.

131 BED JOINT WEEP HOLES
- Form: Open 10 mm diameter hole.
- Locations: Through outer leaf immediately above base of cavity at cavity trays, stepped dpcs and external openings. 75 mm above top of cavity fill at base of cavity.
- Provision: At not greater than 1000 mm centres and not less than two over each opening.

151 PARTIAL FILL CAVITY INSULATION - EXPANDED POLYSTYRENE (EPS)
- Insulation: Expanded polystyrene boards to BS EN 13163.
- Manufacturer: Polyfoam .
  - Product reference: Cavity Board.
- Face size (length x width): as specified in schedules.
- Thickness: Various as specified in schedules.
- Placement: Secure against face of inner leaf.
  - Residual cavity: Clear and unobstructed.
- Joints between boards, at closures and penetrations: No gaps and free from mortar and debris.

160 AIR BRICKS IN EXTERNAL WALLING
- Standard: To BS 493, class 1.
- Manufacturer: Contractors choice.
- Apertures: Square hole or Louvre pattern.
- Work sizes: as specified in schedules.
- Material/ colour: terracotta.
- Placement: Built in with no gaps at joints.

165 GRATINGS/ VENTILATORS IN INTERNAL WALLING
- Standard: To BS 493, class 1.
- Manufacturer: Contractors choice .
- Apertures: Square hole or Louvre pattern.
- Work sizes: as specified in schedules.
- Material/ colour: Galvanised.
- Placement: Built in with no gaps at joints.

REINFORCING/ FIXING ACCESSORIES

210 WALL TIES
- Standard: To BS 1243.
  - Type: Various as specified in schedules.
- Material/ finish: Various as specified in schedules.
- Sizes: Various as specified in schedules.

225 FIXING TIES IN MASONRY CAVITY WALLS
- Embedment in mortar beds (minimum): 50 mm.
- Placement: Sloping slightly downwards towards outer leaf, without bending. Drip centred in the cavity and pointing downwards.
- Spacing: Staggered in alternate courses.
  - Horizontal centres: 900mm.
  - Vertical centres: 450mm.
- Additional ties: Provide within 225 mm of reveals of unbonded openings.
  - Spacing: 300mm.

233 FIXING TIES IN MASONRY CAVITY WALLS WITH PARTIAL FILL CAVITY INSULATION
- Embedment in mortar beds (minimum): 50 mm.
- Placement: Sloping slightly downwards towards outer leaf, without bending. Drip centred in the cavity and pointing downwards.
- Spacing: Evenly space in non staggered horizontal and vertical rows.
  - Horizontal centres: 900mm.
  - Vertical centres: 450mm.
- Spacing centres of top (eaves) row of ties: .
- Provision of additional ties: Within 225 mm of reveals of unbonded openings.
  - Spacing: 300mm.

270 MESHWORK JOINT REINFORCEMENT FOR BRICKWORK
- Manufacturer: Contractors choice .
- Material: Stainless steel .
- Width: Approximately 40-50 mm less in width than wall or leaf.
- Placement: Lay on an even bed of mortar in a continuous strip with 225 mm laps at joints and full laps at angles. Keep back 20 mm from face of external work, 12 mm back from face of internal work and finish joint to normal thickness.

FLEXIBLE DAMP PROOF COURSES/ CAVITY TRAYS

310 DAMP PROOF COURSE - BITUMEN BASED
- Standard: To BS 6398.
  - Class: Various as specified in schedule.
- Manufacturer: Contractors choice.
  - Product reference: Various as specified in schedule.

320 DAMP PROOF COURSE - POLYETHYLENE
- Standard: To BS 6515.
- Manufacturer: Contractors choice.

330 DAMP PROOF COURSE - SLATE
- Manufacturer: Contractors choice.

385 PREFORMED DPC/ CAVITY TRAY JUNCTION CLOAKS/ STOP ENDS
- Manufacturer: Contractors choice.
  - Product references and locations: As shown on drawings.
- Placement: Seal laps with dpcs and/ or cavity trays.

390 SITE FORMED DPC/ CAVITY TRAY JUNCTIONS/ STOP ENDS
- Three dimensional changes in shape: Form to provide a free draining and watertight installation. Seal laps.
- Alternative use of preformed junction cloaks/ stop ends: Submit proposals.

INSTALLATION OF DPCS/ CAVITY TRAYS

415 HORIZONTAL DPCS
- Placement: In continuous lengths on full even bed of fresh mortar, with 100 mm laps at joints and full laps at angles.
- Width: At least full width of leaf unless otherwise specified. Edges of dpc not covered with mortar or projecting into cavity.
- Overlying construction: Immediately cover with full even bed of mortar to receive next masonry course.
- Overall finished joint thickness: As close to normal as practicable.

425 GROUND LEVEL DPCS
- Joint with damp proof membrane: Continuous and effectively sealed.

435 STEPPED DPCS IN EXTERNAL WALLS
- External walls on sloping ground: Install dpcs not less than 150 mm above adjoining finished ground level.

445 SILL DPCS
- Form and placement: In one piece and turned up at back when sill is in contact with inner leaf.

455 COPING/ CAPPING DPCS
- Placement: Bed in one operation to ensure maximum bond between masonry units, mortar and dpc.
- Dpcs crossing cavity: Provide rigid support to prevent sagging.

475 SITE FORMED CAVITY TRAYS
- Requirements to prevent downward ingress of water:
  - Profiles: To match those shown on drawings. Firmly secured.
  - Joint treatment: Use unjointed wherever possible, otherwise lap at least 100 mm and seal to produce a free draining and watertight installation.
  - Horizontal cavity trays: Support using cavity closer.
  - Sloping cavity trays: Prevent sagging.
  - Cleanliness: Free from debris and mortar droppings.

485 CAVITY TRAYS OVER OPENINGS AND OTHER CAVITY BRIDGINGS
- Length: To extend not less than 150 mm beyond ends of lintels/ bridgings.

515 DPC/ CAVITY TRAY LEADING EDGE IN FACEWORK - FLUSH
- Treatment at face of masonry: Finish flush and clear of mortar at the following locations: as specified by client.

525 DPC/ CAVITY TRAY LEADING EDGE IN FACEWORK - SET BACK
- Treatment at face of masonry: Set back 5 mm from face of wall with recessed mortar joint to expose edge at the following locations: as specified by client.

535 DPC/ CAVITY TRAY LEADING EDGE IN FACEWORK - PROJECTING
- Treatment at face of masonry: Projecting 5 mm from face of wall at the following locations: as specified by client.

560 VERTICAL DPCS GENERALLY
- Form: In one piece wherever possible.
  - Joints: Upper part overlapping lower not less then 100 mm.

570 JAMB DPCS AT OPENINGS
- Joint with cavity tray/ lintel at head: Full underlap.
- Joint with sill/ horizontal dpc at base: Full overlap.
- Projection into cavity: Not less than 25 mm.
- Relationship with frame: In full contact.

580 JAMB DPCS TO BUILT IN TIMBER FRAMES
- Fixing: Securely fastened to back of frame.
- Fasteners: Galvanized clout nails or staples.

JOINTS

610 MOVEMENT JOINTS WITH SEALANT
- Joint preparation and sealant application: As section Z22.
- Filler: Bitumen bonded fibreboard.
  - Thickness: To match design width of joint.
  - Manufacturer: Contractors choice or as specified in schedule.
    Product reference: Contractors choice or as specified in schedule.
  - Placement: Build in as work proceeds with no projections into cavities and to correct depth to receive sealant system.
- Sealant: Polysulphide
  - Manufacturer Contractors choice or as specified in schedule.
  - Colour: as appropriate.

630 UNEXPOSED CONTRACTION JOINTS
- Formation: Close butt as work proceeds.

650 POINTING IN FLASHINGS
- Joint preparation: Free of debris and lightly wetted.
- Pointing mortar: As for adjacent walling.
- Placement: Fill joint and finish flush.

655 POINTING IN ASPHALT SKIRTINGS
- Joint preparation: Free of debris and lightly wetted.
- Pointing mortar: 1:4 cement:sand incorporating a bonding agent.
  - Colour: Match adjacent work.
- Placement: Fill joint and finish flush.

660 PINNING UP TO SOFFITS
- Top joint of loadbearing walls: Fill and consolidate with mortar.

PROPRIETARY SILLS/ LINTELS/ COPINGS/ DRESSINGS

720 SILLS
- Standard: To BS 5642-1.
- Material: Precast Concrete.
- Manufacturer: Contractors choice.
- Dimensions: As described in schedules.
- Finish: fair.
- Mortar for bedding/ jointing: Cement gauged as section Z21.
  - Standard: BS4721.
  - Mix: 1:3.
- Joints: Flush.
- Bedding one piece sills: Leave bed joints open except under end bearings and masonry mullions. On completion, point to match adjacent work.

730 PRECAST CONCRETE LINTELS
- Standard: To BS 5977-2.
- Manufacturer: Contractors choice.
- Placement: Bed on mortar used for adjacent work with bearing of not less than 150 mm. Use slate packing pieces.

750 PREFABRICATED STEEL LINTELS
- Standard: To BS 5977-2.
- Manufacturer: as specified in schedules.
- Placement: Bed on mortar used for adjacent work with bearing of not less than 150 mm.

760 COPING UNITS
- Standard: To BS 5642-2.
- Material: Precast Concrete.
- Manufacturer: Contractors choice.
- Dimensions: As detailed in schedules
- Mortar for bedding/ jointing: Cement gauged as section Z21.
  - Standard: BS4721.
  - Mix: 1:3.
- Joints: Full and finished flush.
- Placement: Lay on a full bed of mortar to line and level.

MISCELLANEOUS ITEMS

810 TILE SILLS
- Tiles: Plain clay to BS EN 1304.
  - Manufacturer: Contractors choice.
  - Size: As detailed in schedules.
- Placement: Two courses, broken jointed, true to line and level on full bed of 1:3 cement:sand mortar as section Z21.
- Joints: Full and finished flush.

820 TILE CREASING
- Tiles: Plain clay to BS EN 1304.
  - Manufacturer: Contractors choice.
  - Size: As detailed in schedules.
- Placement: Two courses, broken jointed, on full bed of mortar as used for adjacent work.
- Joints: Full and finished flush.

830 BUILDING IN FRAMES
- Preparation: Remove horns and provide support.
- Fixing cramps: Fully bed in mortar.

840 OPENINGS FOR FRAMES
- Formation: Use accurate, rigid templates to required size.

850 WALL PLATES
- Placement: On full bed of mortar to correct horizontal level.

F31 PRECAST CONCRETE SILLS/ LINTELS/ COPINGS/ FEATURES

100 METHODS OF MEASUREMENT
- Dimensions given refer to concrete sizes and not the overall sizes including projecting reinforcement.

102 CONTRARY TO SMM
- Thicknesses in stages have been given in lieu of stated dimensions for duct covers (SMM E50.1.3.1).
- Sectional area stages have been given in lieu of stated dimensions for lintels, sills and copings (SMM F31.1.*.*).
- Pier caps have been measured in m2 (SMM F31.1.*.*).

105 PRICES ALSO TO INCLUDE
- All necessary handling reinforcement.
- Bending projecting reinforcement.
- Filling lifting hook recesses with similar material to that used in the units or as indicated and finishing to match surrounding surfaces including formwork and cutting hooks.
- Except as otherwise included in descriptions, filling joints and joint recesses with similar material to that used in the units or as indicated and finishing to match the surrounding surfaces and otherwise bedding, jointing, grouting and pointing in cement mortar and all other materials.
- Leaving surfaces rough as necessary to receive applied finish unless stated otherwise.
- Bedding and jointing in mortar as appropriate unless stated otherwise.
- In the case of fair finish work, for flush pointing as the work proceeds unless stated otherwise.
- Any cutting to lengths required on items measured linear as necessary to produce ends, angles and the like.
- For assembling and jointing units supplied in parts.
- Angles, ends and stoolings to copings and cills.

TYPES OF COMPONENT

110 EXTERNAL PRECAST SILLS
- Concrete grade C25
- Finish to visible faces: As detailed in schedules
- Bedding and jointing mortar: 1:3.

120 INTERNAL PRECAST SILLS
- Concrete:
    - Grade: Not less than C25
    - Finish to visible faces: As schedules

140 CONCEALED PRECAST LINTELS
- Concrete: To BS 5328.
    - Mix: Not less than C25.
    - Aggregate nominal maximum size: 20 mm.
- Reinforcement provision for spans up to 1800 mm: As indicated in the following table:

Clear span	Section	Bearing	Reinforcement
Up to 900 mm	140 mm deep x width of wall	150 mm at both ends	1 no. 12 mm mild steel bar for each 105 mm of wall thickness.
900 to 1800 mm	215 mm deep x width of wall	225 mm at both ends	1 no. 16 mm mild steel bar for each 105 mm of wall thickness.

- Cover to reinforcement (nominal): 20 mm minimum.

GENERAL REQUIREMENTS

220 CONCRETE GENERALLY
- Production of concrete: To BS 5328.
- Chloride ion content of mix constituents: Not to exceed 0.4% of weight of cement.
- Admixtures containing calcium chloride: Not allowed.

230 AGGREGATES
- Standard: To BS 882.
- Maximum drying shrinkage when tested to BS 812-120: 0.075%

262 RECORDS
- Records for each type of component: Maintain details including:
  - Unique identification number.
  - Correlation with records of mixes, including batch numbers.
  - Date of each stage of manufacture.
  - Dates and results of all tests, checks and inspections.
  - Dimensions related to specified levels of accuracy.
  - Specific location in the finished work.
  - Damage and making good.
  - Any other pertinent data, e.g. if unit is a production control unit.
- Availability of records for inspection: On request.

FAIR FACED COMPONENTS

310 CONTROL SAMPLES
- Required samples: After finalization of design, one each of the following components: Sills & Lintels.
- Approval of appearance: Obtain before manufacture of remaining units.
- Identification and storage location: Clearly label and retain at factory for comparison with production units.

320 DETAILS OF SAMPLES
- Submittals after approval of appearance and before manufacture of production units:
  - Aggregates: Confirm type, nominal maximum size, grading and source.
  - Mix properties: Evidence of compliance for compressive strength and free-water/ cement ratio.

330 MIXES FOR VISIBLE FACED COMPONENTS
- Constituent materials and mix design for each finish type: To remain constant.
- Colour and appearance of each finish type: To remain constant.
- Aggregates: To BS 882 and consistent colour, free of deleterious materials that may cause popping and staining.
- Origin: Single source for each finish type, having sufficient quantity for whole contract.

350 QUALITY OF FINISHES
- Appearance standard: As established by samples.

361 DURABILITY FOR EXTERNAL COMPONENTS
- Concrete grade: RC50 or C50.
  - Minimum cement content: 400 kg/m^3 with 20 mm aggregate or 440 kg/m^3 with 10 mm aggregate.
  - Maximum free water/ cement ratio: 0.45.
- Nominal cover to reinforcement on visible faces:
  - Single mix: 30 mm minimum.
  - Different mixes for facing and backing work: 50 mm minimum.
- Nominal cover to reinforcement on internal and bedded faces: 20 mm.

370 COVER ON VISIBLE FACES
- Spacers: Not permitted.
- Proposed method statement: Submit.

380 CONSISTENCY OF PRODUCTION METHODS
- Production methods: To remain consistent for each matching type of finish.
- Finish appearance: To remain within the range of variation indicated by the samples submitted.
- Changes to production methods: If variations are proposed for components of the same finish, submit evidence that there will be no difference in appearance.

390 INSPECTION
- Give notice: When completed components are ready to be inspected at factory.

400 DAMAGED COMPONENTS
- Repairs: Submit proposals.

INSTALLATION

420 LAYING
- Bedding components: On full bed of mortar.
  - Packing: If required use slate.
- Removal of marks, stains and extraneous mortar on visible faces: Rubbing not permitted.

430 SUPPORT OF EXISTING WORK OVER NEW LINTELS
- Joint above lintels: Fully fill and compact with semi-dry mortar.

440 ONE PIECE SILLS/ THRESHOLDS
- Bed joints: Leave clear of mortar except at end bearings and beneath masonry mullions.
- On completion: Point with mortar to match adjacent work.

G: STRUCTURAL/CARCASSING/METAL/TIMBER

G12 ISOLATED STRUCTURAL METAL MEMBERS

100 METHODS OF MEASUREMENT
- No allowance has been made for rolling margin or the weight of weld metal

105 PRICES ALSO TO INCLUDE
- Enumerated bolts are to include one nut and maximum two washers per bolt, any type of nut or washer.
- Surface treatment or galvanizing of steel off-site is to include all necessary preparation and protection of surfaces as required by the specification.
- All treatment of steel sliding surfaces and HSFG bolts.
- Repairing surface treatment or galvanizing.
- Carrying out, before steelwork assembly, treatment as measured in M60 Painting/Clear finishing where surfaces will be inaccessible after assembly.

110 FABRICATION OF MEMBERS
- Steel sections: To BS 4-1, BS EN 10055, BS EN 10056 or BS EN 10210, as appropriate.
  - Steel: To BS EN 10025, grade S275.
  - Surface condition: Free from heavy pitting and rust, burrs, sharp edges and flame cutting dross.
- Cuts and holes: Accurate and neat.
- Welding: Metal arc method to BS EN 1011-2.
  - Welded joints: Fully fused, with mechanical properties not less than those of the parent metal.
  - Site welding: Obtain approval.

150 SHOP PRIMING TO BEAMS
- Preparation: Loose scale and rust, burrs, fins, sharp edges and weld spatter removed; crevices cleaned out; surfaces thoroughly degreased, rinsed with clean water and allowed to dry.
- Primer: Zinc phosphate modified alkyd.
  - Application: One full coat within 8 hours of cleaning surfaces.

310 INSTALLATION
- Accuracy: Members positioned true to line and level using, if necessary, steel packs of sufficient area to allow full transfer of loads to bearing surfaces.
- Fixing: Use washers under bolt heads and nuts.
  - Tapered washers: Provide under bolt heads and nuts bearing on sloping surfaces. Match taper to slope angle and align correctly.

G20 CARPENTRY/ TIMBER FRAMING/ FIRST FIXING

100 DEFINITIONS
- "Flat roof" members. Those which do not exceed 15 degrees from the horizontal.
- "Treated". Treatment of timber by any of the processes specified.

105 METHODS OF MEASUREMENT
- Fixing of items generally, no distinction has been made between fixing to new or existing work.
- In amplification of SMM G20. D1, all sawn and regularized timbers are given as basic sizes.
- All wrought timbers are given as nominal sizes. (The limits on planning margins are to be in accordance with BS EN 1313-1 for softwood and BS EN 1313-2 for hardwood).
- The thickness of all manufactured boards (e.g. plywood and chipboard) given in the description is the finished size.
- In amplification of SMM G20.S2, where fixings are left to the discretion of the Contractor, they are deemed nailed, pinned, spiked or stapled, unless otherwise specified.

110 PRICES ALSO TO INCLUDE
- Work creosoted or treated with wood preservative to include additional treatment to holes and cut surfaces as specified .
- Selecting and keeping hardwood clean for clear finish as applicable.
- Enumerated bolts to include one nut and maximum two washers per bolt, any type of nut or washer.
- Selecting, notching, packing and all labours to take up allowable tolerances in sawn timbers.

GENERAL INFORMATION/ REQUIREMENTS

150 STRENGTH GRADING OF TIMBER
- Grader: Any company currently registered under a third party quality assurance scheme operated by a certification body approved by the UK Timber Grading Committee.
- Grading and marking of timber:
  - Timber of a target/ finished thickness less than 100 mm and not specified for wet exposure: Graded at an average moisture content not exceeding 20% with no reading being in excess of 24% and clearly marked as 'DRY' or 'KD' (kiln dried).
  - Timber graded undried (green) and specified for installation at higher moisture contents: Clearly marked as 'WET' or 'GRN'.
  - Structural timber members cut from large graded sections: Regraded to approval and marked accordingly.

TYPES OF TIMBER

210 GRADED SOFTWOOD FOR CARCASSING
- Grading standard: To BS 4978 or BS EN 519 or other national equivalent and so marked.
- Strength class to BS EN 338:
- Treatment: As clause 350.
- Moisture content at time of erection: As clause 450.

230 GRADED SOFTWOOD FOR CARPENTRY
- Species and origin: As specified in schedules.
- Grading standard: To the appropriate standard or rules for the specified grade and so marked.
  - Grade: As specified in schedules.
- Treatment: As clause 350.
- Moisture content at time of erection: As clause 450.

250 GRADED HARDWOOD FOR CARPENTRY
- Grading standard: To BS 5756 and so marked.
- Strength class to BS EN 338:.
- Surface finish: As specified in schedules.
- Treatment: As clause 350.
- Moisture content at time of erection: As clause 450.

270 UNGRADED SOFTWOOD
- Quality of timber: Free from decay, insect attack (except pinhole borers) and with no knots wider than half the width of the section.
- Surface finish: As specified in schedules.
- Treatment: As clause 350.
- Moisture content at time of erection: As clause 450.

275 WROT TIMBER FOR CARPENTRY
- Species: Various.
- Standard: To BS 1186-3.
  - Class: 2.
- Treatment: As clause 350.
- Moisture content at time of fixing: As clause 450.
- Fixing: As specified in schedules.

© AAS 2008 - 2009

310 PLYWOOD
- Standard: To the relevant national standards and quality control procedures specified in BS 5268-2, and so marked.
  - Type: As specified in schedules.
  - Grade: 2/2.
  - Nominal thickness/ number of plies: As specified in schedules.
  - Finish: As specified in schedules.
- Treatment: As clause 350.

320 TEMPERED HARDBOARD
- Standard: To BS EN 622-2.
  - Type: Various.

350 PRESERVATIVE TREATMENT FOR TIMBER
- Clause Z12

355 PRESERVATIVE TREATMENT CERTIFICATE:
- Provide the Contract Administrator with the certificate accompanying each consignment to site.

WORKMANSHIP GENERALLY

401 CROSS SECTION DIMENSIONS OF STRUCTURAL SOFTWOOD AND HARDWOOD TIMBER
- Dimensions: Dimensions in this specification and shown on drawings are target sizes as defined in BS EN 336.
- Tolerances: The tolerance indicators (T1) and (T2) specify the maximums permitted deviations from target sizes as stated in BS EN 336, clause 5.3:
  - Tolerance class 1 (T1) for sawn surfaces.
  - Tolerance class 2 (T2) for further processed surfaces.

402 CROSS SECTION DIMENSIONS OF NONSTRUCTURAL SOFTWOOD TIMBER
- Dimensions: Dimensions in this specification and shown on drawings are finished sizes.
- Maximum permitted deviations from finished sizes: As stated in BS EN 1313-1:
  - Clause 6 for sawn sections.
  - Clause NA.2 for further processed sections.

403 CROSS SECTION DIMENSIONS OF NONSTRUCTURAL HARDWOOD TIMBER
- Dimensions: Dimensions in this specification and shown on drawings are finished sizes.
- Maximum permitted deviations from finished sizes: As stated in BS EN 1313-2:
  - Clause 6 for sawn sections.
  - Clause NA.3 for further processed sections.

420 WARPING OF TIMBER
- Bow, spring, twist and cup: Not greater than the limits set down in BS 4978 or BS EN 519 for softwood, or BS 5756 for hardwood.

430 SELECTION AND USE OF TIMBER
- Timber members damaged, crushed or split beyond the limits permitted by their grading: Do not use.
- Notches and holes: Position in relation to knots or other defects such that the strength of members will not be reduced.
- Scarf joints, finger joints and splice plates: Do not use without approval.

440 PROCESSING TREATED TIMBER
- Cutting and machining: As much as possible before treatment.
- Extensively processed timber: Retreat timber sawn lengthways, thicknessed, planed, ploughed, etc.
- Surfaces exposed by minor cutting and/ or drilling: Treat with two flood coats of a solution recommended by main treatment solution manufacturer.

450 MOISTURE CONTENT
- Moisture content of timber and wood based products at time of installation: Not more than:
  - Covered in generally unheated spaces: 24%.
  - Covered in generally heated spaces: 20%.
  - Internal in continuously heated spaces: 20%.

451 MOISTURE CONTENT TESTING
- Procedure: When instructed, test timber sections with an approved electrical moisture meter.
- Test sample: Test 5% but not less than 10 lengths of each cross-section in the centre of the length.
- Test results: 90% of values obtained to be within the specified range. Provide records of all tests.

510 PROTECTION
- Generally: Keep timber dry and do not overstress, distort or disfigure sections or components during transit, storage, lifting, erection or fixing.
- Timber and components: Store under cover, clear of the ground and with good ventilation. Support on regularly spaced, level bearers on a dry, firm base. Open pile to ensure free movement of air through the stack.
- Trussed rafters: Keep vertical during handling and storage.

530 PAINTED FINISHES
- Structural timber to be painted: Primed as specified before delivery to site.

540 CLEAR FINISHES
- Structural timber to be clear finished: Keep clean and apply first coat of specified finish before delivery to site.

550 EXPOSED TIMBER
- Planed structural timber exposed to view in completed work: Prevent damage to and marking of surfaces and arrises.

JOINTING TIMBER

570 JOINTING/ FIXING GENERALLY
- Generally: Where not specified precisely, select methods of jointing and fixing and types, sizes and spacings of fasteners in compliance with section Z20.

590 METAL PLATE FASTENERS/ GUSSETS
- Manufacturer: Contractors choice.

600 BLACK BOLTS AND NUTS
- Standard: To BS 4190.
- Finish (applied by manufacturer): as clause 670.

620 WASHERS
- Standard: Plain to BS 4320, spring to BS 4464.
- Material and finish: To match bolts.
- Dimensions when seated directly on timber surfaces: Unless specified otherwise:
  - Diameter/ side length: Not less than 3 times bolt diameter.
  - Thickness: Not less than 0.25 times bolt diameter.

630 BOLTED JOINTS
- Holes for bolts: Located accurately and drilled to diameters as close as practical to the nominal bolt diameter and not more than 2 mm larger.
- Washers: Placed under bolt heads and nuts that would otherwise bear directly on timber. Use spring washers in locations which will be hidden or inaccessible in the completed building.
- Bolt tightening: So that washers just bite the surface of the timber. Ensure that at least one complete thread protrudes from the nut.
  - Checking: At agreed regular intervals up to Practical Completion. Tighten as necessary.

640 BOLTED JOINTS WITH CONNECTORS
- Connectors: To BS EN 912.
  - Types and sizes: As shown on drawings.
  - Manufacturer: Contractors choice.
  - Finish: as clause 670.
- Centres of bolt holes: Not more than 2 mm from specified positions.
- Assembly: Do not crush timber, deform washers or overstress bolts.

670 ANTICORROSION FINISHES FOR FASTENERS
- Galvanizing: To BS 7371-6, with internal threads tapped and lightly oiled following treatment.
- Sherardizing: To BS 7371-8, Class 1.
- Zinc plating: To BS EN ISO 4042 and passivated.
- Treatment not specified: Select from the above to suit service conditions.

ERECTION AND INSTALLATION

710 PROPOSALS FOR ERECTING STRUCTURAL TIMBER
- Proposals: Submit details of:
  - Method and sequence of erection.
  - Type of craneage.
  - Temporary guys and bracing proposed for use during erection.
- Latest date for submission: As detailed in contract particulars.

740 PRE-ERECTION CHECKING
- Timing: Not less than 10 days before proposed erection start date.
- Checklist:
  - Foundations and other structures to which timber structure will be attached: Check for accuracy of setting out.
  - Holding down bolts: Check for position, protruding length, condition and slackness.
- Inaccuracies and defects: Report without delay.
- Erection: Obtain permission to commence.

750 MODIFICATIONS/ REPAIRS
- Defects due to detailing or fabrication errors: Report without delay.
- Methods of rectification: Obtain approval of proposals before starting modification or remedial work.
- Defective/ damaged components: Timber members/ components may be rejected if the nature and/ or number of defects would result in an excessive amount of site repair.

760 TEMPORARY BRACING
- Provision: As necessary to maintain structural timber components in position and to ensure complete stability during construction.

770 ADDITIONAL SUPPORTS
- Provision: Where not shown on drawings, position and fix additional studs, noggings or battens for appliances, fixtures, edges of sheets, etc.
- Material properties: Additional studs, noggings and battens to be of adequate size and have the same treatment, if any, as adjacent timber supports.

775 BEARINGS
- Timber surfaces which are to transmit loads: Finished to ensure close contact over the whole of the designed bearing area.
- Packings: Where provided, to cover the whole of the designed bearing area.
  - Crushing strength: Not less than timber being supported.
  - In external locations: Rot and corrosion proof.

780 WALL PLATES
- Position and alignment: To give the correct span and level for trusses, joists, etc.
- Bedding: Fully in fresh mortar.
- Joints: At corners and elsewhere where joints are unavoidable use nailed half lap joints. Do not use short lengths of timber.

784 **JOISTS GENERALLY**
- Centres: Equal, and not exceeding designed spacing.
- Bowed joists: Installed with positive camber.
- End joists: Positioned approximately 50 mm from masonry walls.

786 **JOISTS ON HANGERS**
- Hangers: Bedded directly on and hard against supporting construction. Do not use packs or bed on mortar.
- Joists: Cut to leave not more than 6 mm gap between ends of joists and back of hanger. Rebated to lie flush with underside of hangers.
- Fixing to hangers: A nail in every hole.

790 **JOIST HANGERS**
- Standard: To BS 6178-1.
- Size and type: To suit joist, design load and crushing strength of supporting construction.
- Material/ finish: Galvanized steel.

795 **TRIMMING OPENINGS**
- Trimmers and trimming joists: When not specified otherwise, not less than 25 mm wider than general joists.

800 **TRUSSED RAFTERS**
- Erection: To BS 5268-3, clause 9.3 and TRA site installation guide.
- Trusses generally: Do not modify without approval.
- Damaged trusses: Do not use.
- Fixing: Truss clips and bottom chords of standard trusses and rafters of raised tie trusses bearing fully on wall plates.
- Bottom chords of standard trusses: Do not fix to internal walls until roofing is complete and cisterns are installed and filled.

805 **TRUSS CLIPS**
- Manufacturer: Contractors choice.
- Material/ finish: Galvanized steel.
- Fasteners: 32 x 3.5 mm galvanized or sherardized square twisted nails in every hole.

810 **PERMANENT BRACING OF TRUSSED RAFTERS**
- Bracing and binders: Fixed to every rafter, strut or tie with not less than two 75 x 3.35 mm galvanized round wire nails.
- Lap joints: Extended over and nailed to at least two truss members.

820 **VERTICAL RESTRAINT STRAPS**
- Manufacturer: Contractors choice.
- Material/ finish: Galvanized steel.
- Size:
  - Cross section: Not less than 30 x 5 mm.
  - Length: Various.
- Centres: Not more than 1 m.
- Fixing:
  - To timber members with not less than two 30 x 3.5 galvanized square twist nails.
  - To masonry with not less than four 4.8 x 70mm masonry screws evenly spaced.
  - With at least one screw located within 150 mm of the bottom end of each strap.

830 LATERAL RESTRAINT STRAPS
- Manufacturer: Contractors choice.
- Material/ finish: Galvanized steel.
- Size: Not less than 30 x 5 mm cross section, 150 mm cranked end and 750 mm long.
- Fixing: To top of joists/ rafters/ ties at not more than 1 m centres or as detailed
- Ensure that cranked end is in tight contact with cavity face of wall inner leaf and is not pointing upwards.
- Straps spanning joists/ rafters/ ties running parallel to wall: Fix noggings and packs tightly beneath straps.
  - Size of noggings and packs: Not less than three quarters of joist/ rafters/ tie depth and not less than 38 mm thick.
  - Notching: Notch joists so that straps fit flush with surface. Do not notch rafters/ ties.
- Fasteners: Not less than four 50 mm x 8 gauge sherardized countersunk screws per strap, evenly spread.

840 STRUTTING
- Type: One of the following:
  - Herringbone strutting: At least 38 x 38 mm softwood.
  - Solid strutting: At least 38 mm thick softwood and at least three quarters of joist depth.
  - Fixing: Between joists as follows:
  - Joist spans of 2.5 to 4.5 m: One row at centre span.
  - Joist spans over 4.5 m: Two rows equally spaced.
  - Strutting must not project beyond top and bottom edges of joists.
- Outer joists: Blocked solidly to perimeter walls.

850 INSPECTION
- Structural timberwork: Give reasonable notice before covering up.

860 BOLTED JOINTS
- Inspection: Inspect all accessible bolts at the end of the Defects Liability Period and tighten if necessary.

900 EAVES SOFFIT VENTILATORS
- Manufacturer: As specified in schedules.
  - Type: As specified in schedules.
  - Colour: To suit location.

G32 EDGE REINFORCED WOOD WOOL SLAB DECKING

100 CONTRARY TO SMM
- No distinction has been made between holes made on or off site (SMM G32.3.1.*).

120 WOOD WOOL SLAB DECKING:
- Edge reinforced slabs to BS 1105, type SB.
  Manufacturer and reference: Torvale Building products
  Thickness: 50 mm
  Finish: pre-screeded
  Accessories: Channel reinforced
- Roof supports: As specified by manufacturer
- Method of fixing: As specified by manufacturer

210 STRUCTURE:
- Check that structure is in a suitable state to receive decking before commencing fixing.
- Laying and fixing of slabs will be taken as acceptance by the decking contractor of suitability of the structure.

220 WORKMANSHIP:
- Cut slabs accurately and fix with joints tightly butted and centered on supports.
- Ensure that ends and cut edges are fully supported or reinforced to slab manufacturer's recommendations.
- Protect decking during construction with adequate walkways, crawlboards and all necessary additional supports. Do not apply point or shock loads, or run barrows, etc. on unprotected or unsupported slabs. Do not stack materials on the completed roof deck.

250 SAFE LOAD NOTICE(S): Obtain from the manufacturer the safe working load for each type of decking specified. For each of the locations notified, provide and fix a warning notice displaying the safe working load for the decking in that location.

H: CLADDING/COVERING

General

100 METHODS OF MEASUREMENT
- Not withstanding SMM Clause H20.D1 and H21.D1 all work described as to "roofs" and "dormers" and all "weather boarding is deemed external
- In amplification of SMM H20.D2 and H21.D2 all sawn and regularized timbers are given in basic sizes.
- All wrought timbers are given as nominal sizes. The limits on planing margins are to be in accordance with BS EN 1313-1 for softwood and BS EN 1313-2 for hardwood.
- The thickness of all manufactured boards given in the description is the finished size.
- The amplification of SMM H20.S12 and H21.S12 where fixings are left to the discretion of the Contractor they are deemed nailed, pinned, spiked or stapled unless otherwise stated.
- In amplification of H.***, valleys are measured linear and include for cutting roof coverings both sides.
- Contrary to SMM roofing has not been given a stated pitch but described as flat (SMM J21.3.*.*, J40.1-2* and J41.1-2*)
- Surface finishes have been measured as "extra over" roofing (SMM J21.S4)

105 PRICES ALSO TO INCLUDE
- Work creosoted or treated with wood preservative is to include additional treatment to holes and cut surfaces as specified.
- Manufactured boards and sheeting are to include any necessary balancing veneer when only one face veneer is specified.
- Selecting and keeping hardwood clean for clear finish as applicable
- Punching fixings below exposed surfaces.
- Dressing all joinery in contact with plaster with linseed oil before plastering and subsequently cleaning off.
- Tongueing, crosstongueing, housing and like labours are to include grooves in building board, timber and faced timber.
- Timber flooring and boarding is to include splayed heading joints.
- Holes through existing work are to include holes through any associated membranes, underlays or vapour barriers.
- Accessories to corrugated roofing are to include extra fixings.
- Sheet metal flashings and gutters are to include notching and bending.
- Coverings and flashings etc. generally to include all wedges, clips and tacks.
- Lining to sumps and similar enumerated items to include all dressing over fillets and similar labours but excluding cutting and fitting to outlets.
- All dressing to fillets and similar labours include burned, brazed, welded or soldered angles and seams but excluding holes and jointing to outlets.
- Collars and spigot outlets to include all bends, splay cutting to ends and forming holes in associated new sheet metal roofing for pipe or members passing through.
- Holes through coverings are to include holes and other perforations through any isolating membranes, underlays or vapour barriers included in the description of the covering.

H20 RIGID SHEET CLADDING

TYPES OF SHEET CLADDING

150 SHEET CLADDING
- Board/ Sheet: as specified on schedules
- Breather membrane: As clause 250.

GENERAL REQUIREMENTS

230 TIMBER BATTENS
- General: Regularized softwood free from decay, insect attack (except ambrosia beetle damage) and with no knots wider than half the width of the section.
- Preservative treatment: CCA as section Z12 and British Wood Preserving and Damp-proofing Association Commodity Specification C8.

© AAS 2008- 2009

- Moisture content at time of fixing (maximum): 19%.

240 TREATED TIMBER
- Exposed cut and drilled surfaces: Treat with two flood coats of a solution recommended for the purpose by main treatment solution manufacturer.

250 BREATHER MEMBRANE
- Manufacturer: Contractors choice.
- Continuity: No breaks. Minimize joints.
  - Penetrations and abutments: Attach to breather membrane with tape. Achieve full bond.
  - Laps: Not less than 150 mm, bond with tape. Achieve full bond.
- Tape: As recommended by breather membrane manufacturer.
- Repairs: Lapped patch of breather membrane material secured with continuous band of tape on all edges.
- Junctions at flashings, sills, gutters etc: Overlap and allow free drainage to exterior.

260 FIXING SHEETS
- General: Secure to supports without producing distortion.
- Fasteners: Evenly spaced in straight lines, in pairs across joints and sufficient distance from edge of sheet to prevent damage.

270 COVER STRIPS
- General: Form straight runs in single lengths wherever possible.
- Location and method of forming joints: Submit proposals where not detailed.

H21 TIMBER WEATHERBOARDING

110 TIMBER WEATHERBOARDING
- Breather membrane: As clause 130.
- Counterbattens: As instructed.
- Battens:
  - Size: 25 x 50mm.
  - Centres: 400mm.
  - Fixing: Contractors discretion.
- Boarding:
  - Quality of timber (exposed surfaces): To BS 1186-3, Section 4, Class: 1.
  - Species: Contractors choice.
  - Finished face dimension (overall width): 19mm.
  - Finished thickness: as schedules.
  - Treatment: As Z12.
  - Standard: To NBS section Z12 and British Wood Preserving and Damp-proofing

120 CONTROL SAMPLE
- General: Complete an area of boarding in an approved location and obtain approval of appearance before proceeding.

130 BREATHER MEMBRANE
- Manufacturer: Contractors choice.
- Installation: Fix carefully and neatly to provide a complete barrier to water, snow and wind blown dust. Extend membrane below lowest timber member and into reveals of openings.
  - Laps: Horizontal: 100 mm. Vertical: 150 mm and staggered, to shed water away from substrate.
  - Fasteners: Galvanized, sherardized or stainless steel large head nails or stainless steel staples.

135 TIMBER BATTENS/ COUNTERBATTENS
- Quality of timber: Regularized softwood free from decay, insect attack (except ambrosia beetle damage) and with no knots wider than half the section width.
- Treatment: Preservative impregnation.
  - Standard: To NBS section Z12 and British Wood Preserving and Damp-proofing Association Commodity Specification C8.
  - Type: as instructed.
- Moisture content: Not exceeding 20% at time of fixing.

140 FIXING BATTENS/ COUNTERBATTENS TO MASONRY
- Setting out: In straight, vertical lines.
- Batten/ Counterbatten length (minimum): 1200 mm.
- Installation: Fastener heads to finish flush with or slightly below batten face.

141 FIXING BATTENS/ COUNTERBATTENS TO FRAMING/ SHEATHING
- Setting out: In straight, vertical lines at centres coincident with vertical framing members.
- Batten/ Counterbatten length (minimum): 1200 mm.
- Installation: Where sheathing is provided, fix through sheathing into framing. Fastener heads to finish flush with or slightly below batten face.

142 FIXING BATTENS TO COUNTERBATTENS
- Setting out: In straight, horizontal lines. Align on adjacent areas.
- Batten/ Counterbatten length (minimum): 1200 mm.
- Joints: Square cut, butted centrally on counterbattens and not occurring more than once in any group of four battens on any one counterbatten.
- Installation: Fix each batten to each counterbatten. Use splay fixings at joints. Fastener heads to finish flush with or slightly below batten face.

145 TREATED TIMBER
- Surfaces exposed by minor cutting and/ or drilling: Treat with two flood coats of a solution recommended for the purpose by main treatment solution manufacturer.

150 SURFACE TREATMENT
- Finishing system: Before fixing boards, apply first coat of specified system to all surfaces. Apply liberally to end grain.

160 FIXING BOARDING
- General: Fix boards securely to give flat, true surfaces free from undulations, lipping, splits, hammer marks and protruding fasteners.
- Movement: Allow for movement of boards and fixings to prevent cupping, springing, excessive opening of joints or other defects.
- Heading joints: Position centrally over supports and at least two board widths apart on any one support.
- Nail heads: Punch below surfaces that will be seen in the completed work.

H30 FIBRE CEMENT PROFILED SHEET CLADDING/ COVERING

TYPES OF CONSTRUCTION

120 FIBRE CEMENT CLADDING
- External sheets: To BS EN 494, Type NT (nonasbestos).
  - Manufacturer: Eternit
    Product reference: 2000
- Fittings: As clause 223.
- Primary cladding sheet fasteners: 105mm topfix
  - Fix through Crown of profile
  - Number of fasteners per sheet width:
    Eaves and end laps: 2
    Intermediate supports: 2
- End laps size (minimum): 105mm.
  - Sealing laps: 8mm dia butyl strip
- Side laps size: 70mm.
  - Sealing laps: Unsealed

FIXING CLADDING

215 PAINTING STRUCTURE
- Sequence: Paint outer surface of supporting structure if required before fixing cladding.

219 FASTENERS
- Non specified fasteners: Recommended for the purpose by the cladding manufacturer.

221 FITTINGS AND ACCESSORIES
- Non specified fittings and accessories: Recommended for the purpose by the cladding manufacturer.

223 FIBRE CEMENT FITTINGS:
- Standard: To BS EN 494, Type NT (nonasbestos).
- Manufacturer: Eternit
  - Product types/ reference: As specified in schedules
- Finish/ Colour: to match cladding
- Fixing method:
  - To structure: In conjunction with adjacent sheeting. Fasteners as sheeting.
  - To sheeting: As manufacturers recommendations

360 PROFILE FILLERS GENERALLY
- Manufacturer: Eternit.
  - Product references: Eaves filler piece.
- Colour: As cladding
- Fixing method: As manufacturers instructions.
  - Requirement: To close cavities/ regulate air paths within the external envelope. Tight fit with no unintended gaps.

410 FIXING SHEETS GENERALLY
- Cut sheets: To give clean true lines.
- Sheet orientation: Lay with smooth surface uppermost and with exposed joints of side laps away from prevailing wind unless shown otherwise on drawings.
- Verge sheets: Terminate outside edge with crown.
- Sheet ends, laps and raking cut edges: Fully supported and with fixings at top of lap.
- Fasteners: Drill holes. Position at regular intervals in straight lines, centred on support bearings.
  - Hook/ crook bolt and non wing type fasteners: Position centrally within oversized drilled holes.
  - Fasteners torque: Sufficient to correctly compress washer.
- Debris: Remove dust and other foreign matter before fixing sheets.
- Completion: Check fixings to ensure watertightness and that sheets are secure.

415 FIXING FIBRE CEMENT SHEETS
- Primary fixings: Holes to be 2 mm larger than diameter of fastener (unless using self-drilling type with wing device to create suitable clearance) and located not less than 40 mm from edges of sheets and fittings.
- Mitre joints: Open butt, 3 - 6 mm wide.
  - Box/ check joints: Not permitted.

460 THERMAL MOVEMENT JOINTS
- Type: Proprietary fibre cement movement joint fittings.
- Sheet location: Middle trough.
- Width of gap: 25 mm.
- Requirement: Weathertight.

470 STRUCTURAL MOVEMENT JOINTS
- Type: Proprietary fibre cement movement joint fittings.
- Location: Coincident with structural movement joint.
- Width of gap: To match structural movement requirement.
- Requirement: Weathertight.

540 ABUTMENTS
- Junctions with flashings: Weathertight and neatly dressed down.

550 SEALING LAPS ON EXTERNAL SHEETS
- Sealant tape: Types recommended by sheet manufacturer.
- Position of tape: Below fixing positions in straight unbroken lines, parallel to and slightly back from edge of sheet.
- Seal quality: Effective, continuous and not overcompressed.
- Mitre junctions: Where sealant tape passes through mitre, fill joints at crowns. Dress tails of sealant down mitre; do not leave projecting to edge of sheet.
- End laps: Sealant tape positions:
  - Single line tape: Immediately below line of fasteners.
  - Second line tape (where specified): Slightly set back from edge of external sheet.

560 WARNING NOTICES
- Fixing locations of signs: As indicated by CA.
- Manufacturer: Contractor's choice.
- Signs description: As BS 5378-3 reference A.2.4 with supplementary text sign, lettering 'DANGER Fragile roof'.
Mandatory sign as BS 5378-3 reference A.3.1 with supplementary text sign, lettering 'Use crawling boards'.

H31 METAL PROFILED/ FLAT SHEET CLADDING/ COVERING

TYPES OF CLADDING/ COVERING SYSTEM

110 METAL CLADDING
- External sheets: Plastic coated galvanised steel corrugated sheeting.
  - Thickness (nominal): 0.7mm.
  - Profile: R60.
- Sealing laps:
  - End laps: 150mm
  - Side laps: one and a half corrugations.
- Spacers: to manufacturer's recommendations.
  - Depth of spacer: To be such that insulation is not compressed between external and lining sheets.
  - Fasteners: as manufacturers recommendations.
- Breather membrane: As clause 280.

GENERAL REQUIREMENTS

170 DESIGN
- Cladding/ covering system: Complete detailed design and submit before commencement of fabrication.
  - Standard: To BS 5427-1.
- Related works: Coordinate in detailed design.

172 THERMAL PERFORMANCE/ BRIDGING
- Requirement: Complete thermal design of the cladding/ covering system to avoid excessive thermal bridging.
  - Standard: MCRMA Technical Paper No 14.

FIXING CLADDING/ COVERING

215 PAINTING STRUCTURE
- Sequence: Paint outer surface of supporting structure if required before fixing cladding/ covering.

219 FASTENERS
- Non specified fasteners: Recommended for the purpose by the cladding/ covering manufacturer.

221   FITTINGS AND ACCESSORIES
- Non specified fittings and accessories: Recommended for the purpose by the cladding/ covering manufacturer.

410   FIXING SHEETS GENERALLY
- Cut sheets and flashings: To give clean true lines, with no burrs.
- Penetrations: Cut openings to minimum size necessary.
  - Edge reinforcement: to CA approval.
- Sheet orientation: Lay with exposed joints of side laps away from prevailing wind unless shown otherwise on drawings.
- Sheet ends, laps and raking cut edges: Fully supported and with fixings at top of lap.
- Fasteners: Drill holes. Position at regular intervals in straight lines, centred on support bearings.
  - Oversized drilled holes: Position fasteners centrally.
  - Fasteners torque: Sufficient to correctly compress washers.
- Debris: Remove dust and other foreign matter before finally fixing sheets.
- Completion: Check fixings to ensure watertightness and that sheets are secure.
- Cut edges: Paint to match face finish.

470   STRUCTURAL MOVEMENT JOINTS
- Type: Cover flashing fixed on one side over gap between sheets.
- Location: Coincident with structural movement joint.
- Width of gap: To match structural movement joint requirements.
- Requirement: Weathertight.

480   FLASHINGS/ TRIMS GENERALLY
- Lap joint treatment:
  - Vertical and sloping flashings/ trims: End laps to be same as for adjacent sheeting.
  - Horizontal flashings/ trims: End laps to be 150 mm, sealed and where possible arranged with laps away from prevailing wind.
- Method of fixing: To structure in conjunction with adjacent sheeting. Otherwise to sheeting.
  - Fasteners: to CA approval.

540   ABUTMENTS
- Junctions with flashings: Weathertight and neatly dressed down.

550   SEALING LAPS ON EXTERNAL SHEETS
- Sealant tape: Types recommended by sheet manufacturer.
  - Position: Below fixing positions in straight unbroken lines, parallel to and slightly back from edge of sheet.
- Seal quality: Effective, continuous and not overcompressed.
- End laps: Sealant tape positions:
  - Single line tape: Immediately below line of fasteners.
  - Second line tape (where specified): Slightly set back from edge of external sheet.
- Side laps: Sealant tape positions:
  - Single line tape: Outside line of fasteners.
  - Second line tape (where specified): On other side of fasteners.

554   WATER VAPOUR SEALING AT LAPS AND PENETRATIONS IN METAL LININGS
- Sealant tape: Contractor's choice.
  - Position : Below fixing positions in straight unbroken lines, parallel to and slightly back from edge of sheet.
- Seal quality: Effective, continuous and not overcompressed.

560   SAFETY SIGNS
- Fixing locations of signs: As indicated by CA.
- Manufacturer: Contractor's choice.
- Signs description:
Hazard sign as BS 5499-5, reference number 8.C.0072 with supplementary text sign to BS 5499-1, wording 'Danger fragile roof'.
Mandatory sign to BS 5499-5, registration number BS 5499-5.0103 with supplementary text sign to BS 5499-1, wording 'Use crawling boards'.

H41 GLASSFIBRE REINFORCED PLASTICS PANEL CLADDING/ FEATURES

TYPES OF CLADDING/ FEATURES

110 GRP SINGLE SKIN CLADDING
- Panels: Glass fibre reinforced corrugated translucent sheeting to BS4154
  - Profiles: 3 and 6
- Sealing laps:
  - End laps: 150mm
  - Side laps: one and a half corrugations.
- Spacers: to manufacturer's recommendations.
  - Depth of spacer: To be such that insulation is not compressed between external and lining sheets.
  - Fasteners: as manufacturers recommendations.
- Fixings and fasteners: As clause 480.

INSTALLATION

605 PREPARATION
- Prefabrication: Complete products and attach fixings in workshop wherever possible.
- Identification: Mark or tag products. Do not mark surfaces visible in the complete installation.
- Electrolytic corrosion: Isolate dissimilar metals.

615 SUITABILITY OF STRUCTURE
- Contractor's survey:
  - Scope: Geometric survey of supporting structure, checking line, level and fixing points.
  - Coordinate: With surveys for adjacent cladding.
  - Give notice: If structure will not allow required accuracy or security of erection.
- Setting out: Establish erection datum points, lines and levels for a complete elevation at a time unless otherwise agreed.

625 INSTALLATION OF INTERFACES
- General: Locate flashings, closers etc. correctly with neat overlaps to form weathertight junctions.

640 FINAL FIXING
- Torque figures: Tighten threaded fastenings to figures recommended by manufacturer. Do not overtighten restraint fixings intended to permit lateral movement.

645 METALWORK
- Material standards and fabrication: As section Z11.

660 OPEN DRAINED JOINTS
- Rear seals: Airtight.
- Baffles and flashings: Installed securely and accurately to function as intended.

680 CLEANING DOWN
- General: Return to site at Practical Completion or when instructed and thoroughly clean down the entire area of the GRP work. Cleaning agents for the purpose must be approved by the GRP manufacturer.

H60 PLAIN ROOF TILING

TYPES OF TILING

105 CLAY ROOF TILING
- Underlay: reinforced felt to BS747 type 1F.
  - Laying: As clause 240, parallel to eaves.
- Battens: As clause 245.
  - Size: 19 x 38mm.
  - Fixing: As clause 265.

- Tiles: To BS EN 1304.
  - Manufacturer: to CA approval.
  - Headlap (minimum): 65 mm.
  - Fixing: As clauses 275 and 280:

120 CONCRETE ROOF TILING
- Underlay: reinforced felt to BS747 type 1F.
  - Laying: As clause 240, parallel to eaves.
- Battens: As clause 245.
  - Size: 19 x 38mm.
  - Fixing: As clause 265
- Tiles: To BS EN 490.
  - Manufacturer: to CA approval.
  - Headlap (minimum): 65 mm.
  - Fixing: As clauses 275 and 280.

145 CLAY VERTICAL TILING
- Underlay: reinforced felt to BS747 type 1F.
  - Fixing: As clause 240, parallel to bottom edge.
- Battens: As clause 245.
  - Size: 19 x 38mm.
  - Fixing: As clause 245.
- Tiles: To BS EN 1304.
  - Manufacturer: to CA approval.
  - Headlap (minimum): 65mm.
  - Fixing: As clause 275, two nails per tile in every course.

155 CONCRETE VERTICAL TILING
- Underlay: reinforced felt to BS747 type 1F.
  - Fixing: As clause 240, parallel to bottom edge.
  - Horizontal lap (minimum): 65mm.
- Battens: As clause 245.
  - Size: 19 x 38mm.
  - Fixing: As clause 265.
- Tiles: To BS EN 490, noninterlocking.
  - Manufacturer: to CA approval.
  - Fixing: As clause 275, two nails per tile in every course.

TILING GENERALLY

210 BASIC WORKMANSHIP
- General: Fix tiling and accessories to make the whole sound and weathertight at earliest opportunity.
- Setting out: To true lines and regular appearance, with neat fit at edges, junctions and features.
- Fixings for tiling accessories: As recommended by tile or accessory manufacturer.
- Gutters and pipes: Keep free of debris. Clean out at completion.

220 REMOVING EXISTING TILING
- General: Carefully remove tiles, battens, underlay, etc. with minimum disturbance of adjacent retained tiling.
- Undamaged tiles: Set aside for reuse.

230 MINERAL WOOL INSULATION ON STRUCTURAL METAL LINER TRAYS
- Manufacturer: to CA approval.
- Installation: Fix securely with closely butted joints, leaving no gaps.
  - Fasteners: Use where necessary to prevent slumping.
- Ventilation paths: Do not obstruct.

240 UNDERLAY
- Handling: Do not tear or puncture.
- Laying: Maintain consistent tautness.
- Vertical laps (minimum): 100 mm wide, coinciding with supports and securely fixed.
- Fixing: Galvanized steel, copper or aluminium 20 x 3 mm extra large clout head nails.
- Eaves: Where exposed, underlay must be BS 747, type 5U, or equivalent UV durable type.
- Penetrations: Use proprietary underlay seals or cut underlay to give a tight, watershedding fit around pipes and components.
- Ventilation paths: Do not obstruct.

245 BATTENS/ COUNTERBATTENS
- Timber: Sawn softwood.
  - Standard: BS 5534, clause 4.12.1.
  - Permissible characteristics and defects: Not to exceed limits in BS 5534, annex C.
  - Moisture content at time of fixing and covering (maximum): 22%.
- Preservative treatment: As section Z12 and British Wood Preserving and Damp-proofing Association Commodity Specification C8.

255 COUNTERBATTENS ON RIGID SARKING
- Fixing: Through rigid sarking into rafters at not more than 300 mm centres.

259 COUNTERBATTENS ON RAFTERS
- Fixing: Into rafters at not more than 300 mm centres.

265 BATTEN FIXING
- Setting out: In straight horizontal lines. Align on adjacent areas.
- Batten length (minimum): Sufficient to span over three supports.
- Joints in length: Square cut. Butt centrally on supports. Joints must not occur more than once in any group of four battens on one support.
- Additional battens: Provide where unsupported laps in underlay occur between battens.
- Fixing: Each batten to each support. Splay fix at joints in length.

270 BATTENS FIXED TO MASONRY
- Setting out: In straight horizontal lines. Align on adjacent areas.
- Batten length (minimum): 3 m.
- Fixing centres (maximum): 400 mm.

272 TIMBER FOR TILING SUBSTRATE WORK
- Timber: Sawn softwood, free from wane, pitch pockets, decay and insect attack (ambrosia beetle excepted).
  - Moisture content at time of fixing and covering (maximum): 22%.
- Preservative treatment: As section Z12 and British Wood Preserving and Damp-proofing Association Commodity Specification C8.

275 TILE FIXING
- Setting out: Lay tiles to a half lap bond with joints slightly open. Align tails.
- Ends of courses: Use tile and a half tiles to maintain bond and to ensure that cut tiles are as large as possible.
- Top and bottom courses: Use eaves/ tops tiles to maintain gauge.
- Perimeter tiles:
  - Verges, abutments and each side of valleys and hips: Twice nail end tile in every course.
  - Eaves and top edges: Twice nail two courses of tiles or clip as appropriate.
- Fixings for tiles: Nails/ clips recommended by tile manufacturer.

290 MORTAR BEDDING/ POINTING
- Mortar: As section Z21, 1:3 cement:sand, with plasticizing admixtures permitted.
  - Bond strength providing resistance to uplift: To BS 5534.
- Weather: Do not use in wet or frosty conditions or when imminent.
- Preparation of tiles and accessories to be bedded: Wet and drain surface water before fixing.
- Appearance: Finish neatly as work proceeds and remove residue.

ROOF TILING EDGES/ JUNCTIONS/ FEATURES

305 GENERALLY
- Fittings and accessories: As recommended by tile manufacturer, do not improvise.
  - Exposed fittings and accessories: To match tile colour and finish.
- Cut tiles: Cut only where necessary, to give straight, clean edges.
- Flashings: Fix with or immediately after tiling. Form neatly.

325 FIRE SEPARATING WALLS
- Separating walls: Completely fill space between top of wall and underside of tiles with mineral wool quilt to provide fire stopping.
- Boxed eaves: Completely seal air paths in plane of separating wall with wire reinforced mineral wool, not less than 50 mm thick, fixed to rafters and carefully cut to shape to provide fire stopping.

365 EAVES
- Underlay support: Contractor's choice.
  - Continuous to prevent water retaining troughs.
- Gutter: Dress underlay or underlay support tray to form drip into gutter.
- Undercourse and first course tiles: Fix with tails projecting 50 mm over gutter or to centre of gutter, whichever dimension is the lesser.

445 MORTAR BEDDED VERGES WITH BEDDED UNDERCLOAK
- Underlay: Carry 50 mm onto outer leaf of gable wall and bed on mortar.
- Undercloak: Matching plain tiles.
  - Position: Over underlay, level with underside of tiling battens, sloping towards verge.
  - Projection beyond face of wall: 38-50 mm.
  - Bedding: On mortar identical to that used in gable walling.
- Tiling battens: Carry onto undercloak and finish 100 mm from verge edge.
- Verge tiles:
  - Bedding: Flush with undercloak on 75 mm wide bed of mortar as clause 290.
  - Fixing: Nails. Do not displace or crack mortar.

455 MORTAR BEDDED VERGES WITH NAILED UNDERCLOAK
- Underlay: Carry over full width of verge.
- Undercloak: Fibre cement sheet.
  - Position: Over underlay, level with underside of tiling battens, sloping towards verge.
  - Projection: 38-50 mm beyond face of building.
  - Fixing: Nails.
- Tiling battens: Carry onto undercloak and finish 100 mm from verge edge.
- Verge tiles:
  - Bedding: Flush with undercloak on 75 mm wide bed of mortar as clause 290.
  - Fixing: Nails. Do not displace or crack mortar.

505 DRY HIPS
- Underlay: Lay courses over hip.
  - Overlaps (minimum): 150 mm.
- Dry hip fixing battens: as required.
- Roof tiles: Cut and fix closely at hip.
- Dry hip tiles:

525 MITRED HIPS
- Underlay: Lay courses over hip.
  - Overlaps (minimum): 150 mm.
- Mitred tiles: Cut tile and a half tiles and fix to form a straight, close mitred junction.
- Soakers: Interleave with mitred tiles. Fix by turning down over head of mitred tiles.

555 MORTAR BEDDED HIPS
- Underlay: Lay courses over hip.
  - Overlaps (minimum): 150 mm.
- Roof tiles: Cut and fix closely at hip.

© AAS 2008 - 2009

- Hip irons: Galvanized steel to BS 5534, clause 4.16.1.
  - Fixing: To hip rafter or hip batten with not less than two zinc coated steel screws.
- Hip tiles:
  - Manufacturer: to CA approval.
  - Bedding: Continuous to edges, solid to joints, in mortar as clause 290.
  - Fixing: Where rigid masonry walls support or abut hip, secure hip tiles within 900 mm of such walls to hip rafters or supplementary hip battens with nails/ wire ties or screws.
  - Bottom hip tiles: Shape neatly to align with corner of eaves and fill ends with mortar and slips of tile finished flush.

595 BONNET HIPS
- Underlay: Lay courses over hip.
  - Overlaps (minimum): 150 mm.
- Bonnet hip tiles:
  - Bedding: In mortar as clause 290, neatly struck back about 13 mm from edge of tiles. Course in with roof tiling.
  - Fixing: Nail to hip rafters or supplementary hip battens.
  - Bottom hip tiles: Fill ends with mortar and tile slips finished flush.
- Roof tiles: Cut adjacent tiles to fit neatly.

605 GRP VALLEYS
- Underlay: Lay as recommended by GRP valley manufacturer.
- GRP valleys:
  - Manufacturer: Danelaw.
- Roof tiles: Cut adjacent tiles to fit neatly.
  - Bedding. On mortar as clause 290 on GRP valleys.
  - Valley width between tiles: to CA approval.

615 METAL VALLEYS
- Underlay: Cut to rake. Dress over tilting fillets to lap onto metal valley. Do not lay under metal.
- Roof tiles: Cut adjacent tiles to fit neatly.
  - Bedding: On mortar as clause 290 on fibre cement undercloaks laid loose each side of valleys.
  - Valley width between tiles: to CA approval.

625 CURVED PLAIN TILE VALLEYS
- Underlay: Lay strips not less than 600 mm wide centred on valleys. Overlap with general roof underlay.
- Curved valley tiles:
- Roof tiles: Cut adjacent tiles to fit neatly.

635 MITRED VALLEYS
- Underlay: Lay strips not less than 600 mm wide centred on valleys. Overlap with general roof underlay.
- Mitred tiles: Cut tile and a half tiles and fix to form a straight, close mitred junction.
- Soakers: Interleave with mitred tiles. Fix by turning down over head of mitred tiles.

660 SIDE ABUTMENTS
- Underlay: Turn up not less than 100 mm at abutments.
- Abutment tiles: Cut as necessary. Fix close to abutments.
- Soakers: Interleave with abutment tiles. Fix by turning down over head of abutment tiles.

670 TOP EDGE ABUTMENTS
- Underlay: Turn up not less than 100 mm at abutments.
- Top course tiles: Fix close to abutments.

700 DRY VENTILATED RIDGES
- Underlay: Provide air gap at apex.
- Dry ridge fixing battens: to manufacturer's recommendations.
- Dry ridge tiles:
  - Manufacturer: to CA approval.

710 DRY RIDGES
- Underlay: Lay courses over ridge.
  - Overlap (minimum): 100 mm.
- Dry ridge fixing battens: to manufacturer's recommendations.
- Dry ridge tiles
  - Manufacturer: to CA approval.
- Ridge terminals:
  - Manufacturer: to CA approval.

740 MORTAR BEDDED RIDGES
- Underlay: Lay courses over ridge.
  - Overlap (minimum): 100 mm.
- Ridge tiles:
  - Manufacturer: to CA approval.
  - Bedding: Continuous to edges, solid to joints, in mortar as clause 290.
  - Fixing: Where rigid masonry walls support or abut ridge, secure ridge tiles within 900 mm of such walls to ridge boards or supplementary ridge battens with nails/ wire ties or screws.
  - Gable end ridge tiles: Fill ends with mortar and slips of tiles finished flush.
- Ridge terminals:
  - Manufacturer: to CA approval.

800 CHANGE OF ROOF PITCH
- Tiling substrate work: Fix timber tilting fillet to support flashing and course of tiles above change of pitch.
- Roof tiles: Fix courses above and below change of pitch and make weathertight with flashing.

VERTICAL TILING EDGES/ JUNCTIONS

910 BOTTOM EDGES
- Tiling substrate work: Fix timber tilting fillet to support bottom course of tiles in correct vertical plane. Fix flashing to tilting fillet.
- Underlay: Dress over flashing.
- Undercourse and bottom course tiles: Fix with tails neatly aligned.

920 TOP EDGES
- Top course tiles: Fix under abutment and make weathertight with flashings dressed down not less than 150 mm.

930 SIDE ABUTMENTS
- Tiling substrate work: Chase abutment wall and insert stepped flashing.
  - Flashing: Return not less than 75 mm behind tiling, overlapping underlay and battens. Turn back to form a vertical welt.
- Abutment tiles: Cut and fix neatly.

940 ANGLES WITH ANGLE TILES
- Angle tiles (purpose made): Fix right and left hand in alternate courses to break bond.
- Adjacent tiles: Cut and fix neatly.

950 ANGLES WITH SOAKERS
- Angle tiles: Cut tile and a half tiles and fix to form a straight, close mitred junction.
- Soakers: Interleave with angle tiles. Fix by nailing to battens at top edge.

960 JUNCTIONS WITH ROOF VERGES
- Tiling substrate work: Fix additional tiling batten parallel to and below verges.
- Course end tiles: 'Winchester cut' tile and a half tiles to angle of verge rake. Fix to additional tiling batten with cut edge parallel to and below verge.

H61　FIBRE CEMENT SLATING

TYPES OF SLATING

105　ROOF SLATING
- Underlay: reinforced felt to BS747 type 1F.
  - Laying: As clause 240, parallel to eaves.
  - Horizontal lap (minimum): 150mm.
- Battens:
  - Size: 19 x 38mm.
  - Fixing: As clause 265.
- Slates: To BS EN 492, type NT (nonasbestos).
  - Manufacturer: Eternit.
    Product reference: Eternit 2000 slates.
  - Colour: Blue/black matt.
  - Size: 600 x 300mm.
  - Headlap (minimum): 100mm.
  - Fixing: As clause 275.

110　ROOF SLATING WITH COUNTERBATTENS
- Underlay: reinforced felt to BS747 type 1F.
  - Laying: As clause 240.
  - Horizontal lap (minimum): 150mm .
- Counterbattens:
  - Size: 19 x 38mm.
- Battens:
  - Size: 19 x 38mm.
  - Fixing: As clause 265.
- Slates: To BS EN 492, type NT (nonasbestos).
  - Manufacturer: Eternit.
    Product reference: Eternit 2000 slates.
  - Colour: Blue/black matt.
  - Size: 600 x 300mm.
  - Headlap (minimum): 100mm.
  - Fixing: As clause 275.

120　VERTICAL SLATING
- Underlay: reinforced felt to BS747 type 1F.
  - Fixing: As clause 240, parallel to bottom edge.
  - Horizontal lap (minimum): 100 mm.
- Battens:
  - Size: 19 x 38mm.
  - Fixing: As manufacturer's recommendations
- Slates: To BS EN 492, type NT (nonasbestos).
  - Manufacturer: Eternit.
    Product reference: Eternit 2000 slates.
  - Colour: Blue/black matt.
  - Size: 600 x 300mm.
  - Headlap (minimum): 100mm.
  - Fixing: As clause 275.

SLATING GENERALLY

210　BASIC WORKMANSHIP
- General: Fix slating and accessories to make the whole sound and weathertight at earliest opportunity.
- Setting out: To true lines and regular appearance, with neat fit at edges, junctions and features.
- Fixings for slating accessories: As recommended by slate or accessory manufacturer.
- Gutters and pipes: Keep free of debris. Clean out at completion.

220 REMOVING EXISTING SLATING
- General: Carefully remove slates, battens, underlay, etc. with minimum disturbance of adjacent retained slating.
- Undamaged slates: Set aside for reuse.

240 UNDERLAY
- Handling: Do not tear or puncture.
- Laying: Maintain consistent tautness.
- Vertical laps (minimum): 100 mm wide, coinciding with supports and securely fixed.
- Fixing: Galvanized steel, copper or aluminium 20 x 3 mm extra large clout head nails.
- Eaves: Where exposed, underlay must be BS 747, type 5U, or equivalent UV durable type.
- Penetrations: Use proprietary underlay seals or cut underlay to give a tight, watershedding fit around pipes and components.
- Ventilation paths: Do not obstruct.

245 BATTENS/ COUNTERBATTENS - TREATED
- Timber: Sawn softwood.
  - Standard: BS 5534, clause 4.12.1.
  - Permissible characteristics and defects: Not to exceed limits in BS 5534, annex C.
  - Moisture content at time of fixing and covering (maximum): 22%.
- Preservative treatment: As section Z12 and British Wood Preserving and Damp-proofing Association Commodity Specification C8.

255 COUNTERBATTENS ON RIGID SARKING
- Fixing: Through rigid sarking into rafters at not more than 300 mm centres.

259 COUNTERBATTENS ON RAFTERS
- Fixing: Into rafters at not more than 300 mm centres.

265 BATTEN FIXING
- Setting out: In straight horizontal lines. Align on adjacent areas.
- Batten length (minimum): Sufficient to span over three supports.
- Joints in length: Square cut. Butt centrally on supports. Joints must not occur more than once in any group of four battens on one support.
- Additional battens: Provide where unsupported laps in underlay occur between battens.
- Fixing: Each batten to each support. Splay fix at joints in length.

270 BATTENS FIXED TO MASONRY
- Setting out: In straight horizontal lines. Align on adjacent areas.
- Batten length (minimum): 3 m.
- Fixing centres (maximum): 400 mm.

275 SLATE FIXING
- Setting out: Lay slates to a half lap bond with not more than 5 mm gaps. Align tails.
- Ends of courses: Use extra width slates to maintain bond and to ensure that cut slates are as large as possible. Do not use half slates.
- Extra width slates: Use additional fixings as recommended by slate manufacturer.
- Top courses: Cut top two slate courses as necessary to maintain gauge. Head-nail top course.
- Fixings for slates: Nails/ rivets recommended by slate manufacturer.

290 MORTAR BEDDING/ POINTING
- Mortar: As section Z21, 1:3 cement:sand, with plasticizing admixtures permitted.
  - Bond strength providing resistance to uplift: To BS 5534.
- Weather: Do not use in wet or frosty conditions or when imminent.
- Preparation of slates and accessories to be bedded or pointed: Coat relevant surfaces with a suitable bonding agent.
- Preparation of concrete and clay tile accessories to be bedded: Wet and drain surface water before fixing.
- Appearance: Finish neatly as work proceeds and remove residue.

ROOF SLATING EDGES/ JUNCTIONS/ FEATURES

305 GENERALLY
- Fittings and accessories: As recommended by slate manufacturer, do not improvise.
  - Exposed fittings and accessories: To match slate colour and finish.
- Cut slates: Cut only where necessary, to give straight, clean edges.
- Flashings: Fix with or immediately after slating. Form neatly.

325 FIRE SEPARATING WALLS
- Separating walls: Completely fill space between top of wall and underside of slates with mineral wool quilt to provide fire stopping.
- Boxed eaves: Completely seal air paths in plane of separating wall with wire reinforced mineral wool, not less than 50 mm thick, fixed to rafters and carefully cut to shape to provide fire stopping.

365 EAVES - UNVENTILATED
- Underlay support: to CA approval.
  - Continuous to prevent water retaining troughs.
- Gutter: Dress underlay or underlay support tray to form drip into gutter.
- Undercourse and first course slates: Fix with tails projecting 50 mm over gutter or to centre of gutter, whichever dimension is the lesser.

445 MORTAR BEDDED VERGES WITH BEDDED UNDERCLOAK
- Underlay: Carry 50 mm onto outer leaf of gable wall and bed on mortar.
- Undercloak: Fibre cement slate or sheet.
  - Position: Over underlay, level with underside of slating battens, sloping towards verge.
  - Projection beyond face of wall: 38-50 mm.
  - Bedding: On mortar identical to that used in gable walling.
- Slating battens: Carry onto undercloak and finish 100 mm from verge edge.
- Verge closure battens: Fix between ends of slating battens.
- Verge slates:
  - Bedding: Flush with undercloak on 75 mm wide bed of mortar as clause 290.
  - Pointing: Struck weathered profile, 5 mm back from verge slates.
  - Fixing: Do not displace or crack mortar.

455 MORTAR BEDDED VERGES WITH NAILED UNDERCLOAK
- Underlay: Carry over full width of verge.
- Undercloak: Fibre cement slate or sheet.
  - Position: Over underlay, level with underside of slating battens, sloping towards verge.
  - Projection: 38-50 mm beyond face of building.
  - Fixing: Nails.
- Slating battens: Carry onto undercloak and finish 100 mm from verge edge.
- Verge closure battens: Fix between ends of slating battens.
- Verge Slates:
  - Bedding: Flush with undercloak on 75 mm wide bed of mortar as clause 290.
  - Struck weathered profile, 5 mm back from verge slates.
  - Fixing: Do not displace or crack mortar.

505 DRY CAPPED HIPS
- Underlay: Lay courses over hip.
  - Overlaps (minimum): 150 mm.
- Dry hip fixing battens: as manufacturer's recommendations.
- Roof slates: Cut and fix closely at hip.
- Dry hip cappings:
  - Product reference: to CA approval.
  - Fixing: Screw to hip fixing battens.
  - Joints in length: Apply sealant strip.
  - Bottom hip cappings: Shape neatly to align with corner of eaves.

555 MORTAR BEDDED TILE HIPS
- Underlay: Lay courses over hip.
  - Overlaps (minimum): 150 mm.
- Roof slates: Cut and fix closely at hip.
- Hip irons: Galvanized steel to BS 5534, clause 4.16.1.
  - Fixing: To hip rafter or hip batten with not less than two zinc coated steel screws.
- Hip tiles:
  - Manufacturer: to CA approval.
  - Bedding: Continuous to edges, solid to joints, in mortar as clause 290.
  - Fixing: Where rigid masonry walls support or abut hip, secure hip tiles within 900 mm of such walls to hip rafters or supplementary hip battens with nails/ wire ties or screws.
  - Bottom hip tiles: Shape neatly to align with corner of eaves and fill ends with mortar and slips of tile finished flush.

605 GRP VALLEYS
- Underlay: Lay as recommended by GRP valley manufacturer.
- GRP valleys:
  - Manufacturer: to CA approval.
- Roof slates: Cut double width slates adjacent to valley to fit neatly.
  - Valley width between slates: to CA approval.

615 METAL VALLEYS
- Underlay: Cut to rake. Dress over tilting fillets to lap onto metal valley. Do not lay under metal.
- Roof slates: Cut double width slates adjacent to valley to fit neatly.
  - Valley width between slates: to CA approval.

635 MITRED VALLEYS
- Underlay: Lay strips not less than 600 mm wide centred on valleys. Overlap with general roof underlay.
- Mitred slates: Cut double width slates and fix to form a straight, close mitred junction.
- Soakers: Interleave with mitred slates. Fix by turning down over head of mitred slates.

660 SIDE ABUTMENTS
- Underlay: Turn up not less than 100 mm at abutments.
- Abutment slates: Cut as necessary. Fix close to abutments.
- Soakers: Interleave with abutment slates. Fix by turning down over head of abutment slates.

670 TOP EDGE ABUTMENTS
- Underlay: Turn up not less than 100 mm at abutments.
- Top slate courses: Fix close to abutments.

690 ROOF WINDOWS
- Underlay: Turn up not less than 100 mm at window surrounds under integral flashings/ soakers.
- Roof slates: Cut as necessary and fix closely all round.

750 MORTAR BEDDED AND MECHANICALLY FIXED TILE RIDGES
- Underlay: Lay courses over ridge.
  - Overlap (minimum): 150 mm.
- Ridge tile fixing battens: as manufacturer's recommendations.
- Ridge tiles:
  - Manufacturer: to CA approval.
  - Bedding: Continuous to edges, solid to joints, in mortar as clause 290.
  - Fixing: Secure to ridge tile fixing battens with nails/ wire ties or screws.
  - Gable end ridge tiles: Fill ends with mortar and slips of tiles finished flush.
- Ridge terminals:
  - Manufacturer: to CA approval.

800 CHANGE OF ROOF PITCH
- Slating substrate work: Fix timber tilting fillet to support flashing and courses of slates above change of pitch.
- Roof slates: Make weathertight with flashing.

- Courses above change of pitch: Fix undercourse and first course slates.
- Courses below change of pitch: Cut top two courses as necessary to maintain gauge. Head-nail top course.

VERTICAL SLATING EDGES/ JUNCTIONS

910 BOTTOM EDGES
- Slating substrate work: Fix timber tilting fillet to support bottom course of slates in correct vertical plane. Fix flashing to tilting fillet.
- Underlay: Dress over flashing.
- Undercourse and bottom course slates: Fix with tails neatly aligned.

920 TOP EDGES
- Top slate courses: Fix under abutment and make weathertight with flashings dressed down not less than 150 mm.

930 SIDE ABUTMENTS
- Slating substrate work: Chase abutment wall and insert stepped flashing.
  - Flashing: Return not less than 75 mm behind slating, overlapping underlay and battens. Turn back to form a vertical welt.
- Abutment slates: Cut and fix neatly.

950 ANGLES WITH SOAKERS
- Angle slates: Cut double width slates and fix to form a straight, close mitred junction.
- Soakers: Interleave with angle slates. Fix by nailing to battens at top edge.

960 JUNCTIONS WITH ROOF VERGES
- Slating substrate work: Fix additional slating batten parallel to and below verges.
- Course end slates: Splay cut slate and a half width slates to angle of verge rake. Fix to additional slating batten with cut edge parallel to and below verge.

H62 NATURAL SLATING

TYPES OF SLATING

105 ROOF SLATING.
- Underlay: reinforced felt to BS747 type 1F.
  - Laying: As clause 240, parallel to eaves.
  - Horizontal lap (minimum): 150mm.
- Battens:
  - Size: 19 x 38mm.
  - Fixing: As clause 265.
- Slates:
  - Supplier: Contractor's choice.
    Product reference: to CA approval.
  - Type: Natural slates.
  - Grade: to BS5534 Part 1.
  - Size: 610 x 305mm.
  - Headlap (minimum): 38mm.
  - Fixing: As clause 275.

120 VERTICAL SLATING
- Underlay: reinforced felt to BS747 type 1F.
  - Fixing: As clause 240, parallel to bottom edge.
  - Horizontal lap (minimum): 100 mm.
- Battens:
  - Size: 19 x 38mm.
- Slates:
  - Supplier: Contractor's choice.
    Product reference: to CA approval.
  - Type: Natural slates.

- Grade: to BS5534 Part 1.
- Size: 610 x 305mm.
- Headlap (minimum): 38mm.
- Fixing: As clause 275.

SLATING GENERALLY

210 **BASIC WORKMANSHIP**
- General: Fix slating and accessories to make the whole sound and weathertight at earliest opportunity.
- Setting out: To true lines and regular appearance, with neat fit at edges, junctions and features.
- Fixings for slating accessories: As recommended by manufacturer.
- Gutters and pipes: Keep free of debris. Clean out at completion.

220 **REMOVING EXISTING SLATING**
- General: Carefully remove slates, battens, underlay, etc. with minimum disturbance of adjacent retained slating.
- Undamaged slates: Set aside for reuse.

240 **UNDERLAY**
- Handling: Do not tear or puncture.
- Laying: Maintain consistent tautness.
- Vertical laps (minimum): 100 mm wide, coinciding with supports and securely fixed.
- Fixing: Galvanized steel, copper or aluminium 20 x 3 mm extra large clout head nails.
- Eaves: Where exposed, underlay must be BS 747, type 5U, or equivalent UV durable type.
- Penetrations: Use proprietary underlay seals or cut underlay to give a tight, watershedding fit around pipes and components.
- Ventilation paths: Do not obstruct.

245 **BATTENS/ COUNTERBATTENS - TREATED**
- Timber: Sawn softwood.
  - Standard: To BS 5534, clause 4.12.1.
  - Permissible characteristics and defects: Not to exceed limits in BS 5534, annex C.
  - Moisture content at time of fixing and covering (maximum): 22%.
- Preservative treatment: As section Z12 and British Wood Preserving and Damp-proofing Association Commodity Specification C8.

255 **COUNTERBATTENS ON RIGID SARKING**
- Fixing: Through rigid sarking into rafters at not more than 300 mm centres.

259 **COUNTERBATTENS ON RAFTERS**
- Fixing: Into rafters at not more than 300 mm centres.

265 **BATTEN FIXING**
- Setting out: In straight horizontal lines. Align on adjacent areas.
- Batten length (minimum): Sufficient to span over three supports.
- Joints in length: Square cut. Butt centrally on supports. Joints must not occur more than once in any group of four battens on one support.
- Additional battens: Provide where unsupported laps in underlay occur between battens.
- Fixing: Each batten to each support. Splay fix at joints in length.

270 **BATTENS FIXED TO MASONRY**
- Setting out: In straight horizontal lines. Align on adjacent areas.
- Batten length (minimum): 3 m.
- Fixing centres (maximum): 400 mm.

272 TIMBER FOR SLATING SUBSTRATE WORK
- Timber: Sawn softwood, free from wane, pitch pockets, decay and insect attack (ambrosia beetle excepted).
  - Moisture content at time of fixing and covering (maximum): 22%.
- Preservative treatment: As section Z12 and British Wood Preserving and Damp-proofing Association Commodity Specification C8.

275 SLATE FIXING
- Setting out: Lay slates with an even overall appearance with slightly open (maximum 5 mm) butt joints. Align tails.
- Slate thickness: Consistent in any one course. Lay with thicker end as tail.
- Ends of courses: Use extra wide slates to maintain bond and to ensure that cut slates are as large as possible. Do not use slates less than 150 mm wide.
- Top course: Head-nail short course to maintain gauge.
- Fixing: Centre nail each slate twice through countersunk holes 20-25 mm from side edges.
  - Nails: Copper clout to BS 1202-2 or aluminium clout to BS 1202-3.
  - Nail dimensions: Determine in accordance with BS 5534 to suit site exposure, withdrawal resistance and slate supplier's recommendations.

290 MORTAR BEDDING/ POINTING
- Mortar: As section Z21,1:3 cement:sand, with plasticizing admixtures permitted.
  - Bond strength providing resistance to uplift: To BS 5534.
- Weather: Do not use in wet or frosty conditions or when imminent.
- Preparation of concrete and clay tile accessories to be bedded: Wet and drain surface water before fixing.
- Appearance: Finish neatly as work proceeds and remove residue.

ROOF SLATING EDGES/ JUNCTIONS/ FEATURES

305 GENERALLY
- Fittings and accessories: As recommended by slate supplier, do not improvise.
  - Exposed fittings and accessories: To match slate colour and finish.
- Cut slates: Cut only where necessary, to give straight, clean edges.
- Flashings: Fix with or immediately after slating. Form neatly.

325 FIRE SEPARATING WALLS
- Separating walls: Completely fill space between top of wall and underside of slates with mineral wool quilt to provide fire stopping.
- Boxed eaves: Completely seal air paths in plane of separating wall with wire reinforced mineral wool, not less than 50 mm thick, fixed to rafters and carefully cut to shape to provide fire stopping.

345 VENTILATED EAVES
- Fascia grilles and ventilator trays:
  - Manufacturer: to CA approval.
  - Fix to carry underlay, form drip into gutter and provide free passage of air over insulation.
- Undercourse and first course slates: Fix with tails projecting 50 mm over gutter or to centre of gutter, whichever dimension is the lesser.

365 EAVES
- Underlay support: Contractor's choice.
  - Continuous to prevent water retaining troughs.
- Gutter: Dress underlay or underlay support tray to form drip into gutter.
- Undercourse and first course slates: Fix with tails projecting 50 mm over gutter or to centre of gutter, whichever dimension is the lesser.

445 MORTAR BEDDED VERGES WITH BEDDED UNDERCLOAK
- Underlay: Carry 50 mm onto outer leaf of gable wall and bed on mortar.
- Undercloak: Slates.
  - Position: Over underlay, level with underside of slating battens, sloping towards verge.
  - Projection beyond face of wall: 38-50 mm.

- Bedding: On mortar identical to that used in gable walling.
- Slating battens: Carry onto undercloak and finish 100 mm from verge edge.
- Verge slates:
  - Bedding: Flush with undercloak on 75 mm wide bed of mortar as clause 290.
  - Pointing: Flush profile.
  - Fixing: Do not displace or crack mortar.

455 MORTAR BEDDED VERGES WITH NAILED UNDERCLOAK
- Underlay: Carry over full width of verge.
- Undercloak: Slates.
  - Position: Over underlay, level with underside of slating battens, sloping towards verge.
  - Projection: 38-50 mm beyond face of building.
  - Fixing: Nails.
- Slating battens: Carry onto undercloak and finish 100 mm from verge edge.
- Verge Slates:
  - Bedding: Flush with undercloak on 75 mm wide bed of mortar as clause 290.
  - Pointing: Flush profile.
  - Fixing: Do not displace or crack mortar.

525 MITRED HIPS
- Underlay: Lay courses over hip.
  - Overlaps (minimum): 150 mm.
- Hip slate fixing battens: as manufacturer's recommendations.
- Mitred slates: Cut extra wide slates and fix to form a straight, close mitred junction.
- Soakers: Interleave with mitred slates. Fix by turning down over head of mitred slates.

535 LEAD ROLL HIPS
- Underlay: Lay courses over hip.
  - Overlaps (minimum): 150 mm.
- Hip slates: Cut close to timber roll and fix to form a straight junction.

565 MORTAR BEDDED AND MECHANICALLY FIXED TILE HIPS
- Underlay: Lay courses over hip.
  - Overlaps (minimum): 150 mm.
- Hip tile fixing battens: to manufacturer's recommendations.
- Roof slates: Cut and fix closely at hip.
- Hip tiles:
  - Manufacturer: to CA approval.
  - Bedding: Continuous to edges, solid to joints, in mortar as clause 290.
  - Fixing: Secure to hip tile fixing battens with nails/ wire ties or screws as recommended by hip tile manufacturer.
  - Bottom hip tiles: Shape neatly to align with corner of eaves and fill ends with mortar and tile slips finished flush.

605 GRP VALLEYS
- Underlay: Lay as recommended by GRP valley manufacturer.
- GRP valleys:
  - Manufacturer to CA approval.
- Roof slates: Cut extra wide slates adjacent to valley to fit neatly.
  - Valley width between slates: to CA approval.

615 METAL VALLEYS
- Underlay: Cut to rake. Dress over tilting fillets to lap onto metal valley. Do not lay under metal.
- Roof slates: Cut extra wide slates adjacent to valley to fit neatly.
  - Valley width between slates: to CA approval.

635 MITRED VALLEYS
- Underlay: Lay strips not less than 600 mm wide centred on valleys. Overlap with general roof underlay.
- Mitred slates: Cut extra wide slates and fix to form a straight, close mitred junction.
- Soakers: Interleave with mitred slates. Fix by turning down over head of mitred slates.

660 SIDE ABUTMENTS
- Underlay: Turn up not less than 100 mm at abutments.
- Abutment slates: Cut as necessary. Fix close to abutments.
- Soakers: Interleave with abutment slates. Fix by turning down over head of abutment slates.

670 TOP EDGE ABUTMENTS
- Underlay: Turn up not less than 100 mm at abutments.
- Top slate courses: Fix close to abutments.

690 ROOF WINDOWS
- Underlay: Turn up not less than 100 mm at window surrounds under integral flashings/ soakers.
- Roof slates: Cut as necessary and fix closely all round.

730 LEAD ROLL RIDGES
- Underlay: Lay courses over ridge.
  - Overlap (minimum): 150 mm.
- Top slate courses: Fit close to timber roll.

740 MORTAR BEDDED TILE RIDGES
- Underlay: Lay courses over ridge.
  - Overlap (minimum): 150 mm.
- Ridge tiles:
  - Manufacturer: To approval by CA.
  - Bedding: Continuous to edges, solid to joints, in mortar as clause 290.
  - Fixing: Where rigid masonry walls support or abut ridge, secure ridge tiles within 900 mm of such walls to ridge boards or supplementary ridge battens with nails/ wire ties or screws.
  - Gable end ridge tiles: Fill ends with mortar and slips of tiles finished flush.

800 CHANGE OF ROOF PITCH
- Slating substrate work: Fix timber tilting fillet to support flashing and courses of slates above change of pitch.
- Roof slates: Make weathertight with flashing.
  - Courses above change of pitch: Fix undercourse and first course slates.
  - Courses below change of pitch: Cut top two courses as necessary to maintain gauge. Head-nail top course.

VERTICAL SLATING EDGES/ JUNCTIONS

910 BOTTOM EDGES
- Slating substrate work: Fix timber tilting fillet to support bottom course of slates in correct vertical plane. Fix flashing to tilting fillet.
- Underlay: Dress over flashing.
- Undercourse and bottom course slates: Fix with tails neatly aligned.

920 TOP EDGES
- Top slate courses: Fix under abutment and make weathertight with flashings dressed down not less than 150 mm.

930 SIDE ABUTMENTS
- Slating substrate work: Chase abutment wall and insert stepped flashing.
  - Flashing: Return not less than 75 mm behind slating, overlapping underlay and battens. Turn back to form a vertical welt.
- Abutment slates: Cut and fix neatly.

950 ANGLES WITH SOAKERS
- Angle slates: Cut extra wide slates and fix to form a straight, close mitred junction.
- Soakers: Interleave with angle slates. Fix by nailing to battens at top edge.

960 JUNCTION WITH ROOF VERGES
- Slating substrate work: Fix additional slating batten parallel to and below verges.
- Course end slates: Splay cut slate and a half width slates to angle of verge rake. Fix to additional slating batten with cut edge parallel to and below verge.

H71 LEAD SHEET COVERINGS/ FLASHINGS

TYPES OF LEADWORK

110 ROOFING
- Type of lead: as specified in schedules.

150 DORMERS
- Type of lead: as specified in schedules.

200 GUTTER LINING
- Type of lead: as specified in schedules

235 VALLEY GUTTER LINING TO LEAD ROOFS
- Laying: Dress lead sheet into shallow valley box gutter.
  - Gutter width: 150 mm.
- Lengths: Not more than 150mm
- Fixing: Nail top edge only of each sheet. Dress bottom end neatly into eaves gutter.
- Roofing sheets: Dress over each side of gutter lining, forming laps of not less than 150mm

240 VALLEY GUTTER LINING TO LEAD ROOFS
- Laying: Dress lead sheet into valley.
- Lengths: Not more than 1500mm
- Fixing: Nail top edge only of each sheet. Dress bottom end neatly into eaves gutter.
- Roofing sheets: Dress over each side of gutter lining, forming laps of not less than 150mm
  - Central gutter gap: 150 mm.

310 RIDGE/ HIP ROLLS TO LEAD ROOFS
- Core: Rounded timber.
  - Shape: Tapered to a flat base 30 mm wide.
  - Fixing: To ridge/ hip board with brass or stainless steel screws at not more than 600 mm centres.
- Roof covering: Dress roofing sheets up roll.
  - Fixing: Nail each sheet at underlapping end.
- Lead capping:
  - Thickness: As roof covering.
  - Lengths: Not more than 1500mm
  - Wings: Extend not less than 75 mm on to roof.
  - Laps in length: Not less than 150 mm for ridges, 100 mm for hips.
  - Fixing: Secure wings with one copper or stainless steel clip per roofing bay and at each lap.

315 RIDGE/ HIP ROLLS TO SLATE ROOFS
- Clips: At capping laps and not more than 1500mm
  - Fixing: Nail to top of ridge/ hip board before fixing core. Nail each side not more than 50 mm from edge of capping (drill slates as necessary).
- Core: Rounded timber.
  - Shape: Tapered to a flat base 30 mm wide.
  - Fixing: To ridge/ hip board with brass or stainless steel screws at not more than 600 mm centres, with base not less than 5 mm above slates.
- Lead capping:
  - Thickness: as stated in schedules
  - Lengths: Not more than 1500mm
    Hip capping: Nail head of each length around core.
  - Laps: Not less than 150 mm for ridges, 100 mm for hips.
  - Cover: Wings of capping to extend not less than 150 mm on to roof.

322 SOAKERS FOR MITRED HIPS
- Lead:
  - Thickness: 1.25-1.50 mm (code 3).
- Dimensions:
  - Length: Slate/ tile gauge + lap + 25 mm.
  - Underlaps: Not less than 100mm.

324 SOAKERS FOR VALLEYS GENERALLY
- Lead:
  - Thickness: 1.25-1.50 mm (code 3).
- Dimensions:
  - Length: Slate length at valley mitre + 25 mm.
  - Underlaps: Not less than 150 mm.

410 APRON FLASHINGS
- Lead:
  - Thickness: as stated in schedules
- Dimensions:
  - Lengths: Not more than 1500mm
  - End to end joints: Laps of not less than 100 mm.
  - Upstand: Not less than 75 mm.
  - Cover to abutment: Not less than 100mm

420 COVER FLASHINGS
- Lead:
  - Thickness: as stated in schedules.
- Dimensions:
  - Lengths: Not more than 1500mm.
  - End to end joints: Laps of not less than 100 mm.
  - Cover: Overlap to upstand of not less than 75 mm.

440 SOAKERS AND STEP FLASHINGS
- Lead soakers:
  - Thickness: as stated in schedules.
  - Dimensions:
    Length: Slate/ tile gauge + lap + 25 mm.
    Upstand: Not less than 75 mm.
    Underlap: Not less than 100 mm.
  - Fixing: By roofer.
- Lead step flashings:
  - Thickness: as stated in schedules.
  - Dimensions:
    Lengths: Not more than 1500mm.
    End to end joints: Laps of not less than 100 mm.
    Cover: Overlap to soaker upstands of not less than 65 mm.
  - Fixing: Lead wedges at every course.

450 STEP AND COVER FLASHINGS
- Lead:
  - Thickness: as stated in schedules
- Dimensions:
  - Lengths: Not more than 1500mm
  - End to end joints: Laps of not less than 100 mm.
  - Upstand: Not less than 85 mm.
  - Cover to roof: Not less than 150 mm.
- Fixing: Lead wedges at every course and clips at not more than 500 mm centres along free edge.

456 STEP FLASHINGS WITH SECRET GUTTER
- Lead step flashings:
  - Thickness: as stated in schedules.
  - Dimensions:

Lengths: Not more than 1500mm.
End to end joints: Laps of not less than 100 mm.
Cover: Overlap to gutter lining upstand of not less than 65 mm.
- Fixing: Lead wedges at every course.
- Lead secret gutter lining:
  Thickness: as stated in schedules.
  - Dimensions:
  Lengths: Not more than 1500 mm.
  End to end joints: Laps of not less than 1500mm.
  Upstand: Not less than 65 mm above tiles.
  - Fixing: Dress into secret gutter and form a welted edge at side to be tiled. Nail top edge of each sheet. Dress bottom end neatly into gutter.

458 STEP AND COVER FLASHING WITH SECRET GUTTER
- Lead step and cover flashing:
  - Thickness: as stated in schedules.
  - Dimensions:
  Lengths: Not more than 1500mm.
  End to end joints: Laps of not less than 100 mm.
  Upstand: Not less than 85 mm with overlap to gutter lining upstand of not less than 65 mm.
  Cover to roof: Not less than 150 mm.
  - Fixing: Lead wedges at every course and clips at not more than 500 mm centres along free edge.
- Lead secret gutter lining:
  - Thickness: as stated in schedules.
  - Dimensions:
  Lengths: Not more than 1500 mm.
  End to end joints: Laps of not less than 150mm.
  Upstand: Not less than 65 mm above tiles.
  - Fixing: Dress into secret gutter and form a welted edge at side to be tiled. Nail top edge of each sheet. Dress bottom end neatly into eaves gutter.

460 CHANGE OF ROOF PITCH FLASHINGS
- Lead:
  - Thickness: as stated in schedules.
- Dimensions:
  - Lengths: Not more than 1500mm.
  - End to end joints: Laps of not less than 150 mm.
  - Under course of slates/ tiles above: Not less than 150 mm.
  - Over course of slates/ tiles below: Not less than 150 mm.
- Fixing: Nail top edge at 150 mm centres and welt edge. Clip bottom edge at laps and 500 mm centres.

490 VERTICAL TILING/ SLATING BOTTOM EDGE FLASHINGS
- Lead:
  - Thickness: as stated in schedules.
- Dimensions:
  - Lengths: Not more than 1500 mm.
  - End to end joints: Laps of not less than 100 mm.
  - Width: Adequate for underlap to underlay, dressing over tilting fillet, and welted drip or straight cut bottom edge.

492 VERTICAL TILING/ SLATING TOP EDGE FLASHINGS
- Lead:
  - Thickness: as stated in schedules.
- Dimensions:
  - Lengths: Not more than 1500 mm.
  - End to end joints: Laps of not less than 100 mm.
  - Width: Adequate for underlap to abutment and dressing down over tiles/ slates not less than 150 mm.

494 **VERTICAL TILING/ SLATING SIDE ABUTMENT STEP FLASHINGS**
- Lead:
  - Thickness: as stated in schedules.
- Dimensions:
  - Lengths: Not more than 1500 mm.
  - End to end joints: Laps of not less than 100 mm.
  - Width: Adequate for not less than 75 mm underlap with welted edge to tiles/ slates and not less than 50 mm cover to abutment.

496 **VERTICAL TILING/ SLATING ANGLE SOAKERS**
- Lead:
  - Thickness: 1.25-1.5 mm (code 3).
- Dimensions:
  - Length: Tile/ slate gauge + lap + 25 mm.
  - Underlaps: Not less than 150 mm.

GENERAL REQUIREMENTS/ PREPARATORY WORK

510 **WORKMANSHIP GENERALLY**
- Standard: To BS 6915 and latest edition of 'Rolled lead sheet. The complete manual' published by the Lead Sheet Association.
- Fabrication and fixing: To provide a secure, free draining and weathertight installation.
- Operatives: Trained in the application of lead coverings/ flashings. Submit records of experience on request.
- Measuring, marking, cutting and forming: Prior to assembly wherever possible.
- Marking out: With pencil, chalk or crayon. Do not use scribers or other sharp instruments without approval.
- Bossing and forming: Straight and regular bends, leaving sheets free from ripples, kinks, buckling and cracks.
- Solder: Use only where specified.
- Sharp metal edges: Fold under or remove as work proceeds.
- Finished work: Fully supported, adequately fixed to resist wind uplift but also able to accommodate thermal movement without distortion or stress.
  - Protection: Prevent staining, discolouration and damage by subsequent works.

516 **LEADWELDING**
- In situ welding: Is permitted, subject to completion of a 'hot work permit' form and compliance with its requirements.

520 **LEAD SHEET**
- Production method:
  - Rolled, to BS EN 12588, or
  - Machine cast, Agrément certified and to code thicknesses with a tolerance (by weight) of ±5%.
- Identification: Labelled to show thickness/ code, weight and type.

555 **LAYOUT**
- Setting out of longitudinal and cross joints: Submit proposals.

610 **SUITABILITY OF SUBSTRATES**
- Condition: Dry and free of dust, debris, grease and other deleterious matter.

620 **PREPARATION OF EXISTING TIMBER SUBSTRATES**
- Remedial work: Adjust boards to level and securely fix. Punch in protruding fasteners and plane or sand to achieve an even surface.
- Defective boards: Give notice.
- Moisture content: Not more than 22% at time of covering. Give notice if greater than 16%.

630 **PLYWOOD UNDERLAY**
- Standard: Manufactured to an approved national standard and to BS EN 636, section 7 (plywood for use in humid conditions).
- Sheet size: 2400 or 1200 x 1200 mm and 6 mm thick.

- Moisture content: Not more than 22% at time of covering. Give notice if greater than 16%.
- Laying: Cross joints staggered and a 0.5 to 1 mm gap between boards.
- Fixing: With 25 mm annular ringed shank copper or stainless steel nails, at 300 mm grid centres over the area of each sheet and at 150 mm centres along edges, set in 10 mm from perimeter edges.
  - Nail heads: Set flush or just below the surface.

640 TIMBER FOR USE WITH LEADWORK
- Quality: Planed, free from wane, pitch pockets, decay and insect attack (ambrosia beetle excepted).
- Moisture content: Not more than 22% at time of covering. Give notice if greater than 16%.
- Preservative treatment: Organic solvent as section Z12 and British Wood Preserving and Damp-proofing Association Manual - Commodity Specification C8.

650 UNDERLAY
- Handling: Prevent tears and punctures.
- Laying: Butt or overlap jointed onto a dry substrate.
  - Fixing edges: With copper or stainless steel staples or clout nails.
  - Do not lay over roof edges but do turn up at abutments.
- Wood core rolls: Fixed over underlay.
- Protection: Keep dry and cover with lead at the earliest opportunity.

FIXING LEAD

705 HEAD FIXING LEAD SHEET
- Top edge: Secured with two rows of fixings, 25 mm and 50 mm from top edge of sheet, at 75 mm centres in each row, evenly spaced and staggered.
- Sheets less than 500 mm deep: May be secured with one row of fixings, 25 mm from top edge of sheet and evenly spaced at 50 mm centres.

710 FIXINGS
- Nails to timber substrates: Copper clout nails to BS 1202-2, or stainless steel (austenitic) clout nails to BS 1202-1.
  - Shank type: Annular ringed, helical threaded or serrated.
  - Shank diameter: Not less than 2.65 mm for light duty or 3.35 mm for heavy duty.
  - Length: Not less than 20 mm or equal to substrate thickness.
- Screws to concrete or masonry substrates: Brass or stainless steel to BS 1210, tables 3 or 4.
  - Diameter: Not less than 3.35 mm.
  - Length: Not less than 19 mm.
  - Washers and plastic plugs: Compatible with screws.
- Screws to composite metal decks: Self tapping as recommended by the deck and lead manufacturer/ supplier for clips.

715 CLIPS
- Material:
  - Lead clips: Cut from sheets of same thickness/ code as sheet being secured.
  - Copper clips:
    Thickness: to CA approval.
    Temper: BS EN 1172, designation R220 in welts, seams and rolls, R240 elsewhere; dipped in solder if exposed to view.
  - Stainless steel clips:
    Thickness: to CA approval.
    Grade: BS EN 10088, 1.4301(304) terne coated if exposed to view.
- Dimensions:
  - Width: 50 mm where not continuous.
  - Length: To suit detail.
- Fixing clips: Secure each to substrate with two fixings not more than 50 mm from edge of lead sheet.
- Fixing lead sheet: Welt clips around edges and turn over 25 mm.

780 **WEDGE FIXING INTO DAMP PROOF COURSE JOINTS**
- Joint: Rake/ cut out under damp proof course to a depth of not less than 25 mm.
- Lead: Dress lead into joint.
  - Fixing: Lead wedges at not more than 450 mm centres, at every change of direction and with at least two for each piece of lead.
- Sealant: to CA approval.
  - Application: As section Z22.

790 **SCREW FIXING INTO JOINTS/ CHASES**
- Joint/ Chase: Rake out to a depth of not less than 25 mm.
- Lead: Dress into joint/ chase and up back face.
  - Fixing: Into back face with stainless steel screws and washers and plastics plugs at not more than 450 mm centres, at every change of direction, and with at least two fixings for each piece of lead.
- Sealant: to CA approval.
  - Application: As section Z22.

JOINTING LEAD

810 **FORMING DETAILS**
- Method: Bossing or leadwelding except where bossing is specifically required.
- Leadwelded seams: Neatly and consistently formed.
  - Seams: Do not undercut or reduce sheet thickness.
  - Filler strips: Of the same composition as the sheets being joined.
  - Butt joints: Formed to a thickness one third more than the sheets being joined.
  - Lap joints: Formed with 25 mm laps and two loadings to the edge of the overlap.
- Bossing: Carried out without thinning, cutting or otherwise splitting the lead sheet.

830 **STANDING SEAM JOINTS**
- Joint allowance: 100 mm overlap, 75 mm underlap and copper or stainless steel clips at not more than 750 mm centres.
- Forming joint: Welt overlap and clips around underlap, loosely turn over to form a standing seam of consistent cross section.

840 **WOOD CORED ROLL JOINTS WITHOUT SPLASH LAP**
- Core: Timber.
  - Size: 45 x 45 mm round tapering to a flat base 25 mm wide.
  - Fixing to substrate: Brass or stainless steel countersunk screws at not more than 300 mm centres.
- Undercloak: Dress half way around core.
- Copper or stainless steel clips: Fix to core at not more than 450 mm centres. Do not restrict thermal movement of the undercloak.
- Overcloak: Dress around core with edge welted around ends of clips, finishing 5 mm clear of main surface.

845 **WOOD CORED ROLL JOINTS WITH SPLASH LAP**
- Core: Timber.
  - Size: 45 x 45 mm round tapering to a flat base 25 mm wide.
  - Fixing to substrate: Brass or stainless steel countersunk screws at not more than 300 mm centres.
- Undercloak: Dress three quarters around core.
  - Fixing: Nail to core at 150 mm centres for one third length of the sheet starting from the head.
- Overcloak: Dress around core and extend on to main surface to form a 40 mm splash lap.

847 **HOLLOW ROLL JOINTS**
- Joint allowance: 125 mm overcloak and 100 mm undercloak.
- Copper or stainless steel clips: Fix to substrate at not more than 450 mm centres.
- Overcloak: Welt with clips around undercloak to form a roll of consistent cross section.

860 DRIPS WITH SPLASH LAPS
- Underlap: Dress into rebate along top edge of drip.
  - Fixing: One row of nails at 50 mm centres on centre line of rebate.
- Overlap: Dress over drip and form a 40 mm splash lap.

862 DRIPS WITH SPLASH LAPS
- Underlap: Dress up full height of drip upstand.
  - Fixing: Two rows of nails to lower level substrate, 25 mm and 50 mm from face of drip. At 75 mm centres in each row, evenly spaced and staggered. Seal over nails with a soldered or leadwelded dot.
- Overlap: Dress over drip and form a 75 mm splash lap.
  - Fixing: Lead clips, leadwelded to underlap, with not less than one per bay.

865 DRIPS WITHOUT SPLASH LAPS
- Underlap: Dress into rebate along top edge of drip.
  - Fixing: One row of nails at 50 mm centres on centre line of rebate.
- Overlap: Dress over drip to just short of lower level.

880 WELTED JOINTS
- Joint allowance: 50 mm overlap and 25 mm underlap.
- Copper or stainless steel clips: Fix to substrate at not more than 450 mm centres.
- Overlap: Welt around underlap and clips and lightly dress down.

950 PLAQUES
- Existing plaques/ inscriptions: Retain and refix in approved locations by leadwelding along edges.

955 PLAQUES
- New plaques: Cast to record the following information and fix in approved locations by leadwelding along edges.

970 FINISHING
- Finishing agent: to CA approval.
- Application: As soon as practical, apply a smear coating evenly in one direction and in dry conditions.

H72 ALUMINIUM STRIP/ SHEET COVERINGS/ FLASHINGS

TYPES OF ALUMINIUM WORK

110 ROOFING

130 CLADDING

230 VALLEY GUTTER LINING TO SLATE/ TILE ROOFS
- Laying: Over and beyond tilting fillets. In lengths not more than 3 m.
  - Cross joints: Double lock welts as clause 890.
- Fixing: Fold edges and fix with clips as clause 750 at not more than 450 mm centres. Fold bottom end neatly into eaves gutter.

240 RECESSED VALLEY GUTTER LINING
- Forming: With a clear width not less than 200 mm, side upstands equal to step in base and top flanges of 25 mm.
- Laying: In lengths not more than 3 m.
  - Cross joints: Double lock welts as clause 890.
- Aluminium underlap clips:
  - Cover: Not less than 100 mm with anti-capillary welt at top edge.
  - Fix to roof slope at 200 mm centres.
- Joint with roof covering: Project roof coverings with underlap clips over each side of gutter and single welt around flanges in gutter lining to form drips.

245 VALLEY GUTTER LINING
- Forming: With a clear width not less than 200 mm.
- Laying: In lengths not more than 3 m.
  - Cross joints: Double lock welts as clause 890.
- Joint with roof covering: Overlap gutter lining with roof coverings and join together with single lock welts as clause 880.

410 APRON FLASHINGS
- Dimensions:
  - Lengths: Not more than 2 m, with end to end joints lapped not less than 100 mm.
  - Upstand: Not less than 100 mm.
  - Cover to abutment: Not less than 100mm.
  Bottom edge welted 15 mm.

420 COVER FLASHINGS
- Dimensions:
  - Lengths: Not more than 2 m, with end to end joints lapped not less than 100 mm.
  - Cover to roofing upstand: Not less than 75 mm, with bottom edge welted 15 mm.

430 STEP FLASHINGS
- Dimensions:
  - Lengths: Not more than 2 m, with end to end joints lapped not less than 100 mm.
  - Cover to roofing upstand: Not less than 75 mm, with bottom edge welted 15 mm.

440 SOAKERS AND STEP FLASHINGS
- Soakers:
  - Cut and folded for fixing by roofer.
  - Length: Slate/ tile gauge + lap + 25 mm.
  - Upstand: Not less than 75 mm.
  - Underlap: Not less than 100 mm.
- Step flashings:
  - Lengths: Not more than 2 m, with end to end joints lapped not less than 100 mm.
  - Cover: Overlap to soaker upstands not less than 60 mm, with bottom edge welted 15 mm.
  - Fixing: Aluminium wedges at every course, clips to bottom edge at laps.

450 STEP AND COVER FLASHINGS
- Dimensions:
  - Lengths: Not more than 2 m, with end to end joints lapped not less than 100 mm.
  - Upstand: Not less than 75 mm.
  - Cover to roof: Not less than 150 mm.
- Fixing: Aluminium wedges every course and clips as clause 750 at laps and not more than 500 mm centres along free edge.

460 CHANGE OF ROOF PITCH FLASHINGS
- Dimensions:
  - Lengths: Not more than 2 m, with end to end joints lapped not less than 100 mm.
  - Under slates/ tiles above: Not less than 150 mm.
  - Over slates/ tiles below: Not less than150 mm.
- Fixing: Welt top and bottom edges and secure with clips as clause 750 at 450 mm centres.

GENERAL REQUIREMENTS/ PREPARATORY WORK

510 WORKMANSHIP GENERALLY
- Standard: Generally to CP 143-15.
- Fabrication and fixing: To provide a secure, free draining and completely weathertight installation.
- Operatives: Trained in the application of aluminium coverings/ flashings. Submit records of experience on request.
- Measuring, marking, cutting and forming: Prior to assembly wherever possible.
- Marking out: With pencil, chalk or crayon. Do not use scribers or other sharp instruments without approval.

© AAS 2008– 2009

- Folding: With mechanical or manual presses to give straight, regular and tight bends, leaving panels free from ripples, kinks, buckling and cracks. Use hand tools only for folding details that cannot be pressed.
- Surface protection: Fully coat surfaces to be embedded in concrete or mortar with high build bitumen based paint, after folding.
- Sharp metal edges: Fold under or remove as work proceeds.
- Joints: Do not use sealants to attain waterproofing.
- Finished aluminium work: Fully supported, adequately fixed to resist wind uplift and able to accommodate thermal movement without distortion or stress.
  - Protection: Prevent staining, discolouration and damage by subsequent works.

516 WELDING
- In situ welding: Permitted subject to completion of a 'hot work permit' form and compliance with its requirements.

520 ALUMINIUM STRIP/ SHEET
- Standard: To BS EN 485, BS EN 507, BS EN 515 and BS EN 573.
  - Stamped or labelled with alloy designation, temper, finish and thickness.
- Manufacturer: to CA approval.

535 INTEGRITY OF ALUMINIUM
- Requirement: Design coverings/ flashings and methods of attachment to prevent loss of weathertightness and permanent deformation due to wind pressure or suction.
- Wind loads: Calculate to BS 6399-2.

555 LAYOUT
- Setting out of longitudinal and cross joints: Submit proposals.

570 EXISTING METAL RETAINED
- Cleaning: Remove dirt without damage to metal or adversely affecting other materials.

580 EXISTING METAL REUSED
- Handling/ Storage: Keep for reuse in the Works.

610 SUITABILITY OF SUBSTRATES
- Condition: Dry and free of dust, debris, grease and other deleterious matter.

620 PREPARATION OF EXISTING TIMBER SUBSTRATES
- Remedial work: Adjust boards to level and securely fix. Punch in any protruding fasteners and plane or sand to achieve an even surface.
- Defective boards: Give notice.
- Moisture content: Not more than 22% at time of covering.

640 TIMBER FOR USE WITH ALUMINIUM WORK
- Quality: Planed, free from wane, splits, pitch pockets, decay and insect attack (ambrosia beetle excepted).
- Moisture content: Not more than 22% at time of covering.
- Preservative treatment: Organic solvent as section Z12, and British Wood Preserving and Damp-proofing Association Manual - Commodity Specification C8.

650 UNDERLAY
- Handling: Prevent tears and punctures.
- Laying: Butt jointed onto a dry substrate.
  - Fixing edges: With aluminium or galvanized steel staples or clout nails.
  - Do not lay over eaves and drip/ step aluminium underlaps.
- Protection: Keep dry and cover with aluminium at the earliest opportunity.

FIXING

710 **FIXINGS FOR CLIPS**
- Nails to timber substrates: Aluminium to BS 1202-3 for aluminium clips. Stainless steel (austenitic) for stainless steel clips.
  - Shank type: Annular ringed or helical threaded.
  - Shank diameter: Not less than 2.65 mm.
  - Head: Flat.
  - Length: Not less than 25 mm or equal to substrate thickness.
- Screws to concrete/ masonry substrates: Sherardized or zinc plated steel to BS 1210, table 2, or aluminium to BS 1210, table 5 for aluminium clips. Stainless steel (austenitic) to BS 1210, table 4 for stainless steel clips.
  - Diameter: Not less than 3.35 mm.
  - Length: Not less than 25 mm.
  - Washers and plastic plugs: Compatible with screws.
- Screws to composite metal decks: Self tapping, as recommended by the deck and aluminium manufacturer/ supplier for aluminium or stainless steel clips.

720 **STANDING SEAM FIXED CLIPS**
- Aluminium clips: Cut from same alloy and thickness of metal as that being secured.
- Stainless steel (austenitic) clips: Cut from same thickness of metal as the aluminium being secured.
- Dimensions:
  - Width: Not less than 50 mm.
  - Base length: Not less than 20 mm.
  - Upstand: To suit standing seam profile.
- Fixing: Secure each clip to substrate with two fixings.

725 **STANDING SEAM SLIDING CLIPS**
- Aluminium clips: Cut from same alloy and thickness of metal as that being secured.
- Stainless steel (austenitic) clips: Cut from same thickness of metal as the aluminium being secured.
- Dimensions:
  - Fixed component:
    Width: Not less than 90 mm.
    Base length: Not less than 20 mm.
    Upstand: 20 mm, with slot for locating sliding component.
  - Sliding component:
  - Width: Not less than 35 mm.
  - Upstand: To suit standing seam profile.
- Fixing: Secure each clip to substrate with three fixings.

730 **BATTEN ROLL CLIPS**
- Material: Cut from same alloy and thickness of metal as that being secured.
- Dimensions:
  - Width: Not less than 50 mm.
  - Length: Sufficient to pass under batten and turn up each side, with not less than 20 mm projection for folding into welt.
- Fixing: Secure each clip to substrate with one fixing.

750 **CLIPS FOR FLASHINGS/ CROSS JOINTS**
- Material: Cut from same alloy and thickness as that being secured.
- Dimensions:
  - Width: Not less than 50 mm.
  - Length: Sufficient to suit detail.
- Fixing: Secure each clip to substrate with two fixings, not more than 50 mm from edge of strip/ sheet being fixed.

760 **CONTINUOUS CLIPS**
- Material: Cut from same alloy and thickness as that being secured.
- Dimensions:

- Width: Sufficient to suit detail.
- Length: Not more than 1.8 m.
- Fixing: To substrate at 150 mm centres. Welt edge of strip/ sheet being fixed to continuous clip and dress down.

770 WEDGE FIXING INTO JOINTS/ CHASES
- Joint/ chase: Rake out to a depth of not less than 25 mm.
- Aluminium: Fold 25 mm into joint/ chase with a waterstop welted end.
- Fixing: Aluminium wedges at not more than 450 mm centres, at every change of direction, and with at least two for each piece of aluminium.
- Sealant: to CA approval.
  - Application: As section Z22.

780 WEDGE FIXING INTO DAMP PROOF COURSE JOINTS
- Joint: Rake/ cut out under damp proof course to a depth of not less than 25 mm.
- Aluminium: Fold 25 mm into joint with a waterstop welted end.
- Fixing: Aluminium wedges at not more than 450 mm centres, at every change of direction, and with at least two for each piece of aluminium.
- Sealant: to CA approval.
  - Application: As section Z22.

790 SCREW FIXING INTO JOINTS/ CHASES
- Joint/ chase: Rake out to a depth of not less than 25 mm.
- Aluminium: Fold into joint/ chase and up back face.
- Fixing: Into back face with sherardized, zinc plated steel or aluminium screws, washers and plastics plugs at not more than 450 mm centres, at every change of direction, and with at least two fixings for each piece of aluminium.
- Sealant: to CA approval.
  - Application: As section Z22.

JOINTING

810 FORMING DETAILS
- Folds and welts: Form without thinning or splitting the strip/ sheet.
- Thermal movement: Form details with appropriate allowance for movement, without impairment of security at full expansion or contraction.

830 STANDING SEAM JOINTS
- Joint allowances: 45 mm overlap, 35 mm underlap and 5 mm gap for thermal movement. Preformed interlocking profiles for overlap and underlap are permitted.
- Forming: Double welt overlap and clips around underlap to form standing seam 25 mm high of consistent cross section.

840 BATTEN ROLL JOINTS
- Core: Timber as clause 640.
  - Size: Not less than 40 mm high x 40 mm wide tapering to 36 mm at apex.
  - Fixings to substrate: Sherardized or zinc plated steel, or aluminium countersunk screws at not more than 600 mm centres.
- Aluminium covering:
  - Joint allowances: Form strips/ sheets each side of core with 5 mm gap for thermal movement and upstands to 10 mm above height of core.
    Over upstands: Welt clips.
    To cappings: Single lock welt upstands.
- Cappings: Aluminium of the same alloy, finish and thickness of metal as the strip/ sheet being jointed, in lengths not more than 1.25 m, with single lock welt end to end joints.

J: WATERPROOFING

J21 MASTIC ASPHALT ROOFING/ FINISHES

GENERALLY

100 DEFINITIONS
- "Flat". Not exceeding 15 degrees pitch.

105 CONTRARY TO SMM
- Roofing has not been given a stated pitch but described as flat (SMM J21.3.*.*).
- Surface finishes have been measured as "extra over" roofing (SMM J21.S4).

110 PRICES ALSO TO INCLUDE
- Extra material in keys, grooves and open joints in brickwork and the like over the thickness given.
- Work on existing surfaces to include any extra coat or thickness of asphalt required.
- Arrises to include rounded and chamfered angles.
  - Asphalt in direct contact with metal surfaces to include cleaning, scoring, wirebrushing and priming with bitumen to form key for asphalt.
  - Isolating membranes underlays and vapour barriers included in description of covering to include all holes associated with pipe and other perforations through the covering.
  - Vapour barriers to include returning at all edges against abutments and around perforations etc for the full depth of any insulating underlay separating it from the finish.

TYPES OF COATING/ PAVING

180 MASTIC ASPHALT
- Substrate: As specified in schedules.
- Separating layer: Sheathing felt to BS747 type 4A loose laid with 50mm minimum laps.
- Coating: Mastic asphalt to BS 6925.
  - Type: R988/T25.
  - Nominal thickness: As specified in schedules.
  - Fillet profile: 45° angle, width on face (minimum): 50 mm.
- Surface protection: As specified in schedules.

PRODUCTS

310 ANCILLARY PRODUCTS AND ACCESSORIES
- Types: Recommended by mastic asphalt manufacturer.

325 BONDING COMPOUND
- Type: Oxydised bitumen.
- Manufacturer: Contractors choice.

345 EDGE TRIMS
- Type: As specified in schedules.
- Manufacturer: Paptrim.
  - Product reference: Various.
- Colour: Various.
- Length (maximum): 3 m.

455 SAND FOR RUBBING
- Type: Clean, coarse sand from natural deposits, free from loam.
- Size: Passing a 600 micron sieve and retained on a 212 micron sieve.

475 CHIPPINGS AND DRESSING COMPOUND
- Type: Limestone.
- Size (minimum): 10mm.
- Dressing compound: To BS 3690-1 and BS EN 12591.

EXECUTION GENERALLY

510 ADVERSE WEATHER
- Unfinished areas of the roof: Keep dry.
- Work in severe or continuously wet weather: Suspend or provide and maintain effective temporary cover over the working area.
- Unavoidable wetting: Minimise and make good damage.

520 INCOMPLETE WORK
- Daywork joints in warm roofs and edges of phased roofing: Adequately protected and fully weathertight.

525 PREPARING EDGES OF EXISTING MASTIC ASPHALT
- Single coat applications:
  - Cut edges: Soften and clean.
- Two coat applications:
  - Cut edges: Soften and remove half depth of softened material for width of not less than 75 mm.
  - Jointing: Lapped between new and existing material at prepared edges.
- General:
  - Heating: Torching not permitted.
  - Timing: Immediately prior to laying mastic asphalt.

540 APPLYING BONDING COMPOUNDS
- Temperature of compound: Suitable to achieve bond over the whole surface. Do not overheat.
- Heat sensitive insulation materials: Use cold bituminous adhesive recommended by the insulation manufacturer.

SUBSTRATES/ VAPOUR CONTROL LAYERS/ WARM ROOF INSULATION

610 SUITABILITY OF SUBSTRATES
- Generally:
  - Secure, even textured, clean and dry, and free from frost.
  - Horizontal surfaces: To even falls within tolerances.
- Preliminary work: Complete, including:
  - Formation of upstands and kerbs.
  - Penetrations/ Outlets.
  - Chases: Not less than 25 x 25 mm.
  - External angles: Chamfered where required to maintain full thickness of mastic asphalt.
  - Movement joints.
- Moisture content and stability: Must not impair integrity of roof.

620 REMOVING EXISTING MASTIC ASPHALT
- Existing roof: Do not damage.
- Timing: Only remove sufficient mastic asphalt as will be replaced and waterproofed on same day.

630 MAKING GOOD EXISTING MASTIC ASPHALT
- Dust, dirt, debris, moss, plants and grease: Remove.
- Defective areas of mastic asphalt: Soften and carefully cut out.
  - Hammers, chisels, etc: Do not use to cut cold mastic asphalt.
  - Substrate: Clean and dry.
  - Separating membrane: Make good.
  - Mastic asphalt: Patch level with existing surface in two coats, the top coat lapped 75 mm (minimum) on to existing asphalt and to half its depth.

640 INSTALLING TIMBER FOR TRIMS
- Fasteners: Sherardized steel screws.
- Fixing centres (maximum): 300mm.

642 KEYING TO VERTICAL/ SLOPING DENSE CONCRETE
- Surface preparation: Remove mould oil, clean and apply proprietary high bond primer or proprietary keying mix of cement:sand slurry incorporating a bonding agent.

644 KEYING TO NEW BRICKWORK/ DENSE BLOCKWORK
- Joints: Flush pointed.
- Surface preparation: Apply proprietary high bond primer.

646 KEYING TO EXISTING BRICKWORK/ DENSE BLOCKWORK
- Joints: Sound and flush pointed.
- Surface preparation: Clean and apply proprietary high bond primer.

648 KEYING TO METAL SURFACES
- Surface preparation: Clean and apply proprietary high bond primer.

649 APPLYING STEEL LATHING TO VERTICAL/ SLOPING
- Placing: Long way of mesh is horizontal and pitch of horizontal strands is sloping upwards away from background.
- Jointing: Wire tie butt joints between sheets at 75 mm centres.
- Method of fixing: Contractors discretion.
  - Perimeter edges: 75 mm centres.
  - General areas (maximum): 150 mm vertical and horizontal centres.

660 JOINTS IN RIGID BOARD SUBSTRATES
- Cover strip: Lay centrally over substrate joints before laying vapour control layers or coverings. Adhere to substrate with bonding compound along edges only.

670 LAYING VAPOUR CONTROL LAYER
- Attachment: Secure.
  - Bond: Continuous over whole surface, with no air pockets.
  - Appearance on completion: Flat with no wrinkles.
- Laps: Seal using materials and method recommended by membrane manufacturer.
- Joints in second layer (if any): Stagger by half a sheet.
- Upstands, kerbs and other penetrations: Enclose edges of insulation. Fully seal at abutment by bonding or taping.

695 SEPARATING LAYER
- Give notice: Where it is or becomes apparent that a separating layer is required.

ASPHALTING

720 DELIVERY
- Condition of mastic asphalt as delivered to site:
  - Hot-prepared, do not remelt on site, or
  - Blocks: Remelt on site, temperature of material not to exceed 230°C.

730 TRANSPORTING
- Transport distances: Minimize to avoid excessive cooling of molten mastic asphalt.
- Buckets, barrows or dumpers used for mastic asphalt: Line with minimum quantity of fine inert dust. Use silica or similar acid resisting dust where acid resisting mastic asphalt is being used.

735 LOCALIZED HEATING
- Blowlamps and gas torches: Use only types with controlled gradual heating during laying, removal and repair of mastic asphalt.

740 APPLICATION
- Laying mastic asphalt:
  - Standard: To BS 8218.
- Mastic asphalt coats:
  - Apply in bays to even thickness. Do not use re-heated asphalt.
  - Full thickness: Maintain around external angles, at junctions and tuck-ins.

- Fillets at internal angles: Solid in two layers fully fused to asphalt coating.
- Previously laid coats: Protect whilst exposed.
- Successive coats: Apply without delay and within same working period.
  Stagger joints between bays in consecutive coats (minimum): 75 mm.
- Finish: Even surface, smooth and free from imperfections.
- Completion: During final floating operation, whilst asphalt is still warm, apply sand to horizontal surfaces and rub-in well using wooden float.
- Blowing: Pierce and make good affected areas while mastic asphalt is still at working temperature.
- Condition of contact edges of previously laid bays: Warm and clean.
- Condition at completion: Fully sealed, waterproof and free draining.

750 MASTIC ASPHALT SKIRTINGS AND VERTICAL WORK
- Top edge: Tuck into chase or groove and, maintaining full thickness of asphalt, form splayed top to shed water away from substrate.

785 INSTALLING EDGE TRIMS
- Separating layer: To be terminated at trim. Do not carry under or over trim.
- Trim:
  - Clear distance from wall or fascia: 3 mm (minimum).
  - Fasteners: Countersunk wood screws to BS 1210.
    Type: Aluminium.
    Spacing: 30 mm from ends of trims and 300 mm (maximum) centres.
  - Jointing sleeves: Fix one side only.
  - Corners: Use purpose made corner pieces.

SURFACING

870 LAYING CHIPPINGS
- Preparation:
  - Gravel guards: Fit to outlets.
  - Condition of substrate: Clean.
- Dressing compound: Hot or cold application. Evenly pour at 1.5 kg/m².
- Chippings application (approximately): 16 kg/m².
- Excess loose chippings: On completion, remove without exposing asphalt.

COMPLETION

910 INSPECTION
- Interim and final roof inspections: Submit reports.

940 COMPLETION
- Roof areas: Clean.
  - Outlets: Clear.
- Work necessary to provide a weathertight finish: Complete.
- Storage of materials on finished surface: Not permitted.
- Completed mastic asphalt: Do not damage. Protect from petroleum based solvents and other chemicals, traffic and adjacent or high level working.

J30 LIQUID APPLIED TANKING/ DAMP PROOFING

TYPES OF TANKING/ DAMP PROOFING

110 COLD APPLIED TANKING
- Substrate: Various.
- Coating material:
  - Manufacturer: Ruberoid Building Products.
    Product reference: Synthaprufe.
  - Number of coats: 2 or 3 As specified.
  - Coverage per coat (minimum): As recommended by manufacturer.
- Blinding: sand.

EXECUTION

205 SUITABILITY OF SUBSTRATE
- Preliminary work: Complete including:
  - Chases.
  - External angles.
  - Formation of upstands and kerbs.
  - Movement joints.
  - Penetrations/ Outlets.
- Surface: Sound, even textured, clean and dry.

COATINGS

210 COATING APPLICATION
- Adjacent surfaces exposed to view in finished work: Protect.
- Coatings:
  - Uniform, continuous coverage.
  - Firmly adhered to substrate and free from imperfections.
  - Prevent damage to finished coating.
- Penetrations: Completely impervious.
- Final covering: Apply as soon as possible after coating has hardened.

270 BLINDING
- Coatings: Blind whilst still tacky.
- Surplus material: Remove when coatings are completely dry.

J31 LIQUID APPLIED WATERPROOF ROOF COATINGS

TYPES OF COATING

110 ROOF COATING
- Substrate: Various.
  - Preparation: As specified.
- Waterproof coating: As specified in schedules.
  - System manufacturer: As specified in schedules.
  - Application: As manufacturers recommendations.
  - Reinforcement: As specified in schedules.

EXECUTION GENERALLY

410 ADVERSE WEATHER
- Do not apply coatings:
  - In wet conditions or at temperatures below 5°C, unless otherwise permitted by coating manufacturer.
  - In windy conditions (wind speeds in excess of 7 m/s) unless adequate temporary windbreaks are erected.
- Unfinished areas of roof: Keep dry.

420 SUITABILITY OF SUBSTRATE
- Surfaces to be coated:
  - Firmly fixed, clean, dry, smooth, free from frost, contaminants, loose material, voids, protrusions and organic growths.
  - Compatible with coating system.
- Preliminary work: Complete (including formation of upstands, kerbs, box gutters, sumps, grooves, chases, expansion joints, and fixing of battens, fillets, anchoring plugs/ strips, etc.).
- Moisture content and stability: Must not impair integrity of roof.

EXISTING SUBSTRATES

515 EXISTING FLASHINGS
- General: Raise flashings to clean surfaces to receive coatings. Leave raised during coating application and lower only after full curing.
- Damaged lengths: Replace with new flashings (specified in another section).

520 PRELIMINARY POWER WASH
- General: Before renewing existing coverings, water jet clean all areas. Allow to dry.

530 MAKING GOOD EXISTING LIQUID APPLIED ROOF COATINGS
- General: Inspect for adherence and repair defective areas in accordance with proposed coating manufacturer's recommendations.

535 MAKING GOOD EXISTING BITUMEN SHEET
- Blisters: Star cut, dry out and rebond.
- Cracked and defective areas: Cut back to substrate. Dry out substrate and patch level with existing surface with layers of matching bitumen sheet, lapped not less than 100 mm onto existing sheet.

540 MAKING GOOD EXISTING ASPHALT
- Defective areas: Remove by softening. Dry out substrate and patch level with existing surface with coats of matching asphalt, top coat lapped not less than 75 mm onto existing asphalt and to half its depth.

545 MAKING GOOD EXISTING FIBRE CEMENT SHEET
- Loose and soft surfaces: Wet and remove. Allow to dry.
- Cracked and damaged sheets: Replace.
- Bolt heads: Tighten and crop where necessary.
  - Corrosion and oxidation: Abrade back to bright metal.
- Asbestos based materials: Making good or removal must be carried out by a contractor licensed by the Health and Safety Executive prior to other works starting in their location.

550 MAKING GOOD EXISTING METAL SHEET
- Loose coatings: Remove.
- Corrosion and oxidation: Remove back to bright metal.
- Structurally unsound sheets: Replace.

555 MAKING GOOD EXISTING CEMENTITIOUS SLABS/ SCREEDS
- Loose and hollow surfaces, sharp edges and projections: Remove.
- Voids and cracks: Fill with cement based repair mortar.

560 EXISTING EDGE TRIMS
- Fasteners: Check security. Replace as necessary.
- Existing coverings: Cut out from trim edge recess sufficient to accommodate coatings.

565 EXISTING GUTTERS/ OUTLETS
- Dirt, debris and build up of previous coverings/ coatings: Clean out to restore free flow of water.

570 EXISTING CRACKS/ GAPS
- General: Rake out, clean and make good with sealants or repair systems recommended by coating manufacturer.

575 FINAL POWER WASH
- General: Water jet clean all areas to provide a contamination free surface for coating application. Allow to dry.

580 STERILIZATION TREATMENT
- Preparation: Complete making good and cleaning down of existing coverings.
- Biocidal solution: Apply to all areas previously subject to organic growth. Allow to dry.

ROOF COATING SYSTEM

720 PRIMER
- Application: Apply by brush or roller

750 PRELIMINARY LOCAL REINFORCEMENT
- Reinforcement strip: Apply to junctions at upstands, penetrations and outlets, also to joints and fixings in discontinuous unit substrates. Bed in a preliminary application of coating. Smooth out wrinkles and press into coating to exclude air. Lap joints in length.

760 APPLICATION OF ROOF COATINGS
- Thickness: Monitor by taking wet/ dry film thickness readings.
- Continuity: Maintain full thickness of coatings around angles, junctions and features.
- Rainwater outlets: Form with watertight joints.
- Drainage systems: Do not allow liquid coatings to enter piped rainwater or foul systems.
- Edge trims: Apply coatings over horizontal leg of trim and into recess.

COMPLETION

910 INSPECTION
- Coating surfaces: Check when cured for pinholes and discontinuities.
  - Defective areas: Apply another layer of coating.

940 COMPLETION
- Roof areas: Clean.
  - Outlets: Clear.
  - Flashings: Dressed into place.
- Work necessary to provide a weathertight finish: Complete.
- Storage of materials on finished surface: Not permitted.
- Completed coatings: Protect against damage from traffic and adjacent or high level working.

J40 FLEXIBLE SHEET TANKING/ DAMP PROOFING

GENERALLY

100 PRICES ALSO TO INCLUDE
- Vapour barriers to include returning at all edges against abutments and around perforations etc for the full depth of any insulating underlay separating it from the finish.

105 DEFINITIONS
- "Flat". Not exceeding 15 degrees pitch.

TYPES OF TANKING/ DAMP PROOFING

120 LOOSE LAID POLYETHYLENE DAMP PROOF MEMBRANE
- Substrate: Various.
- Membrane manufacturer: Contractors choice.
  - Product reference: Agrement certified.
- Thickness/ Gauge: As specified in schedules.
- Joints: Continuous mastic strip between minimum 150mm overlaps, top sheet sealed with jointing tape.
  - Surface to be joined: Clean and dry.

WORKMANSHIP

310 WORKMANSHIP GENERALLY
- Condition of substrate:
  - Clean and even textured, free from voids and sharp protrusions.
  - Moisture content: Compatible with dpm/ tanking.
- Air and surface temperature: Do not apply sheets if below minimum recommended by membrane manufacturer.

- Condition of membrane at completion:
  - Neat, smooth and fully supported, dressed well into abutments and around intrusions.
  - Completely impervious and continuous.
  - Undamaged. Prevent puncturing during following work.
- Permanent overlying construction: Cover membrane as soon as possible.

330 PRIMERS
- Types and application: As recommended for the purpose by membrane manufacturer.
- Curing: Allow to dry thoroughly before covering.

360 JUNCTIONS WITH PROJECTING DPCS/ CAVITY TRAYS
- Condition of adjoining surfaces: Clean and dry.
- Projecting dpcs/ cavity trays: Lap and fully bond/ seal with sheeting.
  - Laps (minimum): 150mm.
  - Bonding/ Sealing: Continuous mastic strip.

365 JUNCTIONS WITH FLUSH DPCS/ CAVITY TRAYS
- Preparation of adjacent dpcs/ cavity trays:
  - Expose edge where concealed.
  - Adjoining surfaces: Clean and dry.
- Joints: Bond/ Seal sheeting to wall.
  - Dressing of sheeting beyond dpc/ cavity tray (minimum): 50 mm.
  - Bonding/ Sealing: Continuous mastic strip.

J41 BITUMEN SHEET ROOF COVERINGS

GENERALLY

100 PRICES ALSO TO INCLUDE
- Vapour barriers to include returning at all edges against abutments and around perforations etc for the full depth of any insulating underlay separating it from the finish.
- Holes through coverings to include holes and other perforations through any other materials included in the description of the covering.
- Holes through existing work to include holes through any associated membranes underlays or vapour barriers.
- Felt roofing to include 50 mm side laps and 75 mm end laps and bonding to the base and bonding layers to each other.
- Proprietary roofing to include the number of layers and extent of laps required by the manufacturers.

105 DEFINITIONS
- "Flat". Not exceeding 15 degrees pitch.

110 CONTRARY TO SMM
- Roofing has not been given a stated pitch but described as flat (SMM J40.1-2.* and J41.1-2.*).

TYPES OF COVERING

110 BUILT-UP BITUMEN SHEET WARM ROOF COVERING
- Waterproof covering:
  - System manufacturer: Contractors choice.
  - First layer: As specified in schedules.
    Attachment: Partial bonding.
  - Intermediate layer: As specified in schedules.
    Attachment: Bitumen pour and roll bonding.
  - Top layer: As specified in schedules.
    Attachment: Bitumen pour and roll bonding.
- Surface protection: As specified in schedules.
- Accessories: Aluminium or GRP edge trims.

115 SINGLE LAYER BITUMEN SHEET WARM ROOF COVERING
- Waterproof covering:
  - Manufacturer: Contractors choice.
  - Attachment: Torch on.
- Surface protection: As specified in schedules.

130 BUILT-UP BITUMEN SHEET COLD ROOF COVERING
- Waterproof covering:
  - System manufacturer: Contractors choice.
  - First layer: As specified in schedules.
    Attachment: Galvanised clout nails at 150mm (maximum) grid centres.
  - Intermediate layer: As specified in schedules.
    Attachment: Bitumen pour and roll bonding.
  - Top layer: As specified in schedules.
    Attachment: Bitumen pour and roll bonding.
- Surface protection: Solar reflective paint or chippings as specified.
- Accessories: Aluminium or GRP edge trims.

PRODUCTS

310 ANCILLARY PRODUCTS AND ACCESSORIES
- Types: Recommended by bitumen sheet manufacturer.

320 PRIMERS
- Type:
  - As recommended by bitumen sheet manufacturer.

325 BONDING COMPOUNDS
- Type:
  - As recommended by bitumen sheet manufacturer.
- Restriction: For heat sensitive insulation materials, use cold bonding compounds.

330 TIMBER FOR TRIMS, ETC
- Quality: Planed. Free from wane, pitch pockets, decay and insect attack (except ambrosia beetle damage).
- Moisture content at time of covering (maximum): 22%.
- Preservative treatment: Clause Z12/160 .

335 ANGLE FILLETS
- Material: Softwood.
  - Size (minimum): 50 x 50 mm, triangular in section.
- Restriction: Fillets under torch-on bitumen sheets to be non-combustible.

345 EDGE TRIMS
- Type: Aluminium.
  - Grade to BS EN 12020-1: EN-AW 6063T4 or EN-AW 6063T6.
- Manufacturer: Paptrim.
  - Product reference: As specified in schedules.
- Lengths (maximum): 3 m.

475 CHIPPINGS AND DRESSING COMPOUND
- Type: Limestone.
- Size (minimum): 10mm.
- Dressing compound: To BS 3690-1 and BS EN 12591.

EXECUTION GENERALLY

515 ADVERSE WEATHER
- Unfinished areas of roof: Keep dry.
- Work in severe or continuously wet weather: Suspend or provide and maintain temporary cover over working area.

- Unavoidable wetting: Minimize and make good damage.

530 APPLYING PRIMERS
- Application rate: As recommended by manufacturer.
- Surface coverage: Even and full.
- Coats: Fully bond. Allow volatiles to dry off thoroughly between coats.

540 APPLYING BONDING COMPOUNDS
- Roof sited boilers: Not permitted.
- Temperature of compound: Suitable to achieve bond over whole surface. Do not overheat.

SUBSTRATES/ VAPOUR CONTROL LAYERS/ WARM ROOF INSULATION

610 SUITABILITY OF SUBSTRATES
- Surfaces to be covered: Secure, clean, dry, smooth, free from frost, contaminants, voids and protrusions.
- Preliminary work: Complete (including formation of upstands, kerbs, box gutters, sumps, grooves, chases, expansion joints, and fixing of battens, fillets and anchoring plugs/ strips).
- Moisture content and stability of substrate: Must not impair roof integrity.

620 RENEWING EXISTING COVERINGS
- Existing roof: Do not damage.
- Timing: Remove only sufficient coverings as will be renewed and waterproofed on the same day.

625 REMOVING EXISTING COVERINGS
- Mechanical stripping: Not permitted unless authorised by CA.
- Exposed substrate: Do not damage.

630 MAKING GOOD EXISTING BITUMEN SHEET ROOF COVERING
- Dust, dirt, debris, moss, plants and grease: Remove.
- New materials and accessories: Compatible with existing.
- Blisters: Star cut, dry out and rebond.
- Defective areas of bitumen sheet: Cut back to substrate. Dry out substrate. Patch level with existing covering with layers of matching bitumen sheet, lapped not less than 100 mm onto existing sheet.
- Cracked and split bitumen sheet: Cut back to substrate 150 mm wide at cracks and splits. Dry out substrate. Insert 150 mm wide strip of matching bitumen sheet, bonded to substrate at edges only. Fully bond a layer of bitumen sheet over strip lapped not less than 100 mm onto existing bitumen sheet at edges.
- Stress failure at edge trims: Cut back bitumen sheet to substrate. Secure ends of edge trims. Patch level with existing surface with layers of matching bitumen sheet.
- Detached bitumen sheet at upstands: Repair, re-adhere and protect with additional layer of matching bitumen sheet if necessary.
- Defects at penetrations: Cut out, clean, prime and reseal.

640 INSTALLING TIMBER FOR TRIMS
- Fasteners: Sherardized steel screws.
- Fixing centres (maximum): 600 mm.

660 JOINTS IN RIGID BOARD SUBSTRATES
- Cover strip: Lay centrally over substrate joints before laying vapour control layers or coverings. Adhere to substrate with bonding compound along edges only.

670 LAYING VAPOUR CONTROL LAYER
- Fixing: Securely bond or nail to substrate.
- Laps: Fully bonded.
  - Side (minimum): 50 mm.
  - End (minimum): 75 mm.
- Joints in second layer (if any): Stagger by half a sheet.
- Penetrations: Fully seal using bonding or taping methods recommended by manufacturer.

- Edges of insulation at roof edges, abutments, upstands, kerbs, penetrations and the like: Enclose, with vapour control layer:
  - Dressed up sufficiently, providing 25 mm (minimum) seal when overlapped by the roof covering; or
  - Turned back 150 mm (minimum) over the insulation and sealed down.

WATERPROOF COVERINGS/ ACCESSORIES

710 LAYING BITUMEN SHEETS GENERALLY
- Direction of laying: Where practicable, lay from lowest point of roof, with bitumen sheets unrolled up the slope.
  - Sheets: Install so that water drains over and not into laps.
- Laps: Seal.
  - Side (minimum): 50 mm.
  - End (minimum): 75 mm.
- Head and side laps: Offset.
  - Side laps offset:
    Two layer coverings: 1/2 sheet width.
    Three layer coverings: 1/3 sheet width.
- Intermediate and top layers: Fully bond.
- Successive layers: Apply without delay. Do not trap moisture.

720 NAILING FIRST LAYER OF BITUMEN SHEETS TO TIMBER SUBSTRATE
- Setting out: Unroll bitumen sheet and work from one end. Minimize wrinkles.
- Nails: Galvanized mild steel extra large round-headed clout nails to BS 1202-1, 20 mm long.
- Fixing centres:
  - Perimeter of roof areas and along centres of laps: 50 mm.
  - Over area of bitumen sheets: 150 mm grid centres.

730 PARTIAL BONDING OF BITUMEN SHEETS
- Venting first layer: Loose lay. Do not carry up angle fillets and vertical surfaces or through details.
  - Long edges: Overlap 50 mm.
  - Ends: Butt together.
- Intermediate layer: Fully bond to first layer and through to substrate.

735 POUR AND ROLL BONDING OF BITUMEN SHEETS
- Bonding compound: Hot and fluid when bitumen sheets are laid. Spread evenly so that a small quantity is squeezed out at each edge.
- Bond: Full over whole surface, with no air pockets.
- Excess compound at laps:
  - First and intermediate layers. Spread out.
  - Top layer: Remove.

740 TORCH-ON BONDING OF BITUMEN SHEETS
- Bond: Full over whole surface, with no air pockets.
- Excess compound at laps of top layer: Leave as a continuous bead.

750 LAYING MINERAL FACED BITUMEN SHEETS
- Lap positions and detailing of ridges, eaves, verges, hips, abutments, etc: Submit proposals.
- Setting out: Neat, with carefully formed junctions.
- Lap bonding: Carry out only at prefinished margins or prepared 'black to black' edges.
- Excess bonding compound at laps: Remove whilst still warm.

755 LAYING METAL FACED BITUMEN SHEETS
- Lap positions and detailing of ridges, eaves, verges, hips, abutments, etc: Submit proposals.
- Setting out: Neat with carefully formed junctions.
- Excess bonding compound at laps: Remove whilst still warm.
- Metal face: Do not mark, crease or stain.

770 **DETAILS GENERALLY**
- Requirement: Waterproof. Form with adequate overlapping, staggering of laps and bonding of successive bitumen sheets.
- Strips of bitumen sheet required for 'linear' details: Cut from length of roll, rather than from width.

775 **SKIRTINGS/ UPSTANDS**
- Angle fillets: Fix by bitumen bonding or nailing.
- Venting first layer of bitumen sheet: Stop at angle fillet, or run under angle fillet. Fully bond in bitumen for 300 mm around perimeters. Overlap onto upstand with strips of BS 747 type 3B bitumen sheet, fully bonded.
- Other layers of bitumen sheet: Carry in staggered formation up upstand, with each layer fully bonded. Where practicable, carry top layer over top of upstand.
- Upstands:
  - At ends of rolls: Form with bitumen sheet carried up without using separate strip.
  - Elsewhere: Form with matching strips of bitumen sheet, maintaining laps.

785 **INSTALLING EDGE TRIMS**
- First/ intermediate layers bitumen sheet: Lay over roof edge upstand. Project free edge 25 mm from wall or fascia.
- Trim setting out: 3 mm (minimum) clear from wall or fascia.
- Fasteners: 50 mm countersunk wood screws to BS 1210.
  - Material: As specified in schedules.
  - Spacing: 30 mm from ends of each trim section and at 300 mm (maximum) centres.
- Joints: Formed with jointing sleeves, fixed one side only and with 3 mm expansion gaps.
- Corners: Formed with purpose made corner pieces.
- Completion of trims:
  - Contact surfaces: Prime.
  - Joints: Cover with 150 mm long pads of bitumen sheet, bonded to trim.
- Completion of bitumen sheet:
  - Top layer: Butt joint to rear edge of trim.
  - Cover strip: Fully bond to trim and top layer of bitumen sheet. Carry over roof edge upstand and lap 75 mm onto roof.
  Cover strip material: As specified in schedules.

SURFACING

870 **LAYING CHIPPINGS**
- Outlets: Fit gravel guards.
- Condition of substrate: Clean.
- Dressing compound: Hot or cold application. Evenly pour at 1.5 kg/m^2.
- Chippings application: Do not pile to excessive heights. Spread evenly at a rate of 16kg/m2.
- Completion: Remove loose excess chippings without exposing dressing compound.

COMPLETION

910 **INSPECTION**
- Interim and final roof inspections: Submit roof covering manufacturer's reports.

K: LININGS/SHEATHING/DRY PARTITIONING

K10 PLASTERBOARD DRY LININGS/ PARTITIONS/ CEILINGS

GENERAL

100 METHODS OF MEASUREMENT
- No distinction has been made between plain and irregular internal and external angles.

105 CONTRARY TO SMM
- Walls include attached columns and ceilings include beams (SMM K10.2.2-3.*).

335 ADDITIONAL SUPPORTS
- Framing: Accurately position and securely fix to fully support:
  - Partition heads: Running parallel with, but offset from main structural supports.
  - Fixtures, fittings and service outlets: To suit fixings. Mark framing positions clearly and accurately on linings.
  - Board edges and lining perimeters: As recommended by board manufacturer to suit type and performance of board.

INSTALLATION

435 DRY LININGS GENERALLY
- General: Use fixing, jointing and finishing materials, components and installation methods recommended by board manufacturer.
- Cutting plasterboards: Neatly and accurately without damaging core or tearing paper facing.
  - Cut edges: Minimize and position at internal angles wherever possible. Mask with bound edges of adjacent boards at external corners.
- Fixings plasterboards: Securely and firmly to suitably prepared and accurately levelled backgrounds.
- Finishing: Neatly to give flush, smooth, flat surfaces free from bowing and abrupt changes of level.

445 CEILINGS
- Sequence: Fix boards to ceilings before dry lined walls and partitions.
- Orientation of boards: Fix with bound edges at right angles to supports and with ends staggered in adjacent rows.
- Two layer boarding: Stagger joints between layers.

505 INSTALLING MINERAL WOOL INSULATION
- Fitting: Accurately and firmly with closely butted joints and no gaps.
- Widths: Lay insulation in the widest practical widths to suit grid member spacings.
- Services: Do not cover electrical cables that are not sized accordingly.
  - Ceilings: Cut insulation carefully around electrical fittings, etc. Do not lay over luminaires.
- Fixing:
  - Partitions/ wall linings: Fasten insulation at head of frame to prevent displacement, unless insulation is self supporting and friction fitted between studs.
  - Ceilings (vertical/ sloping areas): Fasten insulation to prevent displacement.

510 SEALING GAPS AND AIR PATHS
- Application of sealant to dry lining systems: Continuous beads leaving no gaps or air paths.
- Acoustic sealant: Apply to perimeter abutments with walls, floors, ceilings and around openings.
  - Gaps greater than 6 mm wide: After application of sealant, complete sealing with jointing compound.
- Air pressure sealant: Apply to board to board and board to metal frame junctions in pressurised shafts, ducts, etc.

555 FIRE STOPPING AT PERIMETERS OF DRY LINING SYSTEMS
- Material: Tightly packed mineral wool or intumescent mastic/ sealant.
- Application: To perimeter abutments to provide a complete barrier to smoke and flame.

560 JOINTS BETWEEN BOARDS
- Tapered edged plasterboards:
  - Bound edges: Lightly butted.
  - Cut/ unbound edges: 3 mm gap.
- Square edged plasterboards: 3 mm gap.
- Square edged fibre reinforced gypsum boards: 5 mm gap.

565 VERTICAL JOINTS
- Centre joints on studs.
  - Partitions: Stagger joints on opposite sides of studs.
  - Two layer boarding: Stagger joints between layers.

570 HORIZONTAL JOINTS
- Surfaces exposed to view: Horizontal joints not permitted.
  - Exception: Where height of partition/ lining exceeds maximum available length of board. Agree positions of joints.
- Two layer boarding: Stagger joints between layers by at least 600 mm.
- Provide support to edges of boards with additional framing.
  - Two layer boarding: Support edges of outer layer.

580 INSULATION BACKED PLASTERBOARD
- General: Do not damage insulation or cut it away to accommodate services.
- Installation at corners: Carefully cut back insulation or plasterboard as appropriate along edges of boards to give a continuous plasterboard face, with no gaps in insulation.

590 FIXING PLASTERBOARD TO METAL FRAMING/ FURRINGS
- Fixing to framing/ furrings: Securely and firmly working from the centre of each board at the following centres (maximum):
  - Partitions/ wall linings: 300 mm. Reduce to 200 mm at external angles.
  - Ceilings: 230 mm. Reduce to 150 mm at board ends and at lining perimeters.
- Position of screws from edges of boards (minimum): 10 mm.
  - Screw heads: Set in a depression. Do not break paper or gypsum core.

592 FIXING INSULATION BACKED PLASTERBOARD TO METAL FURRINGS
- Fixing to furrings: In addition to screw fixings (clause 590) apply continuous beads of adhesive sealant to furrings.

610 FIXING PLASTERBOARD TO TIMBER
- Fixing to timber: Securely and firmly working from the centre of each board at the following centres (maximum):
  - Nail fixing: 150 mm.
  - Screw fixing to partitions/ wall linings: 300 mm. Reduce to 200 mm at external angles.
  - Screw fixing to ceilings: 230 mm.
- Position of nails/ screws from edges of boards (minimum):
  - Bound edges: 10 mm.
  - Cut/ unbound edges: 13 mm.
- Position of nails/ screws from edges of timber supports (minimum): 6 mm.

FINISHING

650 LEVEL OF DRY LINING ACROSS JOINTS
- Sudden irregularities: Not permitted.
- Joint deviations: Measure from faces of adjacent boards using methods and straightedges (450 mm long with feet/ pads) to BS 8212, clause 3.3.5.
  - Tapered edge joints:
    Permissible deviation (maximum) across joints when measured with feet resting on boards: 3 mm.

- External angles:
  Permissible deviation (maximum) for both faces: 4 mm.
- Internal angles:
  Permissible deviation (maximum) for both faces: 5 mm.

670 SEAMLESS JOINTING TO PLASTERBOARDS
- Cut edges of boards: Lightly sand to remove paper burrs.
- Filling and taping: Fill joints, gaps and internal angles with jointing compound and cover with continuous lengths of paper tape, fully bedded.
- Protection of edges/ corners: Reinforce external angles, stop ends, etc. with specified edge/ angle bead.
- Finishing: Apply jointing compound. Feather out each application beyond previous application to give a flush, smooth, seamless surface.
- Nail/ screw depressions: Fill with jointing compound to give a flush surface.
- Minor imperfections: Lightly sand jointing and spotting to remove any minor imperfections.

680 SKIM COAT PLASTER FINISH
- Manufacturer: Contractors choice.
  Thickness: 2 - 3 mm.
- Joints: Fill and tape except where coincident with metal beads.
- Finishing: Trowel/ float to a tight, matt, smooth surface with no hollows, abrupt changes of level or trowel marks.

695 INSTALLING BEADS/ STOPS
- Cutting: Neatly using mitres at return angles.
- Fixing: Securely using longest possible lengths, plumb, square and true to line and level, ensuring full contact of wings with background.
- Finishing: After joint compounds/ plasters have been applied, remove surplus material while still wet from surfaces of beads which are exposed to view.

725 REPAIRS TO EXISTING PLASTERBOARD
- Filling small areas with broken cores: Cut away paper facing, remove loose core material and fill with jointing compound.
  - Finishing: To give a flush, smooth surface suitable for redecoration.
- Large patch repairs: Cut out damaged area and form neat hole with rectangular sides. Replace with identical piece of plasterboard.
  - Fixing: Use methods to suit type of dry lining, ensuring full support to all edges of existing and new plasterboard.
  - Finishing: Fill joints, tape and apply jointing compound to give a flush, smooth surface suitable for redecoration.

K11 RIGID SHEET FLOORING/ SHEATHING/ DECKING/ SARKING/ LININGS/ CASINGS

GENERALLY

100 METHODS OF MEASUREMENT
- Notwithstanding SMM Clause, K11.D1, all work described as to "roofs" and "dormers" and all "weather boarding" is deemed external.
- In amplification of SMM K11.D2 all sawn and regularized timbers are given as basic sizes.
- All wrought timbers are given as nominal sizes. (The limits on planning margins are to be in accordance with BS EN 1313-1 for softwood and BS EN 1313-2 for hardwood).
- The thickness of all manufactured boards (e.g. plywood and chipboard) given in the description is the finished size.
- In amplification of SMM K11.S12, where fixings are left to the discretion of the Contractor they are deemed nailed, pinned, spiked or stapled unless otherwise stated.

105 PRICES ALSO TO INCLUDE
- Work creosoted or treated with wood preservative is to include additional treatment to holes and cut surfaces as specified.
- Manufactured boards and sheeting are to include any necessary balancing veneer when only one face veneer is specified.

- Selecting and keeping hardwood clean for clear finish as applicable.
- Punching fixings below exposed surfaces.
- Dressing all joinery in contact with plaster with linseed oil before plastering and subsequently cleaning off.
- Tongueing, crosstonguing, housing and like labours are to include grooves in building board, timber and faced timber.
- Timber flooring and boarding is to include splayed heading joints.
- Flooring, boarding, panels of timber, manufactured boards and wood and asbestos products are to include fixing each member at each bearing.
- Holes through existing work are to include holes through any associated membranes underlays or vapour barriers.
- Sanding and the like to existing work is to include punching nails below exposed surfaces.

TYPES OF FLOORING/ SHEATHING/ DECKING/ SARKING/ LINING/ CASINGS

815 PLYWOOD
- Plywood: Manufactured to an approved national standard.
   - Manufacturer/ Supplier: Contractors choice.
   - Bonding quality to BS EN 314-2
   - Appearance class to BS EN 635: Class as specified in schedules.
   - Finish: as specified in schedules.
   - Thickness: as specified in schedules.
   - Treatment: as specified in schedules.
- Fixing to supports:
   - Fasteners: Contractors discretion.
   - Fixing centres (maximum):
     Around board edges: 150 mm.
     Along intermediate supports: 300 mm.
   - Fixing distance from edges (minimum): 10 mm.

820 BLOCKBOARD
- Board:
   - Manufacturer/ Supplier: Contractors choice.
   - Face veneer species: As specified by CA.
   - Thickness/ number of plies: 18mm.
- Fixing to supports:
   - Fasteners: Contractors discretion.
   - Fixing centres (maximum):
     Around board edges: 150mm.
     Along intermediate supports: 300mm.
   - Fixing distance from edges (minimum): 10mm.

855 HARDBOARD
- Hardboard: To BS EN 622-2, Type HB.
   - Manufacturer/ Supplier: Contractors choice.
   - Fire rating: Class 1 surface spread of flame.
   - Thickness: as specified in schedules.
   - Edges: Square.
- Fixing to supports:
   - Fasteners: Contractors discretion.
   - Fixing centres (maximum):
     Around edges of sheets: 150mm.
     Along intermediate supports: 300mm.
   - Fixing distance from edges (minimum): 10mm.

875 MEDIUM DENSITY FIBREBOARD
- Medium density fibreboard: To BS EN 622-5, Type MDF.
   - Manufacturer/ Supplier: Contractors choice.
   - Formaldehyde class to BS EN 622-1: Class A.
   - Fire rating: Class 1 surface spread of flame.
   - Thickness: as specified in schedules.

- Edges: Square.
- Fixing to supports:
  - Fasteners: Contractors discretion.
  - Fixing centres (maximum):
    Around board edges: 150mm.
    Along intermediate supports: 300mm.
  - Fixing distance from edges (minimum): 12 mm.

WORKMANSHIP

910 INSTALLATION GENERALLY
- Timing: Building to be weathertight before fixing boards internally.
- Moisture content of timber supports (maximum): 18%.
- Joints between boards: Accurately aligned, of constant width and parallel to perimeter edges.
- Methods of fixing, and fasteners: As section Z20 where not specified otherwise.

930 ADDITIONAL SUPPORTS
- Additional studs, noggings/ dwangs (Scot) and battens:
  - Provision: In accordance with board manufacturer's recommendations and as follows:
    Tongue and groove jointed rigid board areas: To all unsupported perimeter edges.
    Butt jointed rigid board areas: To all unsupported edges.
  - Size: Not less than 50 mm wide and of adequate thickness.
  - Quality of timber: As for adjacent timber supports.
  - Treatment (where required): As for adjacent timber supports.

940 BOARD MOISTURE CONTENT AND CONDITIONING
- Moisture content of boards at time of fixing: Appropriate to end use.
- Conditioning regime: Submit proposals.

950 MOISTURE CONTENT TESTING
- Test regime and equipment: Submit proposals.
- Test results: Submit record of tests and results.

960 FIXING GENERALLY
- Boards/ Sheets: Fixed securely to each support without distortion and true to line and level.
- Fasteners: Evenly spaced in straight lines and, unless otherwise recommended by board manufacturer, in pairs across joints.
  - Distance from edge of board/ sheet: Sufficient to prevent damage.
- Surplus adhesive: Removed as the work proceeds.

980 OPEN JOINTS
- Perimeter joints, expansion joints and joints between boards: Free from plaster, mortar droppings and other debris.
- Temporary wedges and packings: Removed on completion of board fixing.

990 ACCESS PANELS
- Size and position: Agree before boards are fixed.
- Additional noggings/ dwangs (Scot), battens, etc: Provide and fix as necessary.

K20 TIMBER BOARD FLOORING/ SARKING/ LININGS/ CASINGS

TYPES OF FLOORING/ SARKING/ LININGS/ CASINGS

230 TIMBER BOARD
- Substrate: Timber joists at 400mm centres.
- Boards:
  - Species: Contractors choice.
  - Profile: As specified in schedule.
  - Finished overall width (exposed width after fixing): Contractors choice.
  - Finished thickness: As specified in schedule.
  - Moisture content at time of fixing: As appropriate in accordance with BS-EN942.

- Treatment:
  - Standard: To NBS section Z12 and British Wood Preserving and Damp-proofing
- Fixing: Contractors discretion.

WORKMANSHIP

310 WORKMANSHIP GENERALLY
- Protection during and after installation: Keep boards dry. Protect from dirt, stain and damage until Completion.
- Boards to be used internally: Do not install until building is watertight.
- Methods of fixing, and fasteners: As section Z20.
- Moisture content of timber supports at time of fixing boards: Not more than 18%.

330 MOISTURE CONTENT OF TIMBER
- Conditions during and after installation: Control ambient temperature and humidity conditions to maintain moisture content at average level specified in BS EN 942, table B.1 for the relevant service condition until Completion.
- Test for moisture content: When instructed, using an approved moisture meter.

350 TREATED TIMBER
- Surfaces exposed by minor cutting and/ or drilling: Treat with two flood coats of a solution recommended by main treatment solution manufacturer.

360 ACCESS PANELS
- Size and position: Agree before fixing boards.
- Additional noggings/ dwangs, battens, etc: Provide as necessary.

370 FIXING BOARDS
- Generally: Fix boards securely to each support to give flat, true surfaces free from undulations, lipping, splits and protruding fasteners.
- Wood movement: Position boards and fixings to prevent cupping, springing, excessive opening of joints and other defects.
- Heading joints: Tightly butted, central over supports and at least two board widths apart on any one support.
- Edges: Plane off proud edges.
- Exposed nail heads: Neatly punch below surface.

K40 DEMOUNTABLE SUSPENDED CEILINGS

INSTALLATION

305 SETTING OUT
- General: Accurate, continuous, even, and jointed at regular intervals.
  - Soffits: Free from undulations, lipping and distortions in grid members.
- Infill and access units, integrated services: Fitted correctly and aligned.
- Edge/ perimeter infill units size (minimum): Half standard width or length.
- Corner infill units size (minimum): Half standard width and length.
- Grid: Position to suit infill unit sizes.
  - Permitted deviations from nominal sizes: Allow for.
- Infill joints and exposed suspension members: Straight, aligned and parallel to walls.
- Suitability of construction: Give notice where building elements and features to which the ceiling systems relate are not square, straight or level.

315 PROTECTION
- Loading: Do not apply loads for which the suspension system is not designed.
- Ceiling materials: Remove and replace correctly using special tools and clean gloves, etc. as appropriate.

325 INSTALLING HANGERS
- Wire hangers: Straighten and tension before use.
- Installation: Install vertical or near vertical without bends or kinks. Do not allow hangers to press against fittings, services, insulation covering ducts/ pipes.
- Obstructions: Where obstructions prevent vertical installation, either brace diagonal hangers against lateral movement, or hang ceiling system on an appropriate rigid sub-grid bridging across obstructions and supported to prevent lateral movement.
- Extra hangers: Provide as required to carry additional loads.
- Fixing:
  - Wire hangers: Tie securely at top with tight bends to loops to prevent vertical movement.
  - Angle/ strap hangers: Do not rivet for top fixing.

335 PERIMETER TRIMS
- Jointing: Neat and accurate, without lipping or twisting.
  - External and internal corners: Mitre joints.
  - Intermediate butt joints: Minimize. Use longest available lengths of trim. Align adjacent lengths.
- Fixing: Fix firmly to perimeter wall or other building structure.
  - Fixing centres: To suit background.

375 BOARD CEILING SYSTEMS
- Fixing boards to grids:
  - Cutting: Cut boards neatly and accurately.
  - Fixing: Screw boards securely and firmly to grid members. Provide a flat surface free from bowing and lipping. Set heads of screws below surface of boards and fill flush with surface.
  - Movement joints: Provide as appropriate for the area of ceiling system and/ or to coincide with movement joints in surrounding structure.
  - Boards applied in two or more layers: Stagger joints.
  - Board edges: Fully support. Screw to grid members.

390 OPENINGS IN CEILING MATERIALS
- General: Neat and accurate. Suit sizes and edge details of fittings. Do not distort ceiling system.

395 INTEGRATED SERVICES
- General: Position services accurately, support adequately. Align and level in relation to the ceiling and suspension system. Do not diminish performance of ceiling system.
- Small fittings: Support with rigid backing boards or other suitable means. Do not damage or distort the ceiling.
  - Surface spread of flame rating of additional supporting material: Match ceiling material.
- Services outlets:
  - Supported by ceiling system: Provide additional hangers.
  - Independently supported: Provide flanges to support ceiling system.

415 INSTALLING INSULATION
- Fitting: Fit accurately and firmly with closely butted joints and no gaps.
- Insulation within individual infill units: Fit closely. Secure to prevent displacement when infill units are installed or subsequently lifted. Reseal cut dustproof sleeving.
- Width: Lay insulation in the widest practical widths to suit grid member spacings.
- Services: Do not cover electrical cables not sized accordingly. Cut insulation carefully around electrical fittings, etc. Do not lay insulation over luminaires.
- Sloping and vertical areas of ceiling system: Fasten insulation to prevent displacement.

421 CEILING SYSTEMS INTENDED FOR FIRE PROTECTION
- Junctions of ceiling systems with perimeter abutments and service penetrations: Seal gaps with tightly packed mineral wool or intumescent sealant to prevent penetration of smoke and flame.
- Ceiling system/ wall junctions: Maintain protective value of ceiling system.
  - Fixings and grounds: Noncombustible.
  - Metal trim: Provide for thermal expansion.
- Access and access panels: Maintain continuity of fire protection.

L: WINDOWS/DOORS/STAIRS

GENERAL

100 METHODS OF MEASUREMENT
- In amplification of SMM L10.D1 and L11.D1 all wrought timbers are given as nominal sizes. (The limits on planing margins are to be in accordance with BS EN 1313-1 for softwood and BS EN 1313-2 for hardwood).
- The thickness of all joinery items given in the descriptions are finished sizes.
- In amplification of SMM L10.S8, L11.S8, L12.S8, L20.S8, L21.S8, L30.S8 and L31.S8 where fixings are left to the discretion of the Contractor they are deemed nailed, pinned, spiked or stapled unless otherwise specified.

105 CONTRARY TO SMM
- The number of door frames and lining sets has not been given (SMM L207.*.*.*).
- Non-standard doors and casements have been measured m2 (SMM L.10.*.*.* and L20.*.*.*).
- Stairs have been measured m2 (length between bottom and top stair nosings x width) and m as applicable (SMM L30.1.*.* and L31.1.*.*).

107 PRICES ALSO TO INCLUDE
- Dressing all joinery in contact with plaster with linseed oil before plastering and subsequently cleaning off.
- Punching fixings below exposed surfaces.
- Fixing only windows, doors, units and the like is to include easing and adjusting.
- Surfaces treated, coated and veneered are to include keeping clean, covering and protecting.
- Woodwork is to include its assembly and fixing by the use of adhesives where normal practice demands this.
- Holes, mortices, sinkings and like labours are to include work cross-grain.
- Touching up off-site painting and treatment of metal work ready for decoration.
- Work creosoted or treated with wood preservative is to include additional treatment to holes and cut surfaces as specified.
- Selecting and keeping hardwood clean for clear finish as applicable.

L10 WINDOWS/ ROOFLIGHTS/ SCREENS/ LOUVRES

GENERAL INFORMATION/ REQUIREMENTS

110 EVIDENCE OF PERFORMANCE
- Certification: Provide independently certified evidence that all incorporated components comply with specified performance requirements.

120 SITE DIMENSIONS
- Procedure: Before starting work on designated items take site dimensions, record on shop drawings and use to ensure accurate fabrication.

COMPONENTS

250 WOOD WINDOWS
- Standard: To BS 644.
- Manufacturer: As specified in schedules or a firm currently registered under a third party quality assurance scheme.
- Exposure category to BS 6375-1/ Design wind pressure: As specified for location.
- Operation and strength characteristics: To BS 6375-2.
- Timber: Generally to BS EN 942.
    - Species: As specified in schedules.
    - Appearance class: J10 for glazing beads, drip mouldings and the like. J40 or better for all other members.
    - Moisture content on delivery: 12-19%.
- Preservative treatment: Organic solvent as section Z12 to softwood.
- Finish as delivered: Factory primed for softwood, smooth sanded for hardwood.

© AAS 2008- 2009

- Glazing details: As specified in schedules.
- Ironmongery/ Accessories: Manufacturers standard, factory fitted.
- Fixing: Clause 780.

350 PVC-U WINDOWS
- Manufacturer: Anglian Tradelines .
    - Product reference: KT Range .
    - Colour/ Texture: white .
- Glazing details: As specified in schedules.
- Fixing: Clause 783.

460 ROOFLIGHTS
- Manufacturer: Velux .
    - Product reference: As specified in schedules.
- Type: As specified in schedules.
- Frame: Manufacturers standard.
- Kerb: As specified in schedules.
- Glazing details: Manufacturers standard.
- Fixing: As manufacturers recommendations.

INSTALLATION

710 PROTECTION OF COMPONENTS
- General: Do not deliver to site components that cannot be installed immediately or placed in clean, dry floored and covered storage.
- Stored components: Stack vertical or near vertical on level bearers, separated with spacers to prevent damage by and to projecting ironmongery, beads, etc.

730 PRIMING/ SEALING
- Wood surfaces inaccessible after installation: Prime or seal as specified before fixing components.

750 BUILDING IN
- General: Not permitted unless indicated on drawings.
- Brace and protect components to prevent distortion and damage during construction of adjacent structure.

755 PVC-U WINDOW INSTALLATION
- Standard: In accordance with clause 783 and British Plastics Federation 'Code of practice for the installation of PVC-U windows and doorsets in new domestic dwellings'.

756 PVC-U REPLACEMENT WINDOW INSTALLATION
- Standard: In accordance with clause 783 and British Plastics Federation 'Code of practice for the survey and installation of replacement plastics windows and doorsets'.

760 REPLACEMENT WINDOW INSTALLATION
- Standard: To BS 8213-4.

765 WINDOW INSTALLATION GENERALLY
- Installation: Into prepared openings.
- Gap between frame edge and surrounding construction:
    - Minimum: 5mm.
    - Maximum: 10mm.
- Distortion: Install windows without twist or diagonal racking.

770 DAMP PROOF COURSES IN PREPARED OPENINGS
- Location: Ensure correct positioning in relation to window frames. Do not displace during fixing operations.

780 FIXING OF WOOD FRAMES
- Standard: As section Z20.
- Fasteners: As appropriate.

- Spacing: When not predrilled or specified otherwise, position fasteners not more than 150 mm from ends of each jamb, adjacent to each hanging point of opening lights, and at maximum 450 mm centres.

783 FIXING OF PVC-U FRAMES
- Standard: As section Z20.
- Fasteners: As appropriate.
  - Spacing: When not predrilled or specified otherwise, position fasteners 150-250 mm from ends of each jamb, adjacent to each hanging point of opening lights, but no closer than 150 mm to a transom or mullion centre line, and at maximum 600 mm centres.

810 SEALANT JOINTS
- Sealant: Polysulphide or Silicone as appropriate
  - Manufacturer: Contractors choice.
  - Colour: As appropriate.
  - Application: As section Z22 to prepared joints. Finish triangular fillets to a flat or slightly convex profile.

820 IRONMONGERY
- Fixing: Assemble and fix carefully and accurately using fasteners with matching finish supplied by ironmongery manufacturer. Do not damage ironmongery and adjacent surfaces.
- Checking/ Adjusting/ Lubricating: Carry out at completion and ensure correct functioning.

L20 DOORS/ SHUTTERS/ HATCHES

PRELIMINARY INFORMATION/ REQUIREMENTS

110 EVIDENCE OF PERFORMANCE
- Certification: Provide independently certified evidence that all incorporated components comply with specified performance requirements.

115 FIRE RESISTING DOORS/ DOORSETS/ ASSEMBLIES
- Evidence of fire performance: Provide certified evidence, in the form of a product conformity certificate, directly relevant fire test report or engineering assessment, that each door/ doorset/ assembly supplied will comply with the specified requirements for fire resistance if tested to BS 476-22, BS EN 1634-1 or BS EN 1634-3. Such certification must cover door and frame materials, glass and glazing materials and their installation, essential and ancillary ironmongery, hinges and seals.

150 SITE DIMENSIONS
- Procedure: Before starting work on designated items take site dimensions, record on shop drawings and use to ensure accurate fabrication.
- Designated items:
To be advised by CA.

170 CONTROL SAMPLES
- Procedure:
  - Finalize component details.
  - Fabricate one of each of the following designated items as part of the quantity required for the project.
  - Obtain approval of appearance and quality before proceeding with manufacture of the remaining quantity.
- Designated items:
To be advised by CA.

COMPONENTS

210 MATCHBOARDED DOORS
- Manufacturer: Contractors choice.
- Finish as delivered: Factory primed for softwood, smooth sanded for hardwood.

230 WOOD FLUSH DOORS
- Manufacturer: As specified in schedules.
- Preservative treatment: Organic solvent as section Z12 to softwood.
- Finish as delivered: Factory primed for softwood, smooth sanded for hardwood.
- Glazing details: As specified in schedules.

250 WOOD PANELLED DOORS
- Manufacturer: As specified in schedules.
- Preservative treatment: Organic solvent as section Z12 to softwood.
- Finish as delivered: Factory primed for softwood, smooth sanded for hardwood.
- Glazing/ Panel details: As specified in schedules.

310 WOOD DOOR FRAMES EXTERNAL
- Manufacturer: As specified in schedules.
- Species: As specified in schedules.
- Preservative treatment Organic solvent as section Z12 to softwood.
- Finish as delivered: Factory primed for softwood, smooth sanded for hardwood.
- Fixing: Clause 805 .

INSTALLATION

710 PROTECTION OF COMPONENTS
- General: Do not deliver to site components that cannot be installed immediately or placed in clean, dry, floored and covered storage.
- Stored components: Stacked on level bearers, separated with spacers to prevent damage by and to projecting ironmongery, beads, etc.

730 PRIMING/ SEALING
- Wood surfaces inaccessible after installation: Primed or sealed as specified before fixing components.

750 FIXING DOORSETS
- Timing: After associated rooms have been made weathertight and the work of wet trades is finished and dried out.

760 BUILDING IN
- General: Not permitted unless indicated on drawings.

770 DAMP PROOF COURSES ASSOCIATED WITH BUILT IN WOOD FRAMES
- Method of fixing: To backs of frames using galvanized clout nails.

780 DAMP PROOF COURSES IN PREPARED OPENINGS
- Location: Correctly positioned in relation to door frames. Do not displace during fixing operations.

790 FIXING OF WOOD FRAMES
- Spacing of fixings (frames not predrilled): Maximum 150 mm from ends of each jamb and at 600 mm maximum centres.

800 FIXING OF LOOSE THRESHOLDS
- Spacing of fixings: Maximum 150 mm from each end and at 600 mm maximum centres.

805 FIXING EXTERNAL DOOR FRAMES:
- fix to jambs with strong galvanised mild steel cramps (three per jamb) to ensure slam-proof fixing. Stagger cramps on frame. House 75 x 20 mm diameter galvanised steel dowel into foot, let into floor, bed in mortar.

809 FIRE RESISTING AND/ OR SMOKE CONTROL DOORS/ DOORSETS
- Installation: By a firm currently registered under a third party accredited fire door installer scheme in accordance with instructions supplied with the product conformity certificate, test report or engineering assessment.

810 **FIRE RESISTING AND/ OR SMOKE CONTROL DOORS/ DOORSETS**
- Gaps between frames and supporting construction: Filled as necessary in accordance with requirements for certification and/ or door/ doorset manufacturer's instructions.

820 **SEALANT JOINTS**
- Sealant:
  - Manufacturer: Contractors choice.
  - Colour: as appropriate.
  - Application: As section Z22 to prepared joints. Triangular fillets finished to a flat or slightly convex profile.

830 **FIXING IRONMONGERY GENERALLY**
- Fasteners: Supplied by ironmongery manufacturer.
  - Finish/ Corrosion resistance: To match ironmongery.
- Holes for components: No larger than required for satisfactory fit/ operation.
- Adjacent surfaces: Undamaged.
- Moving parts: Adjusted, lubricated and functioning correctly at completion.

840 **FIXING IRONMONGERY TO FIRE RESISTING DOOR ASSEMBLIES**
- General: All items fixed in accordance with door leaf manufacturer's recommendations ensuring that integrity of the assembly, as established by testing, is not compromised.
- Holes for through fixings and components: Accurately cut.
  - Clearances: Not more than 8 mm unless protected by intumescent paste or similar.
- Lock/ Latch cases for FD 60 doors: Coated with intumescent paint or paste before installation.

850 **LOCATION OF HINGES**
- Primary hinges: Where not specified otherwise, positioned with centre lines 250 mm from top and bottom of door leaf.
- Third hinge: Where specified, positioned in centre
- Hinges for fire resisting doors: Positioned in accordance with door leaf manufacturer's recommendations.

860 **INSTALLATION OF EMERGENCY EXIT DEVICES**
- Standard: Unless specified otherwise, install panic bolts/ latches in accordance with BS EN 1125.

L30 STAIRS/ LADDERS/ WALKWAYS/ HANDRAILS/ BALUSTRADES

PRELIMINARY INFORMATION/ REQUIREMENTS

130 **SITE DIMENSIONS**
- Procedure: Before starting work, take site dimensions, record on shop drawings and use to ensure accurate fabrication.

COMPONENTS

230 **INTERNAL WOOD STAIRS SOFTWOOD**
- Manufacturer: Contractors choice.
- Components: Manufacturer's standard
- Moisture content at time of installation: 9-13%.
- Finish as delivered: Sanded.
- Workmanship: To section Z10.

560 **PROPRIETARY BALUSTRADES**
- Manufacturer: Contractors choice .
- Component material and finish as delivered:
  - Guarding: Galvanised mild steel.
  - Handrails: Galvanised mild steel.
- Other requirements: As specified in schedules .
- Fixing: Casting into concete .

INSTALLATION

610 MOISTURE CONTENT
- Temperature and humidity: Monitor and control internal conditions to achieve specified moisture content in wood components at time of installation.

620 PRIMING/ SEALING/ PAINTING
- Surfaces inaccessible after assembly/ installation: Before fixing components, apply full protective/ decorative treatment/ coating system.

640 FIXING GENERALLY
- Fasteners and methods of fixing: To section Z20.
- Structural members: Do not modify, cut, notch or make holes in structural members, except as indicated on drawings.
- Temporary support: Do not use stairs, walkways or balustrades as temporary support or strutting for other work.

L40 GENERAL GLAZING

GENERAL REQUIREMENTS

100 METHODS OF MEASUREMENT
- Hacking out existing glass and preparing rebates for reglazing is measured separately.

103 PRICES ALSO TO INCLUDE
- Setting blocks, location blocks, distance pieces and the like for fixing.
- Cleaning off protective wax coating in rebates prior to glazing aluminium windows, screens and doors.
- Glazing fillets are to include pinning to wood or clipping to metal unless otherwise stated.
- Glass drilled and screwed is to include insulating sleeves to screws.

105 CONTRARY TO SMM
- The number of panes has not been given (SMM L40.1.1.1).
- Special glass has been measured in m2 (SMM L40.3.1.1

111 PREGLAZING
- Preglazing of components: Permitted.
- Prevention of displacement: Submit details of precautions to be taken to protect glazing and compound/ seals during delivery and installation.
- Defective/ displaced glazing/ compound/ seals: Reglaze components in situ.

130 REMOVAL OF GLASS/ PLASTICS FOR REUSE
- Existing glass/ plastics and glazing compound, beads, etc: Remove carefully, avoiding damage to frame, to leave clean, smooth rebates free from obstructions and debris.
- Deterioration of frame/ surround: Submit report on defects revealed by removal of glazing.
- Affected areas: Do not reglaze until instructed.
- Reusable materials: Clean glass/ plastics, beads and other components that are to be reused.

150 WORKMANSHIP GENERALLY
- Glazing generally: To BS 6262.
- Integrity: Glazing must be wind and watertight under all conditions with full allowance made for deflections and other movements.
- Dimensional tolerances: Panes/ sheets to be within ± 2 mm of specified dimensions.
- Materials:
    - Compatibility: Glass/ plastics, surround materials, sealers, primers and paints/ clear finishes to be used together to be compatible. Avoid contact between glazing panes/ units and alkaline materials such as cement and lime.
    - Protection: Keep materials dry until fixed. Protect insulating glass units and plastics glazing sheets from the sun and other heat sources.

152 PREPARATION
- Surrounds, rebates, grooves and beads: Clean and prepare before installing glazing.
- Seal surfaces of rebates, grooves and beads that will be in contact with putty as follows:
  1. For painted softwood surrounds, one coat of primer.
  2. For painted hardwood surrounds, one coat of primer and one undercoat of paint.
  3. For clear finished hardwood surrounds, one coat of undiluted exterior varnish.
  4. For galvanized steel surrounds, one coat of primer

155 GLASS GENERALLY
- Standards: To BS 952 and relevant parts of:
  - BS EN 572 for basic soda lime silicate glass.
  - BS EN 1096 for coated glass.
  - BS EN 1748-1 for borosilicate glass.
  - BS EN 1748-2 for ceramic glass.
  - BS EN 1863 for heat strengthened soda lime silicate glass.
  - BS EN 12150 for thermally toughened soda lime silicate safety glass.
  - BS EN 12337 for chemically strengthened soda lime silicate glass.
  - BS EN 13024 for thermally toughened borosilicate safety glass.
  - BS EN ISO 12543 for laminated glass and laminated safety glass.
- Panes/ sheets: Clean and free from obvious scratches, bubbles, cracks, rippling, dimples and other defects.
- Edges: Generally undamaged. Shells and chips not more than 2 mm deep and extending not more than 5 mm across the surface are acceptable if ground out.

160 LINEAR PATTERNED/ WIRED GLASS
- Alignment: Vertical/ Horizontal as appropriate, and pattern matched across adjacent panes in close proximity.

170 PLASTICS GLAZING SHEET
- Condition: Free from scratches, edge splits and other defects.
- Preparation for use: Protective coverings carefully peeled back from edges and trimmed off to facilitate glazing. Remainder retained in place until completion unless instructed otherwise.

180 BEAD FIXING WITH PINS
- Pin spacing: Regular at maximum 150 mm centres, and within 50 mm of each corner.
- Exposed pin heads: Punched just below wood surface.

181 BEAD FIXING WITH SCREWS
- Brass cups and screws
- Screw spacing: Regular at maximum 225 mm centres, and within 75 mm of each corner.

190 GLASS TO GLASS JOINTING
- Sealant: Clear silicone to BS 5889, Type A or B as recommended for the purpose by the manufacturer.
- Joints:
  - Width: Consistent and suitable to receive sealant.
  - Gap between panes: Completely filled, leaving no voids or bubbles.
  - Surplus sealant: Removed to leave a clean, neatly finished weathertight joint.

TYPES OF GLAZING

210 PUTTY FRONTED SINGLE GLAZING
- Pane material: As specified in schedules.
- Surround: As specified in schedules.
- Sealer: As clause 152.
- Type of putty:
  1. For softwood, linseed oil putty.
  2. For hardwood and metal (except aluminium), metal casement putty

- Glass installation:
  - Glass: Located centrally in surround using setting and location blocks, and secured with glazing sprigs/ cleats/ clips at 300 mm centres.
  - Finished thickness of back bedding after inserting glazing (minimum): 1.5 mm.
  - Front putty: Finished to a smooth, neat triangular profile stopping 2 mm short of sight line. Surface lightly brushed to seal putty to glass and left smooth with no brush marks.
  - Sealing putty: Seal as soon as sufficiently hard but not within 7 days of glazing. Within 28 days apply the full final finish, suitably protected until completion and cleaned down and made good as necessary.
  - Opening lights: Keep in closed position until putty has set sufficiently to prevent displacement of glazing when opened.

230 BEAD FIXED SINGLE GLAZING
- Pane material: As specified in schedules.
- Surround/ bead: As specified in schedules.
  - Preparation: As clause 152 .
  - Bead location: As specified in schedules.
  - Bead fixing: As clause 181.
- Glazing compound: As specified in schedules.
- Glazing installation:
  - Glass: Located centrally in surround using setting and location blocks and distance pieces.
  - Finished thickness of back bedding after inserting glazing (minimum): 3 mm.
  - Front bedding: Applied to fill voids.
  - Beads: Bedded in glazing compound and fixed securely.
  - Visible edge of glazing compound: Finished internally and externally with a smooth chamfer.

520 FIRE RATING
- Assessment of capability: Submit proposed construction details of designated items to a UKAS/ NAMAS accredited laboratory or other approved authority for assessment of capability of achieving specified fire ratings.
- Specimens of construction (if required): Prepared in accordance with BS 476-20, appendix A3, and submitted for testing to BS 476-22.
- Assessment/ test results and reports: Submit immediately they are available, and before installing glazing.
- Designated items: As specified in schedules.

530 INTERNAL TAPE GLAZING
- Pane material: As specified in schedules.
- Surround/ bead: As specified in schedules.
- Bead fixing: Clause 181 .
- Tape/ Section: Washleather .
- Glazing installation: Beads bedded dry to rebate and glazing tape/ section and fixed securely. Tape trimmed flush with sight line on both sides.

610 WINDOW FILM
- Type: Safety film .
- Manufacturer: Contractors choice .
- Colour: as specified .
- Application: Carried out by a firm approved by the film manufacturer in accordance with manufacturer's recommendations.
- Evidence of applicator's competence and experience: Submit on request.
- Sample area: Complete as part of the finished work, in an approved location and obtain approval of appearance before proceeding.
- Ambient air temperature at time of application: Above 5°C.
- Installed film: Fully adhered to the glass with no peeling, and free from bubbles, wrinkles, cracks or tears.
- Further contact with applied films: Avoid until bonding adhesive has cured.
- Cleaning and maintenance instructions: Submit copies.

M: SURFACE FINISHES

GENERALLY

100 METHODS OF MEASUREMENT
- Hacking surfaces of concrete and the like as key for finishings is measurable only where required and where a bonding coat has not been specified.
- Where required to be stated by the SMM, preparatory work to the base which is directly involved in the application of a finish is deemed to be included.

102 CONTRARY TO SMM
- Contrary to SMM 'Additional Rules - work to existing buildings' clause 1.1.*.* bonding/jointing new to existing is also measureable, under section M, 'Surface finishes', where patching/repairs are undertaken.

105 PRICES ALSO TO INCLUDE
- Curing in accordance with the specification.
- Extra dubbing to in-situ finishings and thicknessing to beds and backings to take up any irregularities and tolerances in both new structures and existing surfaces including the cambers on concrete floors.
- Reinforced beds and backings are to include working around all types of reinforcement, ties and the like.
- Brushing while green to roughen the surface, washing down and cement slurrying the surfaces of concrete and the like on which finishings, beds and backings are to be laid.
- Holes through any associated membranes underlays beds or backings.

M10 CEMENT BASED LEVELLING/ WEARING SCREEDS

TYPES OF SCREED

110 BONDED CEMENT:SAND LEVELLING SCREEDS
- Base: Concrete.
- Screed type: Bonded as clause 260.
- Nominal thickness: As specified in schedule.
  - Thickness (minimum): 25 mm.
  - Thickness (maximum): 45 mm.
- Mix:
  - Aggregates: Standards as clause 305.
  - Proportions (cement:sand): 1:3-4.5.
  - Admixture: As clause 307.
- In situ crushing resistance (ISCR): As clause 335.
  - Depth of indentation (maximum): 5mm.
- Flatness/ Surface regularity: Maximum permissible deviation when measured as clause 355.
- Finish: As specified in schedule.

115 CEMENT:SAND LEVELLING SCREEDS
- Base: Concrete.
- Screed type: Unbonded as clause 280.
  - Reinforcement for crack control: As appropriate.
- Nominal thickness: As specified in schedules.
- Mix:
  - Aggregates: Standards as clause 305.
  - Proportions (cement:sand): 1:3-4.5.
  - Admixture: As clause 307.
- In situ crushing resistance (ISCR): As clause 335.
- Flatness/ Surface regularity: Maximum permissible deviation when measured as clause 355.
- Finish: monolithic
  - Method:
    1. Lay not less than 25 mm thick within 3 hours (or agreed lesser time in hot weather) of laying base concrete.
    2. Bring to a smooth even surface to the levels required.

3. Form joints to coincide with construction or movement joints in base concrete, joints square and plain, abutted closely and level. Compact thoroughly at edges.

180 CONCRETE WEARING SCREEDS (GRANOLITHIC)
- Base: In situ concrete slab.
- Screed type: Bonded as clause 260.
- Nominal thickness: As specified in schedule.
  - Thickness (minimum): 25 mm.
  - Thickness (maximum): 45 mm.
- Mix:
  - Aggregates: Standards as clause 305.
  - Coarse aggregate single size: 10 mm.
  - Proportions (cement:sand:coarse aggregate): 1:1:2.
  - Admixture: As clause 307.
- Flatness/ Surface regularity: Maximum permissible deviation when measured as clause 355.
- Abrasion resistance:
  - Standard: To BS 8204-2 table 4.
- Finish: As specified.
  - Slip resistance value (minimum SRV): As measured in clause 690.

GENERALLY/ PREPARATION

210 SUITABILITY OF BASES
- General:
  - Suitable for specified levels and flatness/ regularity of finished surfaces. Consider permissible minimum and maximum thicknesses of screeds.
  - Sound and free from significant cracks and gaps.
- Cleanliness: Remove plaster, debris and dirt.
- Moisture content: To suit screed type. New concrete slabs to receive fully or partially bonded construction must be dried out by exposure to the air for minimum six weeks.

220 PROPRIETARY LEVELLING/ WEARING SCREEDS
- General: Materials, mix proportions, mixing methods,: minimum/ maximum thickness and workmanship must be in accordance with recommendations of screed manufacturer.

230 CONTROL SAMPLES
- General: Complete areas of finished work and obtain approval of appearance before proceeding.

250 CONDUITS UNDER FLOATING SCREEDS
- Haunching: Before laying insulation for floating screeds, haunch up in 1:4
- cement: :sand on both sides of conduits.

251 CONDUITS CAST INTO OR UNDER SCREEDS
- Reinforcement: Overlay with reinforcement selected from:
  - 500 mm wide strip of steel fabric to BS 4483, reference D49, or
  - Welded mesh manufactured in rolls from mild steel wire minimum 1.5 mm diameter to BS 1052, mesh size 50 x 50 mm.
- Placing reinforcement: Mid depth between top of conduit and the screed surface.
- Screed cover over conduit (minimum): 25 mm.

255 PIPE DUCTS/ TRUNKING
- Preformed access ducts: Before laying screed, fix securely to base and level accurately in relation to finished floor surface.

260 FULLY BONDED CONSTRUCTION
- Preparation: Generally in accordance with BS 8204-1.
- Removing mortar matrix: Shortly before laying screed, expose coarse aggregate over entire area of hardened base.
- Texture of surface: Suitable to accept screed and achieve a full bond over complete area.
- Bonding coat: Neat cement grout .

- Lay not less than 40 mm thick in one layer well bonded to hardened base, or in two layers if specified each not less than 20 mm thick where a finish for linoleum or similar is required, the second layer laid immediately following the first and should be of the same composition.
- Bring to a smooth even surface to the levels required.
- Lay finishes in bays of not more than 15 m2 in area, no side to exceed 4000 mm in length, the bays to be laid in chequer-board pattern. Allow at least 24 hours between placing of adjacent bays. Break joints with building paper.
- Lay screeds where a finish for linoleum or similar is required, in strips not more than 3000 mm wide. Allow at least 24 hours between placing adjacent strips.
- Form joints to coincide with construction or movement joints in base concrete, joints square and plain, abutted closely and level. Compact thoroughly at edges.

280 UNBONDED CONSTRUCTION
- Separation: Lay screed over a suitable sheet dpm or a separating layer.
- Installation of separating layer: Lay on clean base. Turn up for full depth of screed at abutments with walls, columns, etc. Lap 100 mm at joints.

BATCHING/ MIXING

305 AGGREGATE STANDARDS
- Cement: Portland to BS EN 197-1, class 42.5 or Portland blast furnace to BS 146, class 42.5.
- Sand: To BS EN 12620.
- Grading limit: To BS 8204-1, table 1.
- Coarse aggregate: To BS EN 12620.

307 ADMIXTURE STANDARD
- Type: Water reducing to BS EN 934-2.
- Calcium chloride: Do not use in admixtures.

310 BATCHING WITH DENSE AGGREGATES
- Mix proportions: Specified by weight.
- Batching: Select from:
  - Batch by weight.
  - Batch by volume: Permitted on the basis of previously established weight:volume relationships of the particular materials. Use accurate gauge boxes. Allow for bulking of damp sand.

311 BATCHING WITH LIGHTWEIGHT AGGREGATES
- Mix proportions: Specified by volume.
- Batching: Use accurate gauge boxes.

330 MIXING
- Water content: Minimum necessary to achieve full compaction, low enough to prevent excessive water being brought to surface during compaction.
- Mixing: Mix materials thoroughly to uniform consistency. Mixes other than no-fines must be mixed in a suitable forced action mechanical mixer. Do not use a free fall drum type mixer.
- Consistency: Use while sufficiently plastic for full compaction.
- Ready-mixed retarded screed mortar: Use within working time and site temperatures recommended by manufacturer. Do not retemper.

335 IN SITU CRUSHING RESISTANCE (ISCR)
- Standards and category: To BS 8204-1, table 5.
  - For testing of bonded screeds refer to Annex C.
  - For testing of floating screeds refer to Annex D.

Class conditions	Service depth	Indentation
A	Heavy use	3 mm
B	Normal use	4 mm
C	Light use	5 mm

340 ADVERSE WEATHER
- Screeds surface temperature: Maintain above 5° C for a minimum of four days after laying.
- Hot weather: Prevent premature setting or drying out.

LAYING

345 LEVEL OF SCREED SURFACES
- Permissible deviation: ±10 mm from datum (allowing for thickness of coverings).

350 SCREEDING TO FALLS
- Minimum screed cover: Maintain at the lowest point.
- Falls: Gradual and consistent.

351 SCREEDING TO RAMPS
- Screed cover: As specified in schedules.
- Falls: Gradual and consistent.

355 FLATNESS/ SURFACE REGULARITY OF FLOOR SCREEDS
- Sudden irregularities: Not permitted.
- Deviation of surface: Measure from underside of a 3 m straightedge (between points of contact), placed anywhere on surface using a slip gauge to BS 8204-1 or -2 (or equivalent).
- Surface regularity standards:
    - SR1: 3 mm high standard.
    - SR2: 5 mm normal standard.
    - SR3: 10 mm utility standard.

375 COMPACTION OF SCREEDS
- General: Compact thoroughly over entire area.
- Screeds over 50 mm thick: Lay in two layers of approximately equal thickness. Roughen surface of compacted lower layer then immediately lay upper layer.

382 STAIR SCREEDS
- Bonding: As clause 260 to treads, risers and landings.
- Risers: Form using fine finish formwork.
- Wearing screed surfaces: Make good with compatible cement:sand mix. Wood float. When hardened remove laitance.

405 JOINTS IN LEVELLING SCREEDS GENERALLY
- Laying screeds: Lay continuously using 'wet screeds' between strips or bays. Minimize defined joints.
- Daywork joints: Form with vertical edge.

428 HEATED SCREEDS
- Screed bays: Coordinate with heating circuits.
- Heating elements: Secure properly. Prevent displacement.
- Laying: Lay carefully. Compact thoroughly around heating elements. Do not damage them.

435 FORMED JOINTS IN WEARING SCREEDS
- Temporary forms: Square edged with a steel top surface and in good condition.
- Placing screed: Compact thoroughly at edges to give level, closely abutted joints with no lipping.

440 CRACK INDUCING GROOVES IN LEVELLING SCREEDS
- Groove depth: At least half the depth of screed.
- Cutting grooves: Straight, vertical and accurately positioned. Select from the following:
    - Trowel cut as screed is laid.
    - Saw cut sufficiently early after laying to prevent random cracking.

445 CRACK INDUCING GROOVES IN WEARING SCREEDS
- Groove dimensions:
  - Depth: At least half the depth of wearing screed.
  - Width: 6mm.
- Cutting grooves: Straight, vertical and accurately positioned. Saw cut sufficiently early after laying to prevent random cracking.

450 SEALANT FOR SAWN JOINTS IN WEARING SCREEDS
- Type: 2 part polysulphide.
- Preparation and application: As section Z22.

FINISHING/ CURING

510 FINISHING GENERALLY
- Timing: Carry out all finishing operations at optimum times in relation to setting and hardening of screed material.
- Prohibited treatments to screed surfaces:
  - Wetting to assist surface working.
  - Sprinkling cement.

520 WOOD FLOATED FINISH
- Finish: Slightly coarse, even texture with no ridges or steps.

530 SMOOTH FLOATED FINISH
- Finish: Even texture with no ridges or steps.

540 TROWELLED FINISH TO LEVELLING SCREEDS
- Floating: To an even texture with no ridges or steps.
- Trowelling: To a uniform, smooth but not polished surface, free from trowel marks and other blemishes, and suitable to receive specified flooring material.

550 TROWELLED FINISH TO WEARING SCREEDS
- Floating: To an even texture with no ridges or steps.
- Trowelling: Successively trowel at intervals, applying sufficient pressure to close surface and give a uniform smooth finish free from trowel marks and other blemishes.

560 DEWATERED TROWELLED FINISH TO WEARING SCREEDS
- Dewatering: Immediately after compaction of wearing screeds, remove water using a vacuum process.
- Floating: Without delay, power float to an even texture with no ridges or steps.
- Trowelling: Successively trowel at intervals, applying sufficient pressure to close surface and give a uniform smooth finish free from trowel marks and other blemishes.

600 POWER GROUND FINISH TO WEARING SCREEDS
- Floating: To an even surface with no ridges or steps. Immediately commence curing.
- Grinding: When concrete is sufficiently hard for sand particles not to be torn from surface, remove 1-2 mm from surface to give an even glass-paper texture, free from blemishes and trowel marks.
- Cleaning: Remove dust and wash down. Resume curing without delay.

650 CURING
- General: Prevent premature drying. Immediately after laying, protect surface from wind, draughts and strong sunlight. As soon as screed has set sufficiently, closely cover with polyethylene sheeting.
- Curing period: Keep polyethylene sheeting in position for a minimum of: 7 days.
- Drying after curing: Allow screeds to dry gradually. Do not subject screeds to artificial drying conditions that will cause cracking or other shrinkage related problems.

655 SURFACE HARDENER
- Apply sodium silicate to in-situ pavings 7 to 10 days after curing as follows:
  - 1. For first dressing mix well together in solution 1 part to 4 parts, or 6 parts if a porous surface, water by volume and spread evenly over the floor with mop or soft brush. Wipe off excess, allow to dry for at least 24 hours.
  - 2. Wash with clean water and apply second coat in solution 1 part to 3 or 4 parts water. Allow to dry and wash with clean hot water.

670 ROOF SCREEDS
- Protection: Cover screeds during wet weather. When weathertight coverings are laid, screeds must be as dry as practicable.

680 SURFACE SEALER TO WEARING SCREEDS
- Manufacturer: Contractors choice .
- Preparation: Clean cured screed surface to remove dirt, grease, oil and other surface contaminants.
- Moisture content of screed: As recommended by sealer manufacturer. Test relative humidity to BS 8203, Annex A if required to verify suitability to receive sealer.
- Application: Evenly to dry surfaces using sufficient coats to form an effective seal but without a glossy finish.

690 SLIP RESISTANCE TESTING OF WEARING SCREEDS
- Test:
  - To the relevant parts of BS 7976-1 and BS 7976-2 using a TRL Pendulum.
  - Report: Submit. Include slip resistance values in the wet and dry states.

M12 TROWELLED BITUMEN/RESIN/RUBBER-LATEX FLOORING

TYPES OF FLOORING

110 RUBBER LATEX FLOORING
- Substrate: Concrete
- Latex levelling compound:
  - Manufacturer: Contractors choice
  - Nominal thickness: 3mm or 6mm as specified in schedules
  - Surface finish/ treatment: trowelled
- Flatness/ Surface regularity: Free from sudden irregularities.

PREPARATION OF SUBSTRATES

230 PREPARATION OF SUBSTRATES GENERALLY
- Cleanliness: Remove surface contaminants, debris, dirt and dust.
- Texture of surface: Suitable to accept latex flooring and achieve a full bond over the complete area.
- Damp down absorbent surfaces

240 EXISTING SUBSTRATES
- Preparation: Remove surface imperfections, ingrained contaminants, coatings and residues.
- Areas impregnated with contaminants: Submit proposals for removal and repair.

250 EXISTING TILE/ SHEET FLOOR COVERINGS
- Preparation: Remove coverings, residual adhesive, bedding, grouting and pointing.

LAYING FLOORING

310 WORKMANSHIP
- Operatives: Trained/ experienced in the application of latex floorings. Submit evidence of training/ experience on request.
- Curing: Allow appropriate periods between coats, before surface treatments, and before trafficking/ use.

320 CONTROL SAMPLES
- General: Complete areas of finished work in specified locations and obtain approval of appearance before proceeding.

400 ADHESION OF LATEX FLOORING
- Bond strength between substrate and fully cured latex flooring: In accordance with manufacturer's performance data.
- Test for bond strength: To BS 8204-6, clause 11.4 and BS EN 1542.

M20 PLASTERED/ RENDERED/ ROUGHCAST COATINGS

TYPES OF COATING

130 CEMENT:SAND RENDER
- Background: As specified in schedules.
    - Preparation: As appropriate for background
- Undercoats:
    - Sand: To BS 1199, type A.
    - Mix: 1:4
    - Thickness (excluding dubbing out): 8mm
- Final coat:
    - Sand: To BS 1199, type A.
    - Mix: 1:4
    - Thickness: 5mm
    - Finish: as clause 856, 861 or 871 as specified

210 LIGHTWEIGHT GYPSUM PLASTER
- Background: As specified in schedules.
    - Preparation: As appropriate for background.
- Undercoats: To BS 1191-2.
    - Manufacturer: Contractors choice
    - Thickness (excluding dubbing out): As specified in schedules
- Final coat: Finish plaster to BS 1191-1, class B.
    - Manufacturer: Contractors choice
    - Thickness: 2 mm.
    - Finish: Smooth as clause 777.

280 GYPSUM PLASTER SKIM ON PLASTERBOARD
- Background: As specified in schedules.
- Backing: As specified in schedules.
    - Preparation: As appropriate for background.
- Plaster: Board finish/ finish plaster to BS 1191-1, class B.
    - Manufacturer: Contractors choice or as specified in schedules
    - Thickness: As specified in schedules.
    - Finish: Smooth as clause 777.

GENERAL

423 UNIFORMITY OF COLOUR AND TEXTURE
- General: Maintain consistent colour and texture. Obtain materials from one source. Mix different loads if necessary.

438 CEMENTS FOR RENDER MORTARS
- Cement:
    - Standard: To BS EN 197-1.
    - Types: Portland cement, CEM I.
        Portland slag cement, CEM II/ B-S.
        Portland fly ash cement, CEM II/ B-V.
    - Strength class: 42.5 or 52.5.
- Sulphate resisting cement:
    - Standard: To BS 4027.

- Strength class: 42.5 or 52.5.
- Masonry cement:
    - Standard: To BS 5224.
    - Class: MC 12.5 (with air entraining agent).
- Certification for all cements: BSI Kitemark scheme.

449 ADMIXTURES FOR CEMENT GAUGED RENDER MORTARS
- Suitable admixtures: Select from:
    - Air entraining (plasticizing) admixtures: To BS 4887-1 and compatible with other mortar constituents.
    - Other admixtures: Submit proposals.
- Prohibited admixtures: Calcium chloride and any admixture containing calcium chloride.

453 MIXING
- Render mortars (site prepared):
    - Batching: By volume. Use clean and accurate gauge boxes or buckets.
    - Mix proportions: Based on damp sand. Adjust for dry sand.
- Mixes: Of uniform consistence and free from lumps. Do not retemper or reconstitute mixes.
- Contamination: Prevent. Keep plant and banker boards clean.

466 SCAFFOLDING
- General: Prevent putlog holes and other breaks in coatings.

474 COLD WEATHER
- General: Do not use frozen materials or apply coatings to frozen or frost bound backgrounds.
- External work: Avoid when air temperature is at or below 5°C and falling or below 3°C and rising. Maintain temperature of work above freezing until coatings have fully hardened.
- Internal work: Take all necessary precautions to enable internal coating work to proceed without damage when air temperature is below 3°C.

498 SPECIAL PROTECTION OF HISTORIC PLASTERWORK
- General: Minimize damage and disturbance to retained plasterwork.
- Protection methods: Submit proposals.

PREPARING BACKGROUNDS

510 SUITABILITY OF BACKGROUNDS
- General: Suitable to receive coatings/ backings.
    - Soundness: Free from loose areas and significant cracks and gaps.
    - Cutting, chasing, making good, fixing of conduits and services outlets and the like: Completed.
    - Tolerances: Permitting specified flatness/ regularity of finished coatings.
- Cleanliness: Remove dirt, dust, efflorescence and mould, and other contaminants incompatible with coatings.

527 RAKING OUT FOR KEY
- Soft joints in existing masonry: Rake out to a depth of 13 mm (minimum).
    - Dust and debris: Remove from joints.

531 ROUGHENING FOR KEY
- Backgrounds: Roughen thoroughly and evenly.
    - Depth of surface removal: Minimum necessary to provide an effective key.

541 BONDING AGENT
- Manufacturer: Contracors choice
- Application: Evenly across background to achieve effective bond or plaster/ render coat to background. Protect adjacent joinery and other surfaces.

551 EXISTING PLASTER/ RENDER
- Location and extent of renewal: Agree, at least on a provisional basis, before work commences.
- Removal of existing coatings: Use methods that minimize the amount of removal and renewal.

556 REMOVING DEFECTIVE EXISTING RENDER
- Render for removal: Loose, hollow, soft, friable, badly cracked, affected by efflorescence or otherwise damaged.
- Patches: Cut out rectangular areas with straight edges.
  - Horizontal and vertical edges: Square cut or slightly undercut.
  - Bottom edges to external render: Do not undercut.
  - Render with imitation joints: Cut back to joint lines.
- Cracks (other than hairline cracks): Cut out to a width of 75 mm (minimum).
- Dust and loose material: Remove from exposed backgrounds and edges.

566 REMOVING DEFECTIVE EXISTING PLASTER
- Plaster for removal: Loose, soft, friable, badly cracked, affected by efflorescence or otherwise damaged.
- Stained plaster: Seek instructions.
- Removing defective plaster. Cut back to a square, sound edge.
- Faults in background (structural deficiencies, damp, etc.): Seek instructions.
- Dust and loose material: Remove from exposed backgrounds and edges.

568 EXISTING DAMP AFFECTED PLASTER/ RENDER
- Plaster affected by rising damp: Remove to a height of 300 mm above highest point reached by damp or 1 m above dpc, whichever is higher.
- Perished and salt contaminated masonry:
  - Mortar joints: Rake out.
  - Masonry units: Seek instructions.
- Faults in background (structural deficiencies, additional sources of damp, etc.): Seek instructions.
- Drying out backgrounds: Established drying conditions. Leave walls to dry for as long as possible before plastering.
- Dust and loose material: Remove from exposed backgrounds and edges.

BACKINGS/ BEADS/ JOINTS

600 ADDITIONAL FRAMING SUPPORTS FOR BACKINGS
- Fixtures, fittings and service outlets: To suit fixings.
- Board edges and perimeters: As recommended by board manufacturer to suit type and performance of board.

605 GYPSUM PLASTERBOARD BACKINGS
- Type: To BS 1230, type 1.
- Core density (minimum): 650 kg/m^3.
- Exposed surface and edge profiles: Suitable to receive specified plaster finish.

607 PROPRIETARY GYPSUM PLASTERBOARD BACKINGS
- Manufacturer: As specified in schedules
  - Product reference: As specified in schedules
- Exposed surface and edge profiles: Suitable to receive specified plaster finish.

610 FIXING PLASTERBOARD BACKINGS TO TIMBER BACKGROUNDS
- Fixings, accessories and installation methods: As recommended by board manufacturer.
- Fixing: At following centres (maximum):
  - Nail fixing: 150 mm.
  - Screw fixing to partitions/ walls: 300 mm reduced to 200 mm at external angles.
  - Screw fixing to ceilings: 230 mm.
- Position of nails/ screws from edges of boards (minimum):
  - Bound edges: 10 mm.
  - Cut/ unbound edges: 13 mm.
- Position of nails/ screws from edges of supports (minimum): 6 mm.
- Nail/ screw heads: Set in a depression. Do not break paper or gypsum core.

611 FIXING PLASTERBOARD BACKINGS
- Accessories, materials and installation methods: As recommended by the plasterboard manufacturer.

612 JOINTS IN PLASTERBOARD BACKINGS
- Ceilings:
    - Bound edges: At right angles to supports and with ends staggered in adjacent rows.
    - Two layer boarding: Stagger joints between layers.
- Partitions/ walls:
    - Vertical joints: Centre on studs. Stagger joints on opposite sides of studs.
   Two layer boarding: Stagger joints between layers.
    - Horizontal joints:
   Two layer boarding: Stagger joints between layers by at least 600 mm. Support edges of outer layer.
- Joint widths (maximum): 3 mm.

630 BEADS/ STOPS FOR INTERNAL USE
- Material: Galvanized steel to BS 6452-1.

634 BEADS/ STOPS
- Manufacturer: Contractors choice
- Material: As specified in schedules.

636 BEADS/ STOPS FOR EXTERNAL USE
- Stainless steel: To BS EN 10088-1, number 1.4301 (name X5CrNi18-10).

640 BEADS/ STOPS GENERALLY
- Location: External angles and stop ends except where specified otherwise.
- Corners: Neat mitres at return angles.
- Fixing: Secure, using longest possible lengths, plumb, square and true to line and level, ensuring full contact of wings with background.
    - External render: Fix mechanically.
- Finishing: After coatings have been applied, remove surplus material while still wet, from surfaces of beads/ stops which are exposed to view.

643 METAL LATHING
- Lap joints 75 mm and wire at intervals not exceeding 75mm.
- Fix to timber with 32 mm galvanized cut staples.
- Fix to steelwork with 1.2 mm galvanized wire.

646 CRACK CONTROL AT JUNCTIONS BETWEEN DISSIMILAR SOLID BACKGROUNDS
- Locations: Where defined movement joints are not required. Where dissimilar solid background materials are in same plane and rigidly bonded or tied together.
- Crack control materials:
    - Isolating layer: Building paper to BS 1521.
    - Metal lathing: As specified.
- Installation: Fix lathing over isolating layer at staggered centres along both edges.
- Width of installation over single junctions:
    - Isolating layer: 150 mm.
    - Lathing: 300 mm.
- Width of installation across face of dissimilar background material (column, beam, etc. with face width not greater than 450 mm):
    - Isolating layer: Not less than 25 mm beyond junctions with adjacent background.
    - Lathing: Not less than 100 mm beyond edges of isolating layer.

659 PLASTERBOARD JOINTS
- Joint reinforcement: Applied to joints and angles except where coincident with metal beads.
- Installation: Continuous lengths of tape pressed well into filled joints, flat and smooth.

673 PLASTERING OVER CONDUITS/ SERVICE CHASES
- General: Prevent cracking over conduits and other services.
- Services chased into background: Isolate from coating by covering with galvanized metal lathing. Fixed at staggered centres along both edges.

PLASTERING

710 APPLICATION GENERALLY
- General: Apply coatings firmly and in one continuous operation between angles and joints. Achieve good adhesion.
- Appearance of finished surfaces: Even and consistent. Free from rippling, hollows, ridges, cracks and crazing.
- Accuracy: Finish to a true plane, to correct line and level, with angles and corners to a right angle unless specified otherwise, and with walls and reveals plumb and square.
- Drying out: Prevent excessively rapid or localised drying out.

715 FLATNESS/ SURFACE REGULARITY
- Sudden irregularities: Not permitted.
- Deviation of plaster surface: Measure from underside of a straight edge placed anywhere on surface.
- Permissible deviation (maximum) for plaster not less than 13 mm thick: 3 mm in any consecutive length of 1800 mm.

718 JUNCTION OF NEW PLASTERWORK WITH EXISTING
- General: Finish new plasterwork flush with original face of existing plasterwork to form a seamless junction.

720 DUBBING OUT
- General: To correct background inaccuracies.
- Smooth dense concrete and similar surfaces: Dubbing out prohibited unless total plaster thickness is within range recommended by plaster manufacturer.
- Thickness of any one coat (maximum): 10 mm.
- Mix: As undercoat.
- Application: Achieve firm bond. Allow each coat to set sufficiently before the next is applied. Cross scratch surface of each coat.

725 UNDERCOATS GENERALLY
- General: Rule to an even surface. Cross scratch to provide a key for the next coat.
- Undercoats on metal lathing: Work well into interstices to obtain maximum key.
- Undercoats gauged with Portland cement: Do not apply next coat until drying shrinkage is substantially complete.

742 THIN COAT PLASTER
- Preparation for plasters less than 2 mm thick: Fill holes, scratches and voids with finishing plaster.

747 PROJECTION PLASTER
- Application: Evenly and in one continuous operation between angles and joints.
- Finish: A level open textured surface before finishing manually.

777 SMOOTH FINISH
- Appearance: A tight, matt, smooth surface with no hollows, abrupt changes of level or trowel marks. Avoid water brush, excessive trowelling and over polishing.

778 WOOD FLOAT FINISH
- Appearance: An even overall texture. Finish with a dry wood float as soon as wet sheen has disappeared.

782 TEXTURED/ PATTERNED FINISHES
- Appearance: Consistent and even. Carry out work on each surface as one continuous operation.

786 PLASTERING ON TIMBER LATHING
- General: Apply undercoat with sufficient pressure to force it between laths and form continuous keys.

RENDERING

810 APPLICATION GENERALLY
- General: Apply coatings firmly and in one continuous operation between angles and joints. Achieve good adhesion.
- Appearance of finished surfaces: Even and consistent. Free from rippling, hollows, ridges, cracks and crazing.
- Accuracy: Finish to a true plane, to correct line and level, with angles and corners to a right angle unless specified otherwise, and with walls and reveals plumb and square.
- Drying: Prevent excessively rapid or localized drying out.

815 FLATNESS/ SURFACE REGULARITY OF RENDERING TO RECEIVE CERAMIC TILES
- Sudden irregularities: Not permitted.
- Deviation of render surface: Measure from underside of a 2 m straight edge placed anywhere on surface.
- Permissible deviation (maximum): 3 mm.

820 DUBBING OUT FOR RENDERING
- General: To correct background inaccuracies.
- Thickness of any one coat (maximum): 16 mm.
- Total thickness (maximum): 20 mm. Seek instructions where this will be exceeded.
- Mix: As undercoat.
- Application: Achieve firm bond. Allow each coat to set sufficiently before the next is applied. Comb surface of each coat.

830 ANCHORED MESH REINFORCEMENT
- Application of first undercoat: Through and round mesh to fully bond with solid background.

840 UNDERCOATS GENERALLY
- General: Rule to an even surface. Comb to provide a key for the next coat. Do not penetrate the coat.
- Undercoats on metal lathing: Work well into interstices to obtain maximum key.

856 FINAL COAT - PLAIN FLOATED FINISH
- Finish: An even, open texture free from laitance.

861 FINAL COAT - SCRAPED FINISH
- Finish: Scraped to expose aggregate and achieve an even texture.

871 FINAL COAT - DRY DASH FINISH
- Coarse aggregate: To BS 63-2. Well washed.
- Application and finishing: Achieve firm adhesion to an even overall appearance. After throwing aggregate tap particles lightly into coating.

880 CURING AND DRYING
- General: Prevent premature setting and uneven drying of each coat.
- Curing coatings: Keep each coat damp by covering with polyethylene sheet and/ or spraying with water.
- Final coat: Hang sheeting clear of the final coat.
- Drying: Allow each coat to dry thoroughly, with drying shrinkage substantially complete before applying next coat.
- Protection: Protect from frost and rain.

M40 STONE/ CONCRETE/ QUARRY/ CERAMIC TILING/ MOSAIC

GENERALLY

100 METHODS OF MEASUREMENT
- Walls, ceilings, isolated beams, isolated columns and floors are deemed plain unless stated otherwise.

105 PRICES ALSO TO INCLUDE
- Employing stock pattern units as appropriate unless stated otherwise.
- Coved angles on skirtings and rounded internal angles generally to include any extra dubbing to bed or backings behind coves and angles.

TYPES OF TILING/ MOSAIC

110 TILING TO FLOORS
- Tiles: Quarry .
  - Manufacturer/ Supplier: Contractors choice .
  - Colour: as specified in schedules.
  - Finish: as specified in schedules.
  - Size: 152 x 152mm.
  - Thickness: 15mm .
- Background/ Base: as specified in schedules.
  - Preparation: As appropriate for background/base.
- Bedding: Sand/cement.

110 TILING TO WALLS
- Tiles: Ceramic .
  - Manufacturer/ Supplier: Contractors choice .
  - Colour: as specified in schedules.
  - Finish: as specified in schedules.
  - Size: 152 x 152mm.
  - Thickness: 6mm .
- Background/ Base: as specified in schedules.
  - Preparation: As appropriate for background/base.

GENERALLY

210 SUITABILITY OF BACKGROUNDS/ BASES
- Background/ base tolerances: To permit specified flatness/ regularity of finished surfaces given the permissible minimum and maximum thickness of bedding.
- New background drying times (minimum):
  - Concrete walls: 6 weeks.
  - Brick/ block walls: 6 weeks.
  - Rendering: 2 weeks.
  - Gypsum plaster: 4 weeks.
- New base drying times (minimum):
  - Concrete slabs: 6 weeks.
  - Cement:sand screeds: 3 weeks.

215 FALLS IN BASES
- General: Give notice if falls are inadequate.

PREPARATION

310 EXISTING BACKGROUNDS/ BASES GENERALLY
- Efflorescence, laitance, dirt and other loose material: Remove.
- Deposits of oil, grease and other materials incompatible with the bedding: Remove.
- Tile, paint and other nonporous surfaces: Clean.
- Wet backgrounds: Dry before tiling.

320 EXISTING CONCRETE/ SCREEDS
- Loose or hollow portions: Cut out.
- Making good: 1:3 sand/cement.

330 EXISTING PLASTER
- Plaster which is loose, soft, friable, badly cracked or affected by efflorescence: Remove. Cut back to straight horizontal and vertical edges.
- Making good: Use plaster or nonshrinking filler.

340 EXISTING GLAZED BRICK
- Defective areas: Cut out.
- Making good: cement/sand render.

350 EXISTING TILES
- Loose or hollow sounding tiles: Remove.
- Making good: cement/sand render.

355 OLD ADHESIVE RESIDUES ON CONCRETE/ SCREED BASES
- Soft or unsound adhesive residues: Remove without damaging base.

360 EXISTING PAINT
- Paint with unsatisfactory adhesion: Remove so as not to impair bedding adhesion.

370 NEW IN SITU CONCRETE
- Mould oil, surface retarders and other materials incompatible with bedding: Remove.

380 NEW PLASTER
- Plaster: Dry, solidly bedded, free from dust and friable matter.
- Plaster primer: Apply if recommended by adhesive manufacturer.

390 PLASTERBOARD BACKGROUNDS
- Boards: Dry, securely fixed and rigid with no protruding fixings and face to receive decorative finish exposed.

410 HACKING FOR KEY
- Keying: Roughen backgrounds thoroughly and evenly. Remove surface to depth of 3 mm.

420 RAKING OUT FOR KEY
- Soft joints in existing masonry: Rake out to a depth of 13 mm (minimum).

438 PREPARING CONCRETE BASES FOR FULLY BONDED BEDDING
- Surface cement:sand matrix: Remove to expose coarse aggregate.
- Surface preparation: Suitable to achieve a full bond with bedding.
- Keep well wetted for several hours. Remove free water then brush in a slurry bonding coat.

450 PREPARING CONCRETE BASES FOR UNBONDED BEDDING - WITHOUT SEPARATING LAYER
- Surface finish: Smooth.
- Surface preparation: Before laying mortar bed, dampen lightly.

451 PREPARING CONCRETE BASES FOR UNBONDED BEDDING
- Separating layer: Building paper to BS1521 class B.
- Lap at joints: 100 mm.

460 SMOOTHING UNDERLAYMENT
- Type: Recommended by adhesive manufacturer.
- Condition: Allow to dry before tiling.

FIXING

510 FIXING GENERALLY
- Colour/ shade: Unintended variations within tiles for use in each area/ room are not permitted.
- Variegated tiles: Mix thoroughly.
- Adhesive: Compatible with background/ base. Prime if recommended by adhesive manufacturer.
- Cut tiles: Neat and accurate.
- Fixing: Provide adhesion over entire background/ base and tile backs.
- Final appearance: Before bedding material sets, make adjustments necessary to give true, regular appearance to tiles and joints when viewed under final lighting conditions.
- Surplus bedding material: Clean from joints and face of tiles without disturbing tiles.

530 SETTING OUT
- Joints: True to line, continuous and without steps.
  - Joints on walls: Horizontal, vertical and aligned round corners.
  - Joints in floors: Parallel to the main axis of the space or specified features.
- Cut tiles: Minimize number, maximize size and locate unobtrusively.
- Joints in adjoining floors and walls: Align.
- Joints in adjoining floors and skirtings: Align.
- Movement joints: If locations are not indicated, submit proposals.

550 FLATNESS/ REGULARITY OF TILING
- Sudden irregularities: Not permitted.
- Deviation of surface: Measure from underside of a 2 m straightedge placed anywhere on surface. The straightedge should not be obstructed by the tiles and no gap should be greater than 3 mm.

560 LEVEL OF TILING ACROSS JOINTS
- Deviation (maximum) between tile surfaces either side of any type of joint:
  - 1 mm for joints less than 6 mm wide.
  - 2 mm for joints 6 mm or greater in width.

570 MORTAR BEDDING
- Bedding mix:
  - Cement: Portland to BS EN 197-1 type CEM I/42.5.
  - Sand for walls: To BS 1199, type A.
  - Sand for floors: To BS 882.
  - Grading limit: To BS 8204-1, table 1.
- Batching: Select from:
  - Batch by weight.
  - Batch by volume: Permitted on the basis of previously established weight:volume relationships of the particular materials. Use accurate gauge boxes. Allow for bulking of damp sand.
- Mixing: Mix materials thoroughly to uniform consistence. Use a suitable forced action mechanical mixer. Do not use a free fall type mixer.
- Application: At normal temperatures use within two hours. Do not use after initial set. Do not retemper.

578 CRACK CONTROL REINFORCEMENT
- Type to BS 4483.
- Installation: Place centrally in depth of bed. Lap not less than 100 mm and securely tie together with steel wire.
- Corners: Avoid a four layer build at corners.

580 POROUS TILES
- Tiles to be bedded in cement:sand: Soak in clean water for at least 30 minutes. Fix as soon as surface water has drained.

590 COVED TILE SKIRTINGS
- Sequence: Bed solid to wall before laying floor tiles.
- Bedding: 1:3 cement /sand.

© AAS 2008– 2009

600   **SIT-ON TILE SKIRTINGS**
- Sequence: Bed solid to wall after laying floor tiles.
- Bedding: 1:3 cement/sand.

651   **THIN BED ADHESIVE - SOLID (WALLS)**
- Application: Apply floated coat of adhesive to dry background in areas of about 1 m². Comb surface.
- Tiling: Apply thin even coat of adhesive to backs of dry tiles. Press tiles firmly onto float coat.
- Finished adhesive thickness (maximum): 3 mm.

670   **THICK BED ADHESIVE - SOLID (WALLS)**
- Application: Apply floated coat of adhesive to dry background. Comb surface.
- Tiling: Apply thin even coat of adhesive to backs of dry tiles. Press tiles firmly onto float coat.
- Finished adhesive thickness: Within range recommended by manufacturer.

690   **CEMENT:SAND (WALLS)**
- Preparation: Dampen background.
- Application: Apply floated coat: 1:3-4 cement:sand mortar bedding.
  - Thickness (maximum): 10 mm.
  - Finish: Equivalent to wood float. Before tiling allow to stiffen slightly.
- Tiling: Without delay, apply 2 mm thick coat of 1:2 cement:fine sand mortar to backs of tiles, filling keys. Press tiles firmly onto float coat. Tap firmly into position.

710   **THICK BED ADHESIVE - SOLID (FLOORS)**
- Application: Apply floated coat of adhesive to dry base and comb surface.
- Tiling: Apply coat of adhesive to backs of tiles filling depressions or keys. Press tiles firmly onto position.
- Finished adhesive thickness: Within range recommended by manufacturer.

720   **CEMENT:SAND BED (FLOORS)**
- Mortar bedding mix: 1:3-4 cement:sand.
  - Consistency: Stiff plastic.
- Laying: Lay suitably small working areas of screeded bed. Compact thoroughly to level.
  - Finished bed thickness: 15-25 mm.
- Tiling: Within two hours and before bedding sets, evenly coat backs of tiles with adhesive. Press tiles firmly into position.
- Finished adhesive thickness: Within range recommended by manufacturer.

MOVEMENT JOINTS/ GROUTING/ COMPLETION

855   **CEMENT:SAND GROUTING MIX**
- Grout mix:
  - Cement: Portland.
  - Sand:
  Joint widths of 6 mm or greater: To BS 1199 table 1, Type B.
  Joint widths of 3-6 mm: To BS 5385-5 table 2.
  - Proportions (cement:sand): 1:3.
- Mixing: Mix thoroughly. Use the minimum of clean water needed for workability.

875   **GROUTING**
- Sequence: Grout when bed/ adhesive has set sufficiently to prevent disturbance of tiles.
- Joints: 6 mm deep (or depth of tile if less). Free from dust and debris.
- Grouting: Fill joints completely, tool to profile, clean off surface. Leave free from blemishes.
- Polishing: When grout is hard, polish tiling with a dry cloth.

885   **COLOURED GROUT**
- Staining of tiles: Not permitted.
- Evaluating risk of staining: Apply grout to a few tiles in a small trial area. If discoloration occurs apply a protective sealer to tiles and repeat trial.

M50 RUBBER/ PLASTICS/ CORK/ LINO/ CARPET TILING/ SHEETING

TYPES OF COVERING

110 FLOOR TILING
- Base: as specified in schedules.
  - Preparation: As appropriate for base.
- Tiles: Vinyl .
  - Manufacturer: as specified in schedules.
  Product reference: as specified in schedules.
  - BS EN 685 class: As appropriate .
  - Size: 300 x 300mm.
  - Thickness: 2mm.
  - Colour/ pattern: as specified.
- Adhesive (and primer if recommended by manufacturer): Contractors choice.

130 CARPET TILING
- Base: as specified in schedules.
  - Preparation: As appropriate for base.
- Carpet tiles:
  - Manufacturer: As specified.
  - Type: As specified.
- Method of laying: Fixing with adhesive .

155 PVC SHEET FLOORING IN WET AREAS
- Base: As specified in schedules.
  - Preparation: As appropriate for base.
- Flooring roll: PVC to BS EN 13553.
  - Manufacturer: Altro .
  Product reference: As specified in schedules.
  - Width: Contactors choice to minimize joints.
  - Thickness: As specified in schedules mm.
- Adhesive (and primer if recommended by manufacturer): Contractors choice.
- Seam welding: Hot welded joints.

GENERAL REQUIREMENTS

210 WORKMANSHIP GENERALLY
- Base condition after preparation: Rigid, dry, sound, smooth and free from grease, dirt and other contaminants.
- Finished coverings: Accurately fitted, tightly jointed, securely bonded, smooth and free from air bubbles, rippling, adhesive marks and stains.

220 SAMPLES
- Covering samples: Before placing orders, submit representative sample of each type.

230 CONTROL SAMPLES
- General: Complete areas of finished work in approved locations, and obtain approval of appearance before proceeding.

330 COMMENCEMENT
- Required condition of works prior to laying materials:
  - Building is weathertight and well dried out.
  - Wet trades have finished work.
  - Paintwork is finished and dry.
  - Conflicting overhead work is complete.
  - Floor service outlets, duct covers and other fixtures around which materials are to be cut are fixed.
- Notification: Submit not less than 48 hours before commencing laying.

© AAS 2008– 2009

340 CONDITIONING
- Prior to laying: Condition materials by unpacking and separating in spaces where they are to be laid. Maintain resilient flooring rolls in an upright position. Unroll carpet and keep flat on a supporting surface.
- Conditioning time and temperature (minimum): As recommended by manufacturer with time extended by a factor of two for materials stored or transported at a temperature of less than 10°C immediately prior to laying.

350 ENVIRONMENT
- Temperature and humidity: Before, during and after laying, maintain approximately at levels which will prevail after building is occupied.
- Ventilation: Before during and after laying, maintain adequate provision.

360 FLOORS WITH UNDERFLOOR HEATING
- Commencement of laying: Not before a period of 48 hours after heating has been turned off.
- Post laying start up of heating system: Slowly return heating to its operative temperature not less than 48 hours after completing laying.

PREPARING BASES

410 NEW BASES
- Suitability of bases and conditions within any area: Commencement of laying of coverings will be taken as acceptance of suitability.

420 EXISTING BASES
- Notification: Before commencing work, confirm that existing bases will, after preparation, be suitable to receive coverings.
- Suitability of bases and conditions within any area: Commencement of laying of coverings will be taken as acceptance of suitability.

430 NEW WET LAID BASES
- Base drying aids: Not used for at least four days prior to moisture content testing.
- Base moisture content test: Carry out in accordance with BS 5325, Annexe A or BS 8203, Annexe A.
- Locations for readings: In all corners, along edges, and at various points over area being tested.
- Commencement of laying coverings: Not until all readings show 75% relative humidity or less.

440 SUBSTRATES TO RECEIVE THIN COVERINGS
- Trowelled finishes: Uniform, smooth surface free from trowel marks and other blemishes. Abrade suitably to receive specified floor covering material.

460 SMOOTHING/ LEVELLING UNDERLAYMENT COMPOUND
- Selection: As recommended by covering manufacturer.

470 EXISTING FLOOR COVERINGS REMOVED
- Substrate: Clear of covering and as much adhesive as possible. Skim with smoothing underlayment compound to give smooth, even surface.

480 EXISTING FLOOR COVERINGS TO BE OVERLAID
- Substrate: Make good by local resticking and patching or filling with smoothing underlayment compound to give smooth, even surface.

510 WOOD BLOCK FLOORING
- Substrate: Clean and free from wax with all blocks sound and securely bonded. Fill hollows with smoothing underlayment compound to give smooth, even surface.
- Missing and loose blocks: Replace and reset in adhesive to match existing. Sand or plane to make level.

LAYING COVERINGS

610 SETTING OUT TILES
- Method: Set out from centre of area/ room, so that wherever possible:
  - Tiles along opposite edges are of equal size.
  - Edge tiles are more than 50% of full tile width.

620 COLOUR CONSISTENCY
- Finished work in any one area/ room: Free from banding or patchiness.

640 ADHESIVE FIXING GENERALLY
- Adhesive: Type to be as specified, recommended by covering/ underlay manufacturer or as approved.
- Primer: Use and type as recommended by adhesive manufacturer.
- Application: As necessary to achieve good bond.
- Finished surface irregularities: Trowel ridges and high spots caused by particles on the substrate not acceptable.

650 SEAMS
- Patterns: Matched.
- Joints: Tight without gaps.

670 BORDERS/ FEATURE STRIPS IN SHEET MATERIAL
- Application: Cut strips along length of sheet to prevent curl.
  - Corners: Mitre joints.

680 SEAM WELDING COVERINGS
- Commencement: Not before a period of 24 hours after laying, or until adhesive has set.
- Joints: Neat, smooth, strongly bonded, flush with finished surface.

690 SEAM BONDING CARPET
- Carpet types: As specified in schedules.
- Seaming adhesive application: Continuous bead to edges.
- Joints: Securely bonded, free of air bubbles.

700 LOOSE LAID CARPET TILES
- Areas of adhered tiles: Secure using double sided tape or peelable adhesive.
- Joints: Tightly butted.
- Perimeter joints: Accurately cut to match abutment and prevent movement.

720 DOORWAYS
- Joint location: On centre line of door leaf.

740 EDGINGS/ COVER STRIPS
- Manufacturer: As specified in schedules.
- Material/ finish: As specified in schedules.
- Fixing: Secure (using matching fasteners where exposed to view) with edge of covering gripped.

750 STAIR NOSINGS/ TRIM
- Manufacturer: As specified in schedules.
- Material/ finish: As specified in schedules.
- Fixing: Secure, level with mitred joints. Adjusted to suit thickness of covering with continuous packing strips of hardboard or plywood. Nosings and packing strips bedded in gap-filling adhesive recommended by nosing manufacturer.
- Screw fixing with matching plugs: As specified in schedules.

760 STAIR COVERINGS - SEPARATE RISERS AND TREADS
- Fixing: Accurately cut. Fit risers before treads.

761 STAIR COVERINGS - ONE PIECE RISER AND TREAD
- Fixing: Accurately cut. Fit with coved riser/ tread.

© AAS 2008– 2009

770 SKIRTINGS
- Types: As specified in schedules.
- Manufacturer: As specified in schedules.
- Fixing: Secure with top edge straight and parallel with floor.
  - Corners: Mitre joints.

M51 EDGE FIXED CARPETING

GENERAL/ PREPARATION

210 WORKMANSHIP GENERALLY
- Finished carpeting: Tightly seamed, accurately fitted, neatly and securely fixed, smooth and evenly tensioned.

230 CONTROL SAMPLES
- General: Complete areas of finished work in approved locations, and obtain approval of appearance before proceeding:

290 CONDITIONING
- Requirements: As recommended by manufacturer.

310 COMMENCEMENT
- Required condition of works prior to laying carpeting:
  - Building is weathertight and well dried out.
  - Wet trades have finished work.
  - Paintwork is finished and dry.
  - Floor service outlets, duct covers and other fixtures around which carpet is to be cut are fixed.
- Notification: Submit not less than 48 hours before commencing laying.

320 ENVIRONMENT
- Temperature and humidity: Before, during and after laying, maintain approximately at levels which will prevail after building is occupied.

330 BASES
- Suitability of bases and conditions within any area: Commencement of laying carpeting will be taken as acceptance of suitability.

340 NEW WET LAID BASES
- Base drying aids: Not used for at least four days prior to moisture content testing.
- Base moisture content test: Carry out in accordance with BS 5325, Annexe A.
- Locations for readings: In all corners, along edges, and at various points over area being tested.
- Commencement of laying carpeting: Not until all readings show 75% relative humidity or less.

350 TIMBER BOARDING/ STRIP FLOORING
- Substrate: Boards securely fixed and acceptably level with no protruding fasteners. Plane, sand or apply smoothing underlayment compound as necessary to give smooth, even surface.

360 EXISTING TILE/ SHEET/ BLOCK/ SCREED FLOORING TO BE OVERLAID
- Substrate: Make good by local rebedding, sanding or applying smoothing underlayment compound to give a secure, smooth, even surface. Allow to dry before laying carpeting.

LAYING CARPETING

410 CARPET GRIPPER
- Manufacturer: Contractors choice.
- Types and method of fixing: As recommended by gripper manufacturer to suit specified carpet, base and conditions of use.
- Fixing: Secure to form continuous length along all edges adjacent to vertical surfaces leaving a 'gully width' of approximately three quarters the thickness of carpet. Do not place across openings.

© AAS 2008 - 2009

- Gripper strip lengths for adhesive fixing (maximum): 200 mm.

420 INTERLAY
- Placement: Fully cover base, with no wrinkles, folds, overlapping or gaps (other than to allow for expansion). Lay at right angles to direction of boarded floors.

430 CARPET UNDERLAY ON FLOORS
- Setting out: Seams not to coincide with those in carpet.
- Placement: Cut exactly to size, tightly butted to grippers and secured at perimeter by stapling or sticking to base.
- Surface of installed underlay: Flat, smooth and free from wrinkles or bubbles.
- Seams: Tight butt joints secured with staples, adhesive or top-taped with no shadow shown through carpet.

440 CARPET UNDERLAY ON STAIRS
- Extent: Underlay pads to cover tread and riser in one piece to full width of carpet (except where edges will be exposed).
- Placement: Tightly butted to grippers and secured to prevent movement and wrinkling.

450 CARPET SEAMS/ JOINTS
- Placement: Straight, flat, evenly tensioned and tightly butted with no trapped surface pile between edges.
- Method and materials: Compatible with carpet and as recommended by manufacturers.
- Bond strength: Consistent throughout whole length of seam, sufficient to withstand stretching without opening up and to last the life of carpet in all locations.
- Pattern matching: Where applicable, accurately matched for full length of seam.

460 RAW EDGE SEAMS (INCLUDING CROSS SEAMS)
- Treatment prior to seaming: Strengthen with cross straps and make secure by sealing, whipping or binding.

470 LAYING CARPET GENERALLY
- Appearance of laid carpet: Pieces of the same carpet type capable of being seen together to be of consistent appearance with pile lying in the same direction.
- Carpet perimeter: Accurately and closely fitted leaving no gaps with edges turned down and secured to grippers.
- Carpet tension: Even and such that carpet is flat and will not ruck, ripple or become slack.
- Doorways and recesses: Cut carpet in. Do not piece in without prior approval.

480 POWER STRETCHING
- General: Power stretch carpets with any one dimension equal to or above 5 m.

490 DOORWAYS
- Carpet joint: On centre line of door leaf.

510 EDGINGS/ COVER STRIPS
- Manufacturer: As specified in schedules.
- Material/ finish: As specified in schedules.
- Fixing: Secure using matching fasteners where exposed to view with edge of carpet firmly gripped.

530 LAYING STAIR CARPET WITH GRIPPER
- Shifting allowance: Provide a minimum additional length of carpet equivalent to one tread and riser. Conceal by substituting for underlay at top or bottom of stairs.
- Gripper locations: At each crotch (intersection of tread and riser), one on each riser and one on each tread. Along landing and winder edges which abut a wall and exceed 300 mm.
- Pile direction: Towards bottom of stairs and perpendicular to nosings.

540 LAYING STAIR CARPET WITH ADHESIVE
- Placement: Fit carpet after fixing nosings. Bond to base with a suitable permanent bond adhesive. Achieve a smooth flat finish with no trapped air.

550 STAIR NOSINGS/ TRIMS
- Manufacturer: Gradus .
  - Product reference: As specified in schedules.
- Material/ finish: As specified in schedules.
- Fixing: Secured, level with mitred joints. Adjusted to suit thickness of carpet with continuous packing strips of hardboard or plywood. Nosings and packing strips bedded in gap-filling adhesive recommended by the manufacturer.
- Screw fixing with matching plugs: As specified in schedules.

570 COMPLETION
- Debris: Remove stay tacks and cut away partly loose warp and face yarns.
- Surface irregularities and tension: Check and make necessary tension adjustments.

M52 DECORATIVE PAPERS/ FABRICS

TYPES OF COVERING

110 COVERING FOR WALLS AND CEILINGS
- Substrate: As specified in schedules.
  - Preparation: As appropriate for substrate.
- Adhesive: as recommended by paper manufacturer .
- Lining: medium grade .
- Covering: As specified in schedules.

GENERALLY

220 SAMPLES
- General: Submit a representative sample of each type of covering before placing orders.

241 ENVIRONMENT
- Conditions: During hanging and drying of linings/ coverings, maintain working area ambient temperature and humidity levels approximate to those proposed in service.

PREPARATION

310 PREPARATION GENERALLY
- Preparation materials: Types recommended by their manufacturers and covering manufacturer for situation and substrate being prepared.
- Substrates: Sufficiently dry in depth to suit covering to be hung.
- Efflorescence salts: Remove.
- Dirt, grease and oil: Remove. Give notice if contamination of substrates has occurred.
- Substrate irregularities: Fill cracks, joints, holes and other depressions with stoppers/ fillers. Work well in and finish off flush with surface. Abrade to a smooth finish.
- Dust, particles and residues from abrasion: Remove.

330 FIXTURES AND FITTINGS
- Before commencing work: Remove fixtures and fittings as specified.
- On completion of work: Refix when coverings are dry.

340 COATED SUBSTRATES
- Removing coatings: Do not damage substrate and adjacent surfaces or adversely affect subsequent coverings.
- Loose, flaking or otherwise defective areas: Carefully remove to a firm edge.
- Water soluble coatings: Completely remove.
- Significant rot, corrosion or other degradation of substrates: If revealed, give notice.
- Retained coatings:
  - Thoroughly clean to remove dirt, grease and contaminants.
  - Abrade gloss coated substrates to provide a key.
  - Carry out tests for compatibility with adhesives.

350 PAPER/ FABRIC COVERED SUBSTRATES
- Existing coverings: Remove by wet or dry stripping.
- Old adhesive and size: Remove by washing.
- Significant loose or damaged plaster or other degradation of substrates: If revealed, give notice.

360 VINYL COVERED SUBSTRATES
- Existing covering: Remove peelable vinyl surface.
- Paper base to vinyl: May be retained as a lining if in good condition and firmly adhering. Stick down lifting edges and corners.

370 ORGANIC GROWTHS
- Loose growths and infected coatings/ decorations: Scrape off and remove.
- Treatment biocide: Apply appropriate solution to growth areas and surrounding surfaces.
- Dead growth: Scrape off and remove.
- Residual effect biocide: Apply appropriate solution to inhibit re-establishment of growths.
- Biocides: Types listed in current Health and Safety Executive (HSE) 'Pesticides', Part B, as surface biocides.

380 SIZE/ SEALER
- Absorbent substrates: Apply one coat of a dilute solution of adhesive where recommended by covering manufacturer.

HANGING

420 HANGING GENERALLY
- Coatings on adjacent surfaces: Complete and dry before commencement of hanging coverings.
- Sequence of hanging coverings:
    - Apply to ceilings before walls.
    - Commence adjacent to main source of natural light.
    - From centre of feature and isolated walls.
- Surplus adhesive: Carefully remove from face of coverings, adjacent surfaces and fittings whilst still wet.
- Completed coverings: Securely adhered, smooth and free of air bubbles, wrinkles, gaps, tears, adhesive marks and stains. Joints truly vertical/ horizontal and straight.

480 LININGS
- Type and weight: To suit coverings and substrates.
- Hanging lengths: Transverse to direction of coverings, with neat butt joints; do not overlap.
- Drying period: Leave for 24 hours before hanging coverings.

490 COVERINGS
- Selvedged coverings: Trim to a true straight edge before hanging, unless overlap joints are recommended by manufacturer.
- Hanging lengths:
    - Wall coverings: Vertical.
    - Ceiling coverings: Parallel to main window wall.

500 JOINTS IN COVERINGS
- Butt joints: Hang lengths with neat butt joints generally.
- Overlap joints: Hang lengths with neat overlap joints only where recommended by covering manufacturer. Cut through joints when stable to a true straight edge, without damaging substrate, and bond joints.
- Cross joints: Hang lengths in one piece generally. Cross joints are only permitted where single lengths are impractical.

520 SHADING
- Matching: Ensure colour consistency of adjacent lengths.
- Hanging lengths: Use in sequence as cut from roll.
- Alternate lengths: Do not reverse unless recommended by covering manufacturer.
- Shade variation: Check after hanging first three lengths. If variation occurs, give notice before proceeding.

530 PATTERN
- Patterned coverings: Accurately align and match.
- Mismatches: Anticipate and obtain approval for locations.

M60 PAINTING/ CLEAR FINISHING

GENERALLY

100 DEFINITIONS
- Wood includes softwood, plywood, hardwood and the like.
- Soft building board includes fibreboard, chipboard, hardboard and the like where there is a high rate of initial absorption or a sealant is required.
- Hard building board includes plasterboard, laminboard and the like where there is a minimal rate of initial absorption.

103 METHODS OF MEASUREMENT
- Frames will be measured and girthed separately each side irrespective of whether the same treatment is to be applied to both sides.

105 PRICES ALSO TO INCLUDE
- Preparing and priming rebates as specified for glazing.
- Taking off any ironmongery and the like necessary for the proper execution of the work and for refixing.
- Decorating surface conduit and the like in with the surface to which it is attached and for any additional area so resulting.
- Extra preparation on metal trims over that of the general surfaces in which they are decorated.
- Touching up primed and galvanized metalwork with primer within seven days of delivery.
- Tints specially selected by the Contract Administrator who will prepare colour schemes which the Contractor is to follow in all respects.
- Preparation of manufactured boards and wood and fibre cement products is to include surface filling.

107 CONTRARY TO SMM
- Painting to staircase and plantroom areas has not been separately identified, the rates include an allowance for such areas (SMM M60.M1).

COATING SYSTEMS

110 EMULSION PAINT
- Manufacturer: Clause 210.
- Surfaces: As Specified in schedules.
  - Preparation: As appropriate for surface.
- Initial coats:
  - Type: As Specified in schedules.
  - Number of coats: As Specified in schedules.
- Finishing coats:
  - Type: As Specified in schedules.
  - Number of coats: As Specified in schedules.

130 GLOSS PAINT
- Manufacturer: Clause 210.
- Surfaces: As Specified in schedules.
  - Preparation: As appropriate for surface.
- Initial coats:
  - Type: As Specified in schedules.
  - Number of coats: As Specified in schedules.
- Finishing coats:
  - Type: As Specified in schedules.
  - Number of coats: As Specified in schedules.

150 EGGSHELL/ SATIN PAINT
- Manufacturer: Clause 210.
- Surfaces: As appropriate for surface.
  - Preparation: As appropriate for surface.
- Initial coats:
  - Type: As appropriate for surface.
  - Number of coats: As appropriate for surface.
- Finishing coats:
  - Type: As appropriate for surface.
  - Number of coats: As appropriate for surface.

GENERALLY

210 COATING MATERIALS
- Manufacturer: Obtain coatings from one of the following:
  - Leyland Paint
  - Dulux Paints
  - Crown Paints
- Akzo Nobel
- Imperial Chemical Industries Plc
- Ronseal Ltd
- Selected manufacturer: Submit name before commencement of coating work.

215 HANDLING AND STORAGE
- Coating materials: Deliver in sealed containers, labelled clearly with brand name, type of material and manufacturer's batch number.
- Materials from more than one batch: Store separately. Allocate to distinct parts or areas of the work.

217 TIME LIMIT
- do not use coating materials more than 18 months after despatch or re-test but remove from site and arrange for re-test.

220 COMPATIBILITY
- Coating materials selected by contractor:
  - Recommended by their manufacturers for the particular surface and conditions of exposure.
  - Compatible with each other.
  - Compatible with and not inhibiting performance of preservative/ fire retardant pretreatments.

280 PROTECTION
- 'Wet paint' signs and barriers: Provide where necessary to protect other operatives and the general public, and to prevent damage to freshly applied coatings.

300 CONTROL SAMPLES
- General: Carry out sample areas of finished work, including preparation, as required by CA
- Approval of appearance: Obtain before commencement of general coating work.

320 INSPECTION BY COATING MANUFACTURER
- General: Permit manufacturers to inspect the work in progress and take samples of their materials from site if requested.

PREPARATION

400   PREPARATION GENERALLY
- Standard: To BS 6150, Section 4.
- Preparation materials: Types recommended by their manufacturers and the coating manufacturer for the situation and surfaces being prepared.
- Substrates: Sufficiently dry in depth to suit coating.
- Efflorescence salts: Remove.
- Dirt, grease and oil: Remove.
  - Give notice if contamination of surfaces/ substrates has occurred.
- Joints, cracks, holes and other depressions:
  - Fill with stoppers/ fillers. Work well in and finish off flush with surface.
  - Abrade to a smooth finish.
- Water based stoppers and fillers:
  - Apply before priming unless recommended otherwise by manufacturer.
  - If applied after priming: Patch prime.
- Oil based stoppers and fillers: Apply after priming.
- Surface irregularities: Abrade to a smooth finish.
- Dust, particles and residues from abrasion: Remove.
- Doors, opening windows and other moving parts:
  - Ease before coating.
  - Prime resulting bare areas.

425   IRONMONGERY
- Before commencing work: Remove ironmongery from surfaces to be coated.
- Hinges: Do not remove.
- On completion of coating work: Refix.

430   EXISTING IRONMONGERY
- General: Remove old coating marks. Clean and polish.

440   PREVIOUSLY COATED SURFACES GENERALLY
- Preparation standard: To BS 6150, Section 6.
- Removing coatings: Do not damage substrate and adjacent surfaces or adversely affect subsequent coatings.
- Loose, flaking or otherwise defective areas: Carefully remove to a firm edge.
- Alkali affected coatings: Completely remove.
- Discovery: Give notice of:
  - Coatings suspected of containing lead.
  - Substrates suspected of containing asbestos.
  - Significant rot, corrosion or other degradation of substrates.
- Retained coatings:
  - Thoroughly clean to remove dirt, grease and contaminants.
  - Gloss coated surfaces: Abrade to provide a key.
- Partly removed coatings:
  - Additional preparatory coats: Apply to restore original coating thicknesses.
  - Junctions: Abrade to give a flush surface.
- Completely stripped surfaces: Prepare as for uncoated surfaces.

461   PREVIOUSLY COATED TIMBER
- Degraded or weathered surface timber: Abrade to remove.
- Degraded substrate timber: Repair with sound timber of same species.
- Exposed resinous areas and knots: Apply two coats of knotting.

471   PREPRIMED TIMBER
- Defective primer: Abrade back to bare timber.
- Bare areas: Reprime.

481 UNCOATED TIMBER
- General: Abrade to a smooth, even finish with arrises and moulding edges lightly rounded or eased.
- Heads of fasteners: Countersink sufficient to hold stoppers/ fillers.
- Resinous areas and knots: Apply two coats of knotting.

490 PREVIOUSLY COATED STEEL
- Defective paintwork: Remove to leave a firm edge and clean bright metal.
- Sound paintwork: Abrade to provide key for subsequent coats.
- Corrosion and loose scale: Abrade back to bare metal.
- Residual rust: Treat with a proprietary removal solution.
- Bare metal: Apply primer as soon as possible.
- Remaining areas: Degrease.

500 PREPRIMED STEEL
- Defective primer, corrosion and loose scale: Abrade back to bare metal.
- Bare areas: Reprime as soon as possible.

511 GALVANIZED, SHERARDIZED AND ELECTROPLATED STEEL
- White rust: Remove.
- Pretreatment: Apply one of the following:
  - 'T wash'/ mordant solution to blacken whole surface.
  - Etching primer recommended by coating system manufacturer.

521 UNCOATED STEEL - MANUAL CLEANING
- Oil and grease: Remove.
- Corrosion, loose scale, welding slag and spatter: Abrade to remove.
- Residual rust: Treat with a proprietary removal solution.
- Primer: Apply as soon as possible.

541 UNCOATED ALUMINIUM/ COPPER/ LEAD
- Surface corrosion: Remove and lightly abrade.
- Pretreatment: Etching primer if recommended by coating system manufacturer.

552 UNCOATED PVC-U
- Dirt and grease: Remove. Do not abrade.

560 UNCOATED CONCRETE
- Release agents: Remove.

570 UNCOATED MASONRY/ RENDERING
- Loose and flaking material: Remove.

580 UNCOATED PLASTER
- Nibs, trowel marks and plaster splashes: Scrape off.
- Overtrowelled 'polished' areas: Abrade lightly.

590 UNCOATED PLASTERBOARD
- Depressions around fixings: Fill with stoppers/ fillers.

601 UNCOATED PLASTERBOARD - TO RECEIVE TEXTURED COATING
- Joints: Fill, tape and feather out with materials recommended by textured coating manufacturer.

611 WALL COVERINGS
- Retained wall coverings: Check that they are in good condition and well adhered to substrate.
- Previously covered walls: Wash down to remove paper residues, adhesive and size.

622 ORGANIC GROWTHS
- Loose growths and infected coatings: Scrape off and remove.
- Treatment biocide: Apply appropriate solution to growth areas and surrounding surfaces.
- Dead growth: Scrape off and remove.
- Residual effect biocide: Apply appropriate solution to inhibit re-establishment of growths.

631 PREVIOUSLY PAINTED WINDOW FRAMES
- Paint encroaching beyond glass sight line: Remove.
- Loose and defective putty: Remove.
- Putty cavities and junctions between previously painted surfaces and glass: Clean thoroughly.
- Finishing:
    - Patch prime, reputty and allow to set.
    - Seal and coat as soon as sufficiently hard.

640 EXTERNAL POINTING TO EXISTING FRAMES
- Defective sealant pointing: Remove.
- Joint depth: Approximately half joint width; adjust with backing strip if necessary.
- Preparation and application: As section Z22.
- Sealant:
    - Manufacturer: Contractors choice.

645 POINTING TO INTERNAL MOVEMENT JOINTS
- General: Apply to junctions of walls and ceilings with architraves, skirtings and other trims.
- Preparation and application: As section Z22.
- Sealant: Water based acrylic.
    - Manufacturer: Contractors choice.

651 EXISTING GUTTERS
- Dirt and debris: Clean from inside of gutters.
- Defective joints: Clean and seal with suitable jointing material.

APPLICATION

711 COATING GENERALLY
- Application standard: To BS 6150, Section 5.
- Conditions: Maintain suitable temperature, humidity and air quality during application and drying.
- Surfaces: Clean and dry at time of application.
- Thinning and intermixing of coatings: Not permitted unless recommended by manufacturer.
- Overpainting: Do not paint over intumescent strips or silicone mastics.
- Priming coats:
    - Thickness: To suit surface porosity.
    - Application: As soon as possible on same day as preparation is completed.
- Finish:
    - Even, smooth and of uniform colour.
    - Free from brush marks, sags, runs and other defects.
    - Cut in neatly.

720 PRIMING JOINERY
- Preservative treated timber: Retreat cut surfaces before priming.
- End grain: Liberally coat, allow to soak in and then recoat.

730 WORKSHOP COATING OF CONCEALED JOINERY SURFACES
- General: Apply coatings to all surfaces of components.

731 SITE COATING OF CONCEALED JOINERY SURFACES
- General: After priming, apply additional coatings to surfaces that will be concealed when fixed in place.

740 CONCEALED METAL SURFACES
- General: Apply additional coatings to surfaces that will be concealed when fixed in place.

751 STAINING TIMBER
- Primer: Apply if recommended by stain manufacturer.
- Stain:
  - Apply in flowing coats.
  - Brush out excess stain before set.
  - Produce uniform depth of colour.

760 VARNISHING TIMBER
- First coat:
  - Thin with white spirit.
  - Brush well in and lay off avoiding aeration.
- Subsequent coats: Rub down lightly between coats along the grain.

770 EXTERNAL DOORS
- Bottom edges: Prime and coat before hanging doors.

780 BEAD GLAZING TO COATED TIMBER
- Before glazing: Apply first two coats to rebates and beads.

790 PUTTY GLAZING
- Setting: Allow putty to set for seven days.
- Sealing:
  - Within a further 14 days, seal with an oil based primer.
  - Fully protect putty with coating system as soon as it is sufficiently hard.
  - Extend finishing coats on to glass up to sight line.

800 GLAZING
- Etched, sand blasted and ground glass: Treat or mask edges before coating to protect from contamination by oily constituents of coating materials.

810 WATER REPELLENT
Application: Liberally flood surface, giving complete and even coverage.

820 SPRAY PAINTING
- use approved equipment. Do not spray the following:
  - priming paints for wood
  - lead paints within the meaning of the Factories Act 1961
  - first coat of bituminous paint on bare metal
  - first coat of emulsion paint on bare plaster
  - tar base paint and creosote
  - where soiling of adjacent surfaces cannot be prevented

N: FURNITURE/EQUIPMENT

N10   GENERAL FIXTURES/ FURNISHINGS/ EQUIPMENT

GENERAL

100   METHODS OF MEASUREMENT
- Worktops, benches and the like will be measured and valued at the rates analogous to shelving.

105   PRICES ALSO TO INCLUDE
- Compensating veneers to manufactured boards where the Contractor has elected to apply hardwood veneers in lieu of providing hardwood faced boards as described.
- Bedding taps, waste fittings and similar accessories in red and white lead or similar material.
- Disconnecting, setting aside and refixing required by other trades.

PRODUCTS

270   SMALL MIRRORS.
- Material: As specified in schedules.
- Edges: Smoothed and silvered.
- Quality: Free from tarnishing, discoloration, scratches and other defects visible in the designed viewing conditions.
- Dimensions:
    - Sizes: As specified in schedules.
    - Thickness: As specified in schedules.
- Backing: As specified in schedules.
- Fixing: Secure with Plugs and screws.
- Positioned: Accurately with sides vertical.
- Reflection: Undistorted.

290   MATWELL FRAMES
- Manufacturer: Contractors choice unless specified in schedules.
    - Product reference: Various.
- Material: As specified in schedule.
    - Finish: As specified in schedule.
- Angles:
    - Corners: Mitred and welded.
    - Angle size: 25 x 25 3mm.
- Matwell size: As specified in schedule.

300   ENTRANCE MATTING
- Manufacturer: Gradus .
    - Product reference: As specified in schedules.
- Colour: As specified in schedules.
- Size: As specified in schedules.

350   MISCELLANEOUS FITTINGS
- Items: As specified in schedules.
- Fixing: As specified in schedules.

EXECUTION

710   MOISTURE CONTENT
- Temperature and humidity: During delivery, storage, fixing and to handover maintain conditions to suit specified moisture contents of timber components.
- Testing: When instructed, test components with approved moisture meter to manufacturer's recommendations.

720   INSTALLATION GENERALLY
- Fixing and fasteners: As section Z20.

740 TAPS
- Fixing: Secure, watertight seal with the appliance.
- Positioning: Hot tap to left of cold tap as viewed by user of appliance.

750 WASTES/ OVERFLOWS
- Bedding: Waterproof jointing compound.
- Fixing: With resilient washer between appliance and backnut.

760 SEALANT POINTING
- Material: Silicone based to BS 5889, Type B with fungicide.
- Colour: As appropriate.
- Manufacturer: Contractors choice .
- Application: As section Z22.

770 TRIMS
- Lengths: Wherever possible, unjointed between angles or ends of runs.
- Running joints: Where unavoidable, obtain approval of location and method of jointing.
- Angle joints: Mitred.

780 COMPLETION
- Doors and drawers: Accurately aligned, not binding. Adjusted to ensure smooth operation.
- Ironmongery: Checked, adjusted and lubricated to ensure correct functioning.

N13 SANITARY APPLIANCES AND FITTINGS

PRODUCTS

300 WCS AND CISTERNS - As specified in schedules
- WC standard: To DEFRA WC suite performance specification or approved by relevant water company.

311 DOCUMENT M PACKAGES
- Manufacturer: Nichols & Clarke .
    - Product reference: Various .
- Type approval certificate: Submit.

315 URINALS AND AUTO FLUSHING CISTERNS
- Urinals: As specified in schedules

316 URINALS AND FLUSHING VALVES
- Urinals: As specified in schedules

331 SINKS As specified in schedules

335 WASH BASINS As specified in schedules

355 BATHS As specified in schedules

375 SHOWER UNITS As specified in schedules

379 DRINKING FOUNTAINS As specified in schedules

436 HANDRAILS AND GRAB BARS : As specified in schedules

EXECUTION

610 INSTALLATION GENERALLY
- Assembly and fixing: Surfaces designed to falls to drain as intended.
- Fasteners: Nonferrous or stainless steel.
- Supply and discharge pipework: Fix before appliances.
- Fixing: Fix appliances securely to structure. Do not support on pipework.
- Jointing and bedding compounds: Recommended by manufacturers of appliances, accessories and pipes being jointed or bedded.
- Appliances: Do not use. Do not stand on appliances.
- On completion: Components and accessories working correctly with no leaks.
- Labels and stickers: Remove.

613 COMPATIBILITY OF COMPONENTS
- General: Each sanitary assembly must consist of functionally compatible components obtained from a single manufacturer.

620 NOGGINGS AND BEARERS
- Noggings, bearers, etc. to support sanitary appliances and fittings: Position accurately. Fix securely.

625 FRAMING FOR PREPLUMBED PANEL SYSTEM
- General: Position accurately. Fix securely.

630 TILED BACKGROUNDS OTHER THAN SPLASHBACKS
- Timing: Complete before fixing appliances.
- Fixing appliances: Do not overstress tiles.

650 INSTALLING WC PANS
- Floor mounted pans: Screw fix and fit cover caps over screw heads. Do not use mortar or other beddings.
- Seat and cover: Stable when raised.

660 INSTALLING SLAB URINALS
- Waterproofing of walls and floor (specified elsewhere): Completed before fixing urinal components.
- Gap between components: 3 mm.
- Space behind channels and slabs: Grout with 1:5 cement:sand grout.
- Pointing: Rake out joints to 10 mm depth. Point flush with waterproof jointing compound recommended by urinal manufacturer.

670 INSTALLING CISTERNS
- Cistern operating components: Obtain from cistern manufacturer.
  - Float operated valve: Matched to pressure of water supply.
- Overflow pipe: Fixed to falls and located to give visible warning of discharge.
  - Location: Agreed, where not shown on drawings.

710 INSTALLING TAPS
- Fixing: Secure against twisting.
- Seal with appliance: Watertight.
- Positioning: Hot tap to left of cold tap as viewed by user of appliance.

720 INSTALLING WASTES AND OVERFLOWS
- Bedding: Waterproof jointing compound.
- Fixing: With resilient washer between appliance and backnut.

© AAS 2008– 2009

N15 SIGNS/ NOTICES

TYPES OF SIGNS

210 ESCAPE ROUTE SIGNS
- Standard: To BS 5499-1, -4 and -5.
- Manufacturer: Seton.
  - Product reference: Various.
- Base material: As specified in schedules.
  - Thickness: As specified in schedules.
- Fixing: As manufacturers recommendations.

220 PHOTOLUMINESCENT SIGNS
- Standard: To BS 5499-1, -4 and -5 and Photoluminescent Safety Products Association (PSPA) Standard 002 part 1.
- Manufacturer: Seton.
  - Product reference: Various.
- Base material: Rigid plastic.
- Fixing: As manufacturers recommendations.

260 PUBLIC INFORMATION SIGNS
- Standard:
  - Layout and graphics: To BS 8501.
  - Colorimetric and photometric properties: To BS 5378-2.
- Manufacturer: Seton.
  - Product reference: Various.
- Fixing: As manufacturers recommendations.

270 TACTILE SIGNS FOR THE VISUALLY IMPAIRED
- Manufacturer: Seton.
  - Product reference: Various.
- Corners of rectangular rigid signs: Radiused.
- Surface: Nonreflective with maximum gloss factor of 15% when tested to BS 2782-5 or BS EN ISO 2813.
- Characters: Embossed between 1 and 1.5 mm with a stroke width that allows both sides of the character to be felt with the fingers at a single pass.
- Braille: English Standard located 6 mm below the bottom of the text with braille locator at left edge of sign.
- Fixing: As manufacturers recommendations.

FIXING SIGNS

410 FIXING SIGNS GENERALLY
- Fixing signs: Secure, plumb and level, using fixing methods recommended by manufacturer.
- Fasteners/ Adhesives: As section Z20.
- Strength of fasteners: Sufficient to support all live and dead loads.
- Fasteners for external signs: Corrosion resistant material or with a corrosion resistant finish. Isolate dissimilar metals to avoid electrolytic corrosion.
- Fixings showing on surface of sign: Must not detract from the message being displayed.

420 SIGNS FOR THE VISUALLY IMPAIRED
- Protection of users: Ensure that fasteners for tactile signs have no sharp edges or protrusions that would cause confusion or injury to users.

450 SIGNS FIXED TO POSTS
- Protrusion of post top above sign: Not permitted unless supporting a luminaire.
- Position of posts:
  - Single post: Centre of sign.
  - Two posts: Centre of posts 300 mm from sign edges.

460 CONCRETE FOUNDATIONS FOR SIGN POSTS
- Mix: To BS 5328-2, Designated mix not less than GEN 1 or Standard mix not less than ST2.
- Alternative mix for small quantities: 50 kg Portland cement, class 42.5, to 100 kg fine aggregate to 180 kg 20 mm nominal maximum size coarse aggregate, medium workability.
- Admixtures: Submit proposals.
    - Prohibited content: Calcium chloride.
- Blinding to post holes: 50 mm concrete.
- Installation of posts: Plumb and central in holes.
- Concrete fill: Fully compacted with concrete to not less than 150 mm below ground level.
- Duration of support to posts after placing concrete: Not less than three days.
- Backfilling: Not less than 48 hours after placing concrete.

470 SIGN POSTS PLACED IN POCKETS IN STRUCTURES
- Concrete mix: Standard mix ST5 to BS 5328-2.
- Pockets in structure:
    - Cleaning: Fully cleaned.
    - Blinding: 50 mm concrete.
- Installation of posts: Plumb and central in pockets.
- Concrete fill: Fully compacted to finished foundation level.
- Duration of support to posts after placing concrete: Not less than three days.
- Backfilling over pockets: Not less than 48 hours after placing concrete.

480 EXPOSED CONCRETE FOUNDATIONS TO POSTS
- Finish: Compacted until air bubbles cease to appear on the upper surface, then weathered to shed water and trowelled smooth.

P: BUILDING FABRIC SUNDRIES

P10 SUNDRY INSULATION/ PROOFING WORK/ FIRE STOPS

100 CONTRARY TO SMM
- The centres of members have not been stated (SMM P10.2.3.*).

120 MINERAL WOOL INSULATION LAID BETWEEN CEILING/FLOOR JOISTS
- Material: Mineral wool mats to BS 5803-1, Kitemark certified.
- Manufacturer: Contractors choice.
- Thickness: As specified in schedule.
- Installation: To BS 5803-5.
- Installation requirements:
    - Joints: Closely butted, no gaps.
    - Extent of insulation: Over wall plates.
    - Service holes: Sealed, and debris removed before laying insulation.
    - Eaves ventilation: Unobstructed.
    - Electric cables overlaid by insulation: Sized accordingly.

130 MINERAL WOOL INSULATION LAID ACROSS CEILING/FLOOR JOISTS
- Material: Mineral wool mat to BS 5803-1, Kitemark certified.
- Manufacturer: Contractors choice.
- Thickness: As specified in schedule.
- Installation: To BS 5803-5.
- Installation requirements:
    - Insulation widths: Widest practical.
    - Laid direction: At right angles to ties/ joists.
    - Joints: Closely butted, no gaps.
    - Insulation: Fitted neatly around rafter ends and extended over wall plates.
    - Service holes: Sealed, and debris removed before laying insulation.
    - Eaves ventilation: Unobstructed.
    - Electric cables overlaid by insulation: Sized accordingly.

170 LOFT ACCESS HATCHES
- Insulation: Mineral wool mat.
    - Thickness: Same as loft insulation.
    - Laying: Cut to fit, with no gaps and securely fixed.
- Edges of hatch: Sealed with an approved compressible draught excluder.

P20 UNFRAMED ISOLATED TRIMS/ SKIRTINGS/ SUNDRY ITEMS

GENERAL

100 METHODS OF MEASUREMENT
- In amplification of SMM P20.D1 all wrought timbers are given as nominal sizes. (The limits on planing margins are to be in accordance with BS EN 1313-1 for softwood and BS EN 1313-2 for hardwood).
- In amplification of SMM P20.S8, where fixings are left to the discretion of the Contractor, they are deemed nailed, pinned, spiked or stapled unless otherwise specified.

105 PRICES ALSO TO INCLUDE
- Tonguing, crosstonguing, housing and like labours to include grooves in softwood, hardwood, manufactured boards and wood and asbestos products.
- Lippings to include tonguing.
- Dressing all joinery in contact with plaster with linseed oil before plastering and subsequently cleaning off.
- Punching fixings below exposed surfaces.
- Fixing only windows, doors, units and the like is to include easing and adjusting.
- Surfaces treated, coated and veneered are to be kept clean covered and protected.
- Holes, mortices, sinkings and like labours are to include work cross-grain.

- Work creosoted or treated with wood preservative is to include additional treatment to holes and cut surfaces as specified.
- Selecting and keeping hardwood clean for clear finish as applicable.

TRIMS/ CASINGS

110 SOFTWOOD
- Quality of wood and fixing: To BS 1186-3.
  - Species: As appropriate.
  - Class: 1.
- Preservative treatment: Not required.
- Profile: As specified in schedules.
  - Finished size: As specified in schedules.
- Finish as delivered: Natural.
- Fixing: Contractors discretion.

120 HARDWOOD
- Quality of wood and fixing: To BS 1186-3.
  - Species: As specified in schedules.
  - Class: 1.
- Preservative treatment: Not required.
- Profile: As specified in schedules.
  - Finished size: As specified in schedules.
- Finish as delivered: Natural.
- Fixing: Contractors discretion.

BOARD MATERIALS

200 MEDIUM DENSITY FIBREBOARD
- Manufacturer: Contractors choice.
- Standard: To BS EN 622-5.
  - Type: MDF or MDF-H
- Profile: As specified in schedules.
  - Dimensions: As specified in schedules.
- Finish as delivered: Factory applied priming coat.
- Support/ Fixing: Contractors discretion.

240 PLYWOOD
- Thickness: As specified in schedules.
- Face ply species: Birch.
- Appearance class to BS EN 635: I.
- Bond quality to BS EN 314-2: 1.
- Edges: lipping to match veneer.
- Support/ Fixing: As specified.

SUNDRY ITEMS

440 DECORATIVE FACINGS
- Material: Plastics laminated sheet.
- Manufacturer: Formica.
  - Product reference: Formica.
  - Roll width: Contractors choice to minimise waste.
- Fixing: Adhesive.

INSTALLATION

510 INSTALLATION GENERALLY
- Joinery workmanship: As section Z10.
- Metal workmanship: As section Z11.
- Methods of fixing and fasteners: Unless specified as section Z20.
- Straight runs: To be in one piece, or in long lengths with as few joints as possible.

- Running joints: Location and method of forming to be agreed where not detailed.
- Joints at angles: Mitre unless shown otherwise.
- Position and level: To be agreed where not detailed.

P21 DOOR/ WINDOW IRONMONGERY

GENERALLY

100 METHODS OF MEASUREMENT
- When described as to softwood, for fixing to flush doors, plywood, blockboard or any other building board including drilling through any facing other than metal sheet or hardwood veneer.
- Where part of an item is fixed to one type of background and part to another, the whole is described as being to that background involving the more expensive fixing unless otherwise stated.

105 PRICES ALSO TO INCLUDE
- All necessary normal pattern keeps, sockets, striking plates and the like unless stated otherwise.
- Filling floor springs with oil.
- Unless otherwise stated, for providing two keys per lock or latch.
- Two knobs or handles with back plates and spindle as required for each set of lock or latch furniture.
- All necessary connectors to joints in the length and lugs for letting into concrete for bottom tracks to sliding door gear.
- Labelling and handing over keys.
- Testing, easing, adjusting and oiling.
- Unless otherwise stated ironmongery, whether as fixed in its place or fixed only, is to include fixing with countersunk matching screws.
- Fixing by whatever means are necessary to uphold the item including drilling, plugging or tapping as necessary and the provision of all fixing materials.

IRONMONGERY

120 IRONMONGERY RANGE SELECTED BY CONTRACTOR
- Source: Single co-ordinated range.
- Notification: Submit details of selected range, manufacturer and/ or supplier.
- Principal material/ finish: As specified in schedules.
- Items unavailable within selected range: Submit proposals.

122 IRONMONGERY FROM LISTED PROPRIETARY RANGES
- Source: as specified in schedules
- Items unavailable within selected range: Submit proposals.

170 IRONMONGERY FOR FIRE DOORS
- Relevant products: Ironmongery fixed to, or morticed into, the component parts of a fire resisting door assembly.
- Compliance: Ironmongery included in successful tests to BS 476-22 or BS EN 1634-1 on door assemblies similar to those proposed.
- Melting point of components (except decorative non functional parts): 800°C minimum.

180 IRONMONGERY GENERAL
- All ironmongery details including manufacturer and type shall be as specified in the schedules of rates.

190 WORKMANSHIP
- Assemble and fix in accordance with the manufacturer's recommendations.
- Use fastenings with matching finish supplied by the ironmongery manufacturer.

P30 TRENCHES/ PIPEWAYS/ PITS FOR BURIED ENGINEERING SERVICES

GENERALLY

100 METHODS OF MEASUREMENT
- All Preambles in the following Work Sections apply equally to this section where appropriate:
  - D20 - Excavating and filling
  - E10 - In-situ concrete
- Underground ductwork, cover tiles, identification tapes are deemed straight unless stated otherwise.

105 PRICES ALSO TO INCLUDE
- Excavating trenches to include compacting, backfilling in layers and setting aside and replacing top soil as necessary.
- In the case of pipes etc not laid on or encased in concrete, for forming handholes as necessary.
- In the case of pitch fibre pipes not laid on or encased in concrete, for the extra work involved in preparing the bottoms of trenches and in backfilling.
- In the case of cable trenches, for the extra work involved in sifting the backfill and, where cable covers are stated, for backfilling in two stages.
- Locating and exposing existing services to include all extra excavation, earthwork support and backfilling as necessary including any extra allowance for rock and any extra breaking up and making good surface finishes over that allowed for the new trench.
- Underground ducts for buried electrical services to include all necessary draw-in wires and temporary plugs.

110 ROUTES OF SERVICES BELOW GROUND
- Locations of new service runs and pipeducts: Submit proposals.
- Temporary marking: Indicate new service runs and pipeducts with 75 x 75 mm softwood posts painted white and projecting not less than 600 mm above ground level, or with clearly visible waterproof markings on hard surfaces.

115 EXISTING ROADS AND PAVINGS
- Excavation and backfilling: To Highways Authorities and Utilities Committee (Stationery Office) 'Specification for the reinstatement of openings in highways'.

116 EXISTING ROADS AND PAVINGS
- Excavation and backfilling: To Northern Ireland Road Authority and Utilities Committee (Stationery Office) 'Specification for the reinstatement of openings in roads'.

120 IN SITU CONCRETE
- In situ concrete for buried engineering services:
  - Standard: To BS 5328-1, -2, -3 and -4, or to BS 8500-1, -2 and BS EN 206-1.
  - Mix: C20.
  - Equivalent or better mix: Submit proposals.
  - Different mixes may be used for different parts of the work.

125 CUTTING TREE ROOTS IN SERVICE TRENCHES
- Protected area: The larger of the branch spread of the tree or an area with a radius of half the tree's height, measured from the trunk.
- Roots in protected area: Do not cut.
- Roots exceeding 25 mm diameter: Give notice and do not cut without permission.
- Cutting:
  - Use a hand saw to make clean smooth cuts.
  - Minimise wound area and ragged edges.
  - Pare cut surfaces smooth with a sharp knife.
- Unintentionally severed roots: Give notice and form a new clean cut slightly nearer the trunk.
- Backfilling: Original topsoil.

130 TRENCHES
- Width: As small as practicable.
- Trench sides: Vertical.

- Trench bottoms: Remove mud, rock projections, boulders and hard spots. Trim level.
- Give notice: To inspect trench for each section of the work.

BACKFILLING

260 BACKFILLING GENERALLY
- Backfill from top of pipeduct surround: Material excavated from the trench.
- Backfilling: Lay and compact in 300 mm maximum layers. Do not use heavy compactors before backfill is 600 mm deep.

265 BACKFILLING UNDER ROADS AND PAVINGS
- Backfill from top of pipeduct surround: Granular sub-base material to Highways Agency Specification for highway works, clause 803 (Type 1).
- Backfilling: Lay and compact in 150 mm maximum layers.

275 TEMPORARY BRIDGES
- Construction traffic: After backfilling, provide temporary bridges over trenches to prevent damage to pipeducts or services.

280 WARNING MARKER TAPES
- Type: Continuous colour coded, heavy gauge polyethylene identification tapes.
- Installation: During backfilling, lay along the route of all service pipes, ducts etc.
- Location, depth, colour and markings: To requirements of service undertaker.

ACCESS CHAMBERS/ FITTINGS/ DRAWPITS

400 SMALL SURFACE BOXES
- Standard: To BS 5834-2, Grade Light vehicles.
- Manufacturer: Contractors choice.
- Material: Cast iron.
- Sizes: As specified in schedules.
- Frame bedding: As clause 440.

440 BEDDING OF FRAMES FOR ACCESS COVERS AND SURFACE BOXES
- Bedding: Solid in 1:3 cement:sand mortar, centrally over opening, level with surrounding finishes and square with joints in surrounding finishes or with building.
- Frames in soil or grassed areas: Haunch back edge of bedding so that it is not visible.

P31 HOLES/ CHASES/ COVERS/ SUPPORTS FOR SERVICES

110 COORDINATION
- Locations and dimensions of holes and chases for services: Submit details.

185 HOLES, RECESSES AND CHASES IN MASONRY
- Locations: To maintain integrity of strength, stability and sound resistance of construction.
- Sizes: Minimum needed to accommodate services.
  - Holes (maximum): 300 x 300 mm.
- Walls of hollow or cellular blocks: Do not chase.
- Walls of other materials:
  - Vertical chases: No deeper than one third of single leaf thickness, excluding finishes.
  - Horizontal or raking chases: No longer than 1 m. No deeper than one sixth of the single leaf thickness, excluding finishes.
- Chases and recesses: Do not set back to back. Offset by a clear distance at least equal to the wall thickness.
- Cutting: Do not cut until mortar is fully set. Cut carefully and neatly. Avoid spalling, cracking and other damage to surrounding structure.

230 NOTCHES AND HOLES IN STRUCTURAL TIMBER
- General: Avoid if possible.
- Sizes: Minimum needed to accommodate services.
- Position: Do not locate near knots or other defects.

- Notches and holes in same joist: Minimum 100 mm apart horizontally.
- Notches in joists: Locate at top. Form by sawing down to a drilled hole.
    - Depth (maximum): 0.125 x joist depth.
    - Distance from supports: Between 0.07 and 0.25 x span.
- Holes in joists: Locate on neutral axis.
    - Diameter (maximum): 0.25 x joist depth.
    - Centres (minimum): 3 x diameter of largest hole.
    - Distance from supports: Between 0.25 and 0.4 of span.
- Notches in roof rafters, struts and truss members: Not permitted.
- Holes in struts and columns: Locate on neutral axis.
    - Diameter (maximum): 0.25 x minimum width of member.
    - Centres (minimum): 3 x diameter of largest hole.
    - Distance from ends: Between 0.25 and 0.4 of span.

310   PIPE SLEEVES
- Material: PVC .
- Sleeves: Extend through full thickness of wall or floor. Position accurately.
    - Clearance around service (maximum): 20 mm or diameter of service, whichever is the lesser.
    - Installation: Bed solid.
- Space between service and sleeve: Seal with mineral wool.
- Where exposed to view: Finish bedding and sealing neatly.

Q: PAVING/PLANTING/FENCING/SITE FURNITURE

Q10 KERBS/ EDGINGS/ CHANNELS/ PAVING ACCESSORIES

GENERALLY

100 METHODS OF MEASUREMENT
- All Preambles in the following Work Sections apply equally to this section where appropriate:
  - D20 - Excavating and filling
  - E10 - In-situ concrete
  - E30 - Reinforcement for in-situ concrete

TYPES OF KERBS/ EDGINGS/ CHANNELS

110 PRECAST CONCRETE KERBS
- Standard: To BS 7263-3.
- Manufacturer: Contractors choice.
- Type: HB.
- Size (width x height x length): As specified in schedules.
- Special shapes: As specified in schedules.
- Joints: Clause 640 .

110 PRECAST CONCRETE EDGING
- Standard: To BS 7263-3.
- Manufacturer: Contractors choice.
- Type: EB.
- Size (width x height x length): As specified in schedules.
- Special shapes: As specified in schedules.
- Joints: Clause 640 .

250 MATERIAL SAMPLES
- Samples representative of colour and appearance of designated materials: Submit before placing orders.

ROAD/ PAVING ACCESSORIES/ MARKING

305 TREE GRILLES
- Manufacturer: Nicholls and Clarke.
  - Product reference: As specified in schedules.
- Size: As specified in schedules.
- Material: As specified in schedules.

395 ROAD MARKING (THERMOPLASTIC)
- Standard: Road Safety Markings Association standard specification document for road marking and road studs (StanSpec).
- Colour: As specified in schedules.

LAYING

510 LAYING GENERALLY
- Cutting: Neat, accurate and without spalling. Form neat junctions.
- Bedding of units: Positioned true to line and levelled along top and front faces, in a mortar bed on accurately cast foundations or, where clause 547 applies, on a race of fresh concrete.
- Securing of units: After bedding has set, secured with a continuous haunching of concrete or, where clause 547 applies, backing concrete cast monolithically with fresh concrete race.

520 ADVERSE WEATHER
- Conditions: Do not construct if the temperature is below 3°C on a falling thermometer or 1°C on a rising thermometer. Adequately protect foundations, bedding and haunching against frost and rapid drying by sun and wind.

530 CONCRETE FOR FOUNDATIONS, RACES AND HAUNCHING
- Standard: To BS 5328-1, -2, -3 and -4, or to BS 8500-1, -2 and BS EN 206-1.
- Designated mix: Not less than GEN0 or Standard mix ST1.
- Workability: Very low.

540 MORTAR BEDDING
- General: To section Z21.
- Mix: 1:3.
- Bed thickness: 12-40 mm.

547 BEDDING/ BACKING OF UNITS ON FRESH CONCRETE RACES
- Standard: To BS 7533-6, clause 4.2.

550 KERB DOWELS
- Dowels: Steel bar to BS 4482.
  - Size: 12 mm diameter, 150 mm long.
- Installation of dowels: Vertically into foundation while concrete is plastic.
  - Centres: To suit holes in kerbs.
  - Projection: 75 mm.
- Grouting of holes in kerbs: Filled with 1:3 cement:sand mortar, finished flush.

560 HAUNCHING DOWELS
- Dowels: Steel bar to BS 4482.
  - Size: 12 mm diameter, 150 mm long.
- Installation of dowels: Vertically into foundation while concrete is plastic.
  - Centres: 450 mm.
  - Distance from back face of kerb: 50 mm.
  - Projection: 75 mm.
- Haunching: Rectangular cross section, cast against formwork, fully enclosing and protecting dowels.

570 CHANNELS
- Installation: To an even gradient, without ponding or backfall.
- Lowest points of channels: 6 mm above drainage outlets.

580 DRAINAGE CHANNEL SYSTEMS
- Installation: To an even gradient, without ponding or backfall. Commence laying from outlets.
- Silt and debris: Removed from entire system immediately before handover.
- Washing and detritus: Safely disposed without discharging into sewers or watercourses.

610 ANGLE KERBS
- Usage: Internal and external 90° changes of direction.
- Cutting of mitres: Not permitted.

620 ACCURACY
- Deviations (maximum):
  - Level: ± 6 mm.
  - Horizontal and vertical alignment: 3 mm in 3 m.

630 NARROW MORTAR JOINTS
- Jointing: Ends of units buttered with bedding mortar as laying proceeds. Joints completely filled, tightly butted and surplus mortar removed immediately.
  - Joint width: 3 mm.

640 TOOLED MORTAR JOINTS
- Jointing: Ends of units buttered with bedding mortar as laying proceeds. Joints completely filled and tooled to a neat flush profile.
  - Joint width: 6 mm.

Q20 GRANULAR SUB-BASES TO ROADS/ PAVINGS

100 METHODS OF MEASUREMENT
- Filling materials deemed to be obtained off site unless otherwise stated.

120 CHECKING CALIFORNIA BEARING RATIO (CBR) OF SUBGRADES
- Subgrade variation: If material appears to vary from that stated in the site investigation report, or if there are extensive soft spots, test subgrade CBR to BS 1377-4 or BS 1377-9. Submit results and obtain instructions before proceeding.

140 EXCAVATION OF SUBGRADES
- Final excavation to formation/ subformation level: Carry out immediately before compaction of subgrade.
- Soft spots and voids: Give notice.
- Wet conditions: Do not excavate or compact when the subgrade may be damaged or destabilized.

145 PREPARATION/ COMPACTION OF SUBGRADES
- Timing: Immediately before placing sub-base.
- Compaction: Thoroughly, by roller or other suitable means, adequate to resist subsidence or deformation of the subgrade during construction and of the completed roads/ pavings when in use. Take particular care to compact fully at intrusions, perimeters and where local excavation and backfilling has taken place.

150 SUBGRADES FOR VEHICULAR AREAS
- Preparation and treatment: To Highways Agency 'Specification for highway works', clauses 616 and 617.

211 GRANULAR MATERIAL
- Quality: Free from excessive dust, well graded, all pieces less than 75 mm in any direction, minimum 10% fines value of 50 kN when tested in a soaked condition to BS 812-111, and in any one layer only one of the following:
    - Crushed rock (other than argillaceous rock) or quarry waste with not more binding material than is required to help hold the stone together.
    - Crushed concrete, crushed brick or tile, free from plaster, timber and metal.
    - Crushed non-expansive slag.
    - Gravel or hoggin with not more clay content than is required to bind the material together, and with no large lumps of clay.
    - Well-burned non-plastic colliery shale.
    - Natural gravel.
    - Natural sand.
- Filling: Spread and levelled in 150 mm maximum layers, each layer thoroughly compacted.

220 FROST SUSCEPTIBLE GRANULAR MATERIAL
- Definition: To Highways Agency 'Specification for highway works' clause 801.17.
- Limitations: Do not use within 450 mm of the final surface of the paving.
- Testing: Test materials used if required and supply certificates.

230 PLACING GRANULAR MATERIAL GENERALLY
- Preparation: Loose soil, rubbish and standing water removed.
- Structures, membranes and buried services: Ensure stability and avoid damage.

240 LAYING GRANULAR SUB-BASES FOR VEHICULAR AREAS
- General: Spread and levelled in layers. As soon as possible thereafter compact each layer.
- Standard: Highways Agency 'Specification for highway works' clause 801.3-801.15.
- At drainage fittings, inspection covers, perimeters and where local excavation and backfilling has taken place: Take particular care to compact fully.

241 LAYING GRANULAR SUB-BASES FOR VEHICULAR AREAS
- Proposals: Well in advance of starting work submit details of:
    - Maximum depth of each compacted layer.
    - Type of plant.

- Minimum number of passes per layer.
- General: Spread and levelled in layers. As soon as possible thereafter compact each layer.
- At drainage fittings, inspection covers, perimeters and where local excavation and backfilling has taken place: Take particular care to compact fully.
- Sub-base surface after compaction and immediately before overlaying: Uniformly well closed and free from loose material, cracks, ruts or hollows.

250 LAYING GRANULAR SUB-BASES FOR PEDESTRIAN AREAS
- General: Spread and levelled.
- Compaction:
  - Timing: As soon as possible after laying.
  - Method: By roller or other suitable means, adequate to resist subsidence or deformation of the sub-base during construction and of the completed paving when in use. Take particular care to compact fully at intrusions, perimeters and where local excavation and backfilling has taken place.

310 ACCURACY
- Permissible deviation (maximum) from required levels, falls and cambers:

	Roads Parking areas	Footways Recreation areas
Subgrade	+ 20 mm - 30 mm	± 20 mm

320 BLINDING
- Locations: Surfaces to receive sand bedded interlocking brick or block paving to sections Q24 and Q25.
- Material: Sand, fine gravel, PFA or other approved.
- Finish: Close, smooth, compacted surface.

330 COLD WEATHER WORKING
- Frozen materials: Do not use.
- Freezing conditions: Do not place fill on frozen surfaces. Remove material affected by frost. Replace and recompact if not damaged after thawing.

340 PROTECTION
- Sub-bases: As soon as practicable, cover with subsequent layers, specified elsewhere.
- Subgrades and sub-bases: Prevent damage from construction traffic, construction operations and inclement weather.

Q21 **IN SITU CONCRETE ROADS/ PAVINGS/ BASES**

NOTE: The rates in this section allow for ready mixed concrete delivered to a point ne 50m horizontally to placing point and all hoisting by non mechanical means

GENERALLY

100 METHODS OF MEASUREMENT
- All Preambles in the following Work Sections apply equally to this section where appropriate:
  - E10 – Mixing/casting/curing in-situ concrete
  - E20 - Formwork for in-situ concrete
  - E30 - Reinforcement for in-situ concrete
  - E42 - Accessories cast into in-situ concrete

TYPES OF PAVING

110 UNREINFORCED CONCRETE PAVING (DESIGNATED)
- Granular sub-base: As section Q20.
  - Thickness: As specified in schedules.
- Separation membrane: Polyethylene sheet 125 micrometres thick, edges lapped 300 mm.
- Concrete:
  - Designated concrete: C25 to BS 8500-1, -2 and BS EN 206-1.
  - Slab thickness (minimum): 150mm.

- Finish: As specified in schedules.

GENERAL/ PREPARATION

140 **READY-MIXED CONCRETE**
- Production plant: Currently certified by a body accredited by UKAS to BS EN 45011 for product conformity certification of ready-mixed concrete.
- Source of ready-mixed concrete: Obtain from one source if possible. Otherwise, submit proposals.
  - Name and address of depot: Submit before any concrete is delivered.
  - Delivery notes: Retain for inspection.
- Declarations of nonconformity from concrete producer: Notify immediately.

145 **ADMIXTURES**
- Calcium chloride and admixtures containing calcium chloride: Do not use.

240 **SUB-BASE PREPARATION**
- Surface: Sound, free of debris, mud and soft spots, and suitably close textured.
- Levels and falls: Within specified tolerances:
  - Vehicular areas: ±20 mm.
  - Pedestrian areas: ±12 mm.
  - Drainage outlets: +0 to -10 mm of required finished level.
- Kerbs and edgings: Complete, adequately bedded and haunched, and to required levels.

265 **TIMBER PERMANENT FORMWORK**
- Side forms: Softwood board, drilled as required for dowel bars.
  - Size: 150 x 38 mm.
  - Fixing: Galvanized nails to 50 x 50 x 450 mm long, softwood pegs driven into the ground at 1200 mm centres.
- Preservative treatment : As section Z12 and British Wood Preserving and Damp-Proofing Association Commodity Specification C4.

LAYING CONCRETE

310 **TRANSPORTING CONCRETE**
- General: Avoid contamination, segregation, loss of ingredients, excessive evaporation and loss of workability. Protect from heavy rain.
- Entrained air: Anticipate effects of transport and placing methods in order to achieve specified air content.
- Placing: Use suitable walkways and barrow runs for traffic over reinforcement and freshly placed concrete.

320 **LAYING CONCRETE GENERALLY**
- Timing: Place as soon as practicable after mixing and while sufficiently plastic for full compaction. After discharge from the mixer do not add water or retemper.
- Temperature of concrete at point of delivery:
  - In hot weather (maximum): 30°C.
  - In cold weather (minimum): 5°C.
- Cold weather:
  - Do not use frozen materials.
  - Do not place concrete against frozen or frost covered surfaces.
  - Do not place concrete when air temperature is below 3°C on a falling thermometer. Do not resume placing until rising air temperature has reached 3°C.
- Surfaces on which concrete is to be placed: Free from debris and standing water.
- Placing in final position: Place in one continuous operation up to construction joints.
  - Do not place concrete simultaneously on both sides of movement joints.
- Spreading: Spread and strike off with surcharge sufficient to obtain required compacted thickness.
- Adjacent work: Form neat junctions and prevent damage. Keep clean all channels, kerbs, inspection covers, etc.

330 COMPACTING
- General: Fully compact concrete to full depth (until air bubbles cease to appear on the surface) especially around reinforcement, cast-in accessories, into corners and at joints.
- Poker vibrators: Do not use to make concrete flow into position. Do not allow to come into contact with fabric reinforcement.
- Wet formed joint grooves: Rectify any irregularities by means of a vibrating float.
- Finish: A dense, even textured surface free from laitance or excessive water.
- Excess concrete: Remove from top of groove formers.

340 MANHOLE COVER/ GULLY GRATING FRAMES
- General: Set frames in independent concrete slabs placed over, but slightly larger than, exterior of manhole shaft or gully pot and any concrete surround.
- Positioning of joints in main slab: Set out so that manhole/ gully slabs are adjacent to a main transverse joint, wherever possible.
- Joints: Separate the independent slabs from main slabs with 25 mm thick joint filler board. Set board 20 mm below top of slab to form a sealing groove.

350 LEVELS
- Lines and levels of finished surface: Smooth and even, with regular falls to prevent ponding.
- Finished surfaces: Within ±6 mm of required levels (+6-0 mm adjacent to gullies and manholes).

360 SURFACE REGULARITY
- General: Where appropriate in relation to the geometry of the surface, the variation in gap under a 3 m straightedge (with feet) placed anywhere on the surface to be not more than 5 mm.
- Sudden irregularities: Not permitted.

JOINTS

410 JOINTS GENERALLY
- Layout: All joints to be accurately located, straight and well aligned.
- Construction joints made at end of working day: Form as contraction joints.
- Modifications to joint design or location: Submit proposals.
- Temporary support: Prior to concreting, set formwork, dowel bars, tie bars, joint filler boards, sealing groove fillets and the like rigidly in position and support to prevent displacement. Maintain support until concrete has set.
- Keep clean:
  - Do not allow concrete to enter any gaps or voids in the formwork or to render the movement joints ineffective.
  - Do not allow concrete to impregnate or penetrate any materials used as compressible joint fillers.

440 LONGITUDINAL CONSTRUCTION JOINTS
- Definition: Longitudinal joints are those parallel to the main axis of the paving.
- Standard: To Concrete Society Technical report 28.
- Formed groove:
  - Size (minimum width x depth): 15 x 13 mm.
  - Preparation: Repair damaged edges of initially cast slab prior to forming groove.
  - Method: Fix preformed fillet against top edge of the initially cast slab before placing the adjacent slab. Remove when concrete is fully cured.
- Completion: Round upper edges of slabs at joints to 5 mm radius. Do not overwork concrete.

SURFACE FINISH

520 TAMPED FINISH
- Method: Tamp surface with edge of a board or beam to give an even texture of parallel ribs.

540 SMOOTH FLOATED FINISH
- Timing: After compaction allow concrete to stiffen sufficiently to be properly worked.
- Finish: Even smooth surface with no ridges or steps.

550 POWER TROWEL FINISH
- Preparation: Float concrete to an even surface with no ridges or steps, then immediately commence curing.
- Surface finish: Uniform, smooth and free from trowel marks and other blemishes.
- Completion: Resume specified curing without delay.

CURING/ PROTECTION/ FINISHING

610 CURING
- General: Immediately after completion of surface treatment prevent evaporation from surface and exposed edges of slabs for a minimum period of seven days.
- Early curing:
  - Cover with waterproof sheeting held clear of surface. Seal against draughts at edges and junctions.
  - Do not apply sprayed compounds or sheets in direct contact until surface is in a suitable state and will not be marked.
- Coverings for curing: Contractor's choice of:
  - Impervious sheet material.
  - Resin based aluminized curing compound containing a fugitive dye and with an efficiency index of 90% when tested to BS 7542.
  - Sprayed plastics film.

660 PROTECTION
- Prevent damage to concrete:
  - From rain, indentation, physical damage, dirt, staining, rust marks and other disfiguration.
  - From thermal shock.
  - In cold weather, from freezing expansion of water trapped in pockets, etc.
  - By use as a building platform or for storing, mixing or preparing materials.

670 OPENING TO TRAFFIC
- Light vehicles: Seven days after placing concrete.
- Heavy vehicles: 28 days after placing concrete.

Q22 COATED MACADAM/ ASPHALT ROADS/ PAVINGS

GENERALLY

100 METHODS OF MEASUREMENT
- Notwithstanding the provision of SMM Clause S2 the method of application is deemed to be as detailed in the Materials and Workmanship Preambles.
- Notwithstanding the provision of SMM Clause S6, preparatory work to the base which is directly involved in the application of a finish is deemed to be included with the measured item.

TYPES OF PAVING

125 COATED MACADAM PAVING
- Materials and workmanship: To BS 4987-1 and -2.

150 ROLLED ASPHALT PAVING
- Workmanship: To BS 594-2.

PREPARATORY WORK/ REQUIREMENTS

210 TIMBER EDGING
- Softwood board:
  - Size: 38 x 150 mm.
  - Fixing: Galvanized nails into softwood pegs.
- Softwood pegs:
  - Size: As appropriate .
  - Fixing: Drive into ground.

- Centres: 1200mm.
- Preservative treatment:
  - Type: ACQ high pressure vacuum.

220 BITUMINOUS MATERIALS GENERALLY
- Suppliers' names: Submit.
  - Timing (minimum): 2 weeks before starting work.
- Test certificates: At the time of delivery for each manufacturing batch submit certificate:
  - Confirming compliance with this specification and the relevant British Standard.
  - Stating full details of composition of mix.

240 ACCEPTANCE OF SUB-BASE
- Surface: Sound, clean and suitably close textured.
- Levels and falls: To be within the specified tolerances:
  - Vehicular areas: +10 to -30 mm.
  - Pedestrian areas: ±12 mm.
  - Drainage outlets: 0 to -10 mm of the required finished level.
- Kerbs and edgings: Complete, adequately bedded and haunched and to the required levels.

LAYING

310 LAYING GENERALLY
- Preparation: Remove all loose material, rubbish and standing water.
- Adjacent work: Form neat junctions. Do not damage.
- Channels, kerbs, inspection covers etc: Keep clean.
- New paving:
  - Keep traffic free until it has cooled to prevailing atmospheric temperature.
  - Do not allow rollers to stand at any time.
  - Prevent damage.
  - Lines and levels: With regular falls to prevent ponding.
  - Overall texture: Smooth, even and free from dragging, tearing or segregation.
  - State on completion: Clean.

320 ADVERSE WEATHER
- Frozen materials: Do not use.
- Suspend laying:
  - During freezing conditions.
  - If the air temperature reaches 0°C, or in calm dry conditions -3°C, on a falling thermometer.
  - Hot rolled asphalt: During periods of continuous or heavy rain.

330 LEVELS
- Permissible deviation from the required levels, falls and cambers (maximum):
  - Finished surface: ±6 mm.
  - Adjacent to gullies and manholes: 0 to +3 mm.

350 CONTRACTOR'S USE OF PAVEMENTS
- Before use: Fill interstices of open-grained surface with coated grit to BS 4987-1, tables 34-36 with soluble bitumen content for all aggregates in the range of 1.5-3.5%. Remove surplus. Apply binder and chippings as clause 360.
- Final surfacing:
  - Timing: Defer laying until as late as practicable.
  - Immediately before laying final surfacing: Clean and make good the base/ binder course. Allow to dry. Uniformly apply, without puddles, a tack coat of sprayed bitumen emulsion of a suitable grade to BS 434-1 at 1.5-2.0 kg/m². Allow emulsion to break completely before applying surfacing.

351 CONTRACTOR'S USE OF PAVEMENTS
- Final surfacing:
  - Timing: Defer laying until as late as practicable.
  - Immediately before laying final surfacing: Clean and make good the base/ binder course. Allow to dry. Uniformly apply, without puddles, a tack coat of sprayed bitumen emulsion of a

suitable grade to BS 434-1 at 1.5-2.0 kg/m². Allow emulsion to break completely before applying surfacing.

Q24 INTERLOCKING BRICK/ BLOCK ROADS/ PAVINGS

TYPES OF PAVING

110 CONCRETE BLOCK PAVING
- Granular sub-base: As clause Q20.
  - Thickness: As specified in schedules.
- Bedding: Sand.
  - Nominal thickness after compaction: 50mm.
- Blocks: To BS 6717.
  - Manufacturer: Marshalls.
  Product reference: As specified in schedules.
  - Sizes: 200 x 100 x 65mm.
  - Special blocks: As specified.
  - Colour/ Finish: As specified.

GENERAL

211 COLOUR BANDING
- General: Unless premixed by manufacturer, select blocks/ pavers/ setts vertically from at least 5 separate packs in rotation, to avoid colour banding.

220 SAMPLES
- General: Before ordering, submit samples of blocks/ pavers/ setts, that are representative of colour and appearance.

240 ADVERSE WEATHER
- General: Do not use frozen materials or lay bedding on frozen or frost covered sub-bases.

242 ADVERSE WEATHER (SAND BEDDED AND JOINTED PAVING)
- Stockpiled bedding material: Protect from saturation.
- Exposed areas of sand bedding and uncompacted areas of paving: Protect from heavy rainfall.
- Sand bedding that becomes saturated before laying paving: Remove and replace, or allow to dry before proceeding.
- Damp conditions: Brush in as much jointing sand as possible. Minimize site traffic over paving. As soon as paving is dry, top up joints and complete compaction.

248 BEDDING LAYER TOLERANCES AFTER FINAL COMPACTION
- 50 mm nominal thickness: +15 to -20 mm.
- 30 mm nominal thickness: +12 to -0 mm.

260 LEVELS OF PAVING
- Permissible deviation from specified levels:
  - Generally: ±6 mm.
- Height of finished paving above features:
  - At drainage channels and kerbs: +3 to +6 mm.

270 REGULARITY
- General: Where appropriate in relation to the geometry of the surface, variation in gap under a 3 m straight edge placed anywhere on the surface (maximum) to be 10 mm.
- Sudden irregularities: Not permitted.
- Difference in level between adjacent blocks/ pavers/ setts (maximum): 2 mm.

PREPARATION/ LAYING

360 CONDITION OF SUB-BASES/ BASES BEFORE LAYING SAND BEDDING COURSE
- Granular surfaces:
  - Sound, clean, smooth and close-textured enough to prevent migration of sand bedding into the sub-base/ overlay during compaction and use.
  - Free from movement under compaction plant and free from compaction ridges, cracks and loose material.
- Prepared existing and new bound bases (roadbases): Sound, clean, free from rutting or major cracking and cleared of sharp stones, projections or debris.
- Bound base (roadbase) surface tolerance: +0 to -12 mm.
- Levels and falls: Accurate and within specified tolerances.
- Drainage outlets: Within +0 to -10 mm of required finished level.
- Edge restraints, manhole covers, drainage outlets and the like: Complete, to required levels, and adequately bedded and haunched in mortar that has reached sufficient strength.
- Haunching to gullies, manhole covers and inside face of edge restraints: Vertical, so that pavings do not 'ride up' when compacted.

455 SAND FOR BEDDING
- Type: Naturally occurring clean sharp sand from the quaternary geological series or sea dredged, graded to BS 7533-3, Annex.
- Purity: Free from deleterious salts, contaminants and cement.
- Procurement: Obtain from one source and ensure consistent grading.
- Moisture content: Maintain even moisture content that will give maximum compaction. Sand squeezed in the hand should show no free water and bind together when pressure is released.

465 LAYING BEDDING GENERALLY
- Site trial: Determine by trial the depth of loose bedding material needed to ensure specified bedding course thickness after final compaction of paving.
- Bedding materials: Do not deliver to working area over uncompacted paving.
- Bedding course prepared area: 1 m (minimum) to 3 m (maximum) in advance of laying face, and 1 m (maximum) at end of working period.
- Protection of prepared bedding course: Do not allow traffic or leave exposed. Fill, rescreed and recompact areas disturbed by removal of screed rails or trafficking. Lay blocks/ pavers/ setts immediately.

470 LAYING BEDDING (PRECOMPACTED)
- General: Spread bedding material in one loose layer. Compact with a plate compactor and level the surface by screeding.

475 LAYING BEDDING (PARTIALLY PRECOMPACTED)
- General: Spread, in a loose uncompacted layer, approximately the required final thickness of bedding material and compact with a plate compactor. Spread a further loose layer approximately 15 mm thick, and screed to levels.

480 LAYING BEDDING (POSTCOMPACTED)
- General: Spread bedding material in one loose uniform layer and screed to levels.

485 LAYING BLOCKS/ PAVERS/ SETTS
- General: Commence from an edge restraint. Vibrate to produce thoroughly interlocked paving of even overall appearance with regular joints and accurate to line, level and profile.
- Bedding of units: Firm, so that rocking or subsidence does not occur or develop.
- Working face: Lay blocks/ pavers/ setts hand tight. Maintain open noncastellated working face. Do not use mechanical force to obtain tight joints.
- Placing: Squarely with minimum disturbance to bedding.
- Supply to laying face: Deliver over newly laid paving but stack at least 1 m back from laying face. Do not allow plant to traverse areas of uncompacted paving.
- Alignment: Continually check with string lines as work proceeds to ensure maintenance of accurate bond.

- Infill at edge restraints: Complete as work proceeds. Cut units accurately and maintain specified joint width. Avoid very small infill pieces at edges by breaking bond on the next course in from the edge, using cut blocks/ pavers/ setts not less than 1/3 full size.
- Cut edges: Turn inwards where possible; do not position against edge restraints or other features.
- In situ mortar infill: 1:3 sand/cement.

495 OBSTRUCTIONS
- Cutting: After laying full paving units, trim blocks/ pavers/ setts neatly around drainage fittings and other obstructions. Cut units accurately and maintain joint width within specified range. Do not reduce thickness of blocks/ pavers/ setts.
- In situ concrete surrounds:
  - Standard: To BS 5328, Grade C35 air entrained, maximum aggregate size 10 mm.
  - Shape and size: Rectangular, 100 mm (minimum) all round obstruction.
  - Thickness (minimum): Combined depth of blocks/ pavers/ setts and sand bedding.
  - Colour: As specified .

500 CUTTING BLOCKS/ PAVERS/ SETTS
- General: Cleanly and accurately, without spalling.
  - Do not mark or damage visible surfaces.

510 SAND FOR JOINTING
- Type: Clean free flowing dried silica sand graded to BS 7533-3, table D3.
  - Do not use sand that will stain paving units.
- Purity: Free from deleterious salts, contaminants and cement.

520 COMPACTING AND JOINTING
- Timing:
  - After edge restraint/ kerb haunchings have matured.
  - As laying proceeds, but after infilling at edges. Do not leave uncompacted areas of paving at end of working periods.
- Compacting:
  - Standard: To BS 7533-3, Annex F.
  - General: Thoroughly compact blocks/ pavers/ setts with vibrating plate compactor to achieve specified bedding course thickness. Apply same compacting effort throughout. Do not damage kerbs and adjacent work.
- Areas not to be compacted:
  - Within 1 m of working face.
  - Within 1 m of unrestrained edges.
- Surface levels: Check frequently. Lift and relay areas not at specified level.
- Jointing: Where jointing material is specified, brush into joints, revibrate surface and repeat as necessary to completely fill joints.

525 COMPACTING CLAY PAVERS
- Timing: Brush sand into the joints and remove surplus before commencing any compaction.

COMPLETION

615 COMPLETION OF PAVING WITH SAND FILLED JOINTS
- Sand dressing: washed river sand.
- Vacuum cleaning machines: Do not use.

Q25 SLAB/ BRICK/ BLOCK/SETT/ COBBLE PAVINGS

GENERAL

100 METHODS OF MEASUREMENT
- All Preambles in E10 Insitu concrete are to apply equally to this section where applicable.
- Notwithstanding the provision of SMM Clause S8, preparatory work to the base which is directly involved in the application of a finish and is detailed in the specification is deemed to be included with the measured item.

102 PRICES ALSO TO INCLUDE
- Employing stock pattern units as appropriate unless stated otherwise.

125 ADVERSE WEATHER
- Temperature: Do not lay or joint paving if the temperature is below 3°C on a falling thermometer or below 1°C on a rising thermometer.
- Frozen materials: Do not use. Do not lay bedding on frozen or frost covered bases.
- Paving with mortar joints and/ or bedding: Protect from frost damage, rapid drying out and saturation until mortar has hardened.
- Stockpiled bedding sand: Protect from saturation.
- Exposed areas of sand bedding and uncompacted areas of sand bedded paving: Protect from heavy rainfall.
- Saturated sand bedding: Remove and replace, or allow to dry before proceeding.
- Laying dry-sand jointed paving in damp conditions: Brush in as much jointing sand as possible. Minimize site traffic over paving. As soon as paving is dry, top up joints and complete compaction.

130 CEMENTITIOUS BASES AND SUB-BASES
- General: Protect from moisture loss, if not covered by another pavement course within 2 hours of completion.

135 CONDITION OF SUB-BASES/ BASES BEFORE LAYING BEDDING COURSE
- Trenches and excavation of soft or loose spots in subgrade: Filled and compacted as specified.
- Granular surfaces: Sound, clean, smooth and close-textured enough to prevent migration of bedding materials into the sub-base during compaction and use, free from movement under compaction plant and free from compaction ridges, cracks and loose material.
- Prepared existing and new bound bases (roadbases): Sound, clean, free from rutting or major cracking and cleared of sharp stones, projections or debris.
- Sub-base/ Roadbase level tolerances: To BS 7533-7, Annex A.
- Levels and falls: Accurate and within the specified tolerances.
- Drainage outlets: Within 0 to -10 mm of the required finished level.
- Features in sand bedded paving (including mortar bedded restraints and drainage ironwork): Completed, to required levels, and adequately bedded and haunched in mortar that has reached sufficient strength.
- Cementitious sub-bases: Cured for the minimum times specified in BS 7533-4, clause 4.5.3.

160 LIME FOR MORTAR BEDDING AND POINTING
- General: Nonhydraulic to BS 890 or BS EN 459 or, if ready-mixed lime:sand, to BS 4721.

164 MORTAR BEDDED CONCRETE SLABS
- Bedding:
    - Mortar: As section Z21, 1:3 Cement/sand.
    - Thickness: 10 mm minimum to 40 mm maximum.
- Laying: Bed units on foundation, and secure with continuous mortar haunching.
    - Keep exposed faces clean and free from mortar.
- Jointing: 1:3 Cement/sand.

171 TOOLED JOINTS IN MORTAR BEDDED UNITS
- Joints: Completely filled with bedding mortar as work proceeds.
    - Joint width: 5 mm.
    - Finish: Neat flush profile.

172 TOOLED COLOURED JOINTS IN MORTAR BEDDED UNITS
- Joints: Completely filled with bedding mortar as work proceeds.
    - Joint width: 5 mm.
- Pointing: 1:3 cement:sand mortar with pigment, colour As instructed by CA .
    - Depth: 10 mm.

184 LAYING PAVINGS
- Cutting: Cut units cleanly and accurately, without spalling, to give neat junctions with edgings and adjoining finishes.
- Lines and levels of finished surface: Smooth and even, with falls to prevent ponding.

- Bedding of units: Firm so that rocking or subsidence does not occur or develop.
- Slopes: Lay paving units upwards from the bottom of slopes.
- Appearance: Even and regular with even joint widths and free of mortar and sand stains.

188 LEVELS OF PAVING
- Permissible deviation from specified levels:
  - Generally: ± 6 mm.
- Height of finished paving above features:
  - At gullies: +6 to +10 mm.
  - At drainage channels and kerbs: +3 to +6 mm.

192 REGULARITY
- Maximum variation in gap under a 3 m straight edge placed anywhere on the surface (where appropriate in relation to the geometry of the surface): 10 mm.
- Sudden irregularities: Not permitted.
- Difference in level between adjacent blocks/ pavers/ setts (maximum): 2 mm.

196 PROTECTION
- Cleanliness: Keep paving clean and free from mortar droppings, oil and other materials likely to cause staining.
- Materials storage: Do not overload pavings with stacks of materials.
- Handling: Do not damage paving unit corners, arrises, or previously laid paving.
- Mortar bedded pavings: Keep free from traffic after laying:
  - Pedestrian traffic: For 2 days (minimum).
  - Vehicular traffic: For 5 days (minimum).
- Access: Restrict access to paved areas to prevent damage from site traffic and plant.

SLAB/ FLAG PAVING

215 CONCRETE FLAG PAVING
- Granular sub-base: As clause Q20/.
  - Thickness: As specified by CA.
- Bedding: As specified in schedules.
  - Nominal thickness: 50mm.
- Flags: To BS 7263-1.
  - Manufacturer: Marshalls.
  Product reference: Various .
  - Colour/ Finish: As specified by CA.
  - Nominal sizes: As specified in schedules.
- Jointing: As specified in schedules

235 CUTTING CONCRETE FLAGS
- General: Cleanly and accurately, without spalling.
  - Visible surfaces: Do not mark or damage.
- Notches cut from corner of flag exceeding 25% of flag area: Mitre cut remaining shape from internal corner of notch to opposite external corner.
- Diagonal mitre cutting: Cut flags or portions of flags to form a mitre at abrupt changes of level at the ends of ramped footpath crossings and the like.

245 SAND FOR BEDDING
- Type: Naturally occurring clean sharp sand, graded as for laying course sand to BS 7533-4, table 4.
- Purity: Free from deleterious salts, contaminants and cement.
- Procurement: Obtain from one source and ensure consistent grading.
- Moisture content: Maintain even moisture content that will give maximum compaction. Sand squeezed in the hand should show no free water and start to exhibit cracking when pressure is released.

255 SAND BEDDING
- Bedding sand: As clause 245.
- Laying and compacting to specified final thickness: Vibrating plate as BS 7533-4, table 1. Loosen the top 10 mm using a rake; alternatively, screed out a further 10 mm of loose sand.
- Bedding materials: Do not deliver to working area over uncompacted paving.
- Bedding course prepared area: 1 m (minimum) to 3 m (maximum) in advance of laying face, and 1 m (maximum) at end of working period.
- Protection of prepared bedding course: Do not allow traffic or leave exposed. Fill, rescreed and recompact areas disturbed by removal of screed rails or trafficking. Lay paving units immediately.
- Supply of slabs/ flags to laying face: Over newly laid paving but stack at least 1 m back from laying face.

280 NARROW SAND FILLED JOINTS FOR SAND BEDDED PAVING
- Jointing sand: Clean dried sand, graded as for jointing sand to BS 7533-4, table 4 and free from deleterious salts, contaminants and cement.
- Placing slabs/ flags: Squarely with minimum disturbance to bedding. Lay away from previously laid slabs/ flags.
- Joint width: 2-5 mm. Do not use mechanical force to obtain tight joints.
- Jointing:
  - Timing: On the same day as laying and when rain is not expected.
  - Jointing: Brush clean dry sand over joints, then bed down slabs/ flags using plate vibrator to BS 7533-4, clause 6.9.4.
  - Completion: Refill joints with sand and revibrate until joints are completely filled. Repeat if jointing sand settles during early trafficking.
- Vacuum cleaning machines: Do not use.

300 FULL MORTAR BEDDING
- Mortar: As section Z21.
  - Mix: 1:3 .
  - Consistency: firm.
  - Sand: To BS 882, grading limit M or F.
- Spreading and levelling mortar: To give specified average nominal thickness after bedding of slabs/ flags.
- Laying slabs/ flags: On full mortar bed. Bed down to line and level with a maul.

315 MORTAR SPOT BEDDING
- Mortar: As section Z21.
  - Mix: 1:3.
  - Sand: To BS 882, grading limit M or F.
- Bedding: Spread five spots of wet mortar to give specified average nominal thickness after bedding of flags.
- Laying: Bed slabs/ flags down to line and level with a maul, ensuring full contact at all five spots.

335 MORTAR POINTED JOINTS
- Jointing material: 1:3 cement/sand mortar .
- Timing: When the surface of the paving is dry and rain is not expected.
- Jointing: Carefully and thoroughly fill joints. Tool to smooth bucket handle profile 2-3 mm below slab/ flag surface. Clean mortar from face of slabs/ flags before it sets.
- Protection: Protect from rain.

Q30 SEEDING/ TURFING

GENERALLY

100 DEFINITIONS
- "Cultivating". Breaking up the soil with suitable implements to a specified depth and degree of tilth in preparation for seed sowing or turfing.

105 PRICES ALSO TO INCLUDE
- Cultivating to include removing and disposing of weeds.
- Sowing grass seed evenly in two directions at right angles.

- Working grass to any levels, ground contour or plan shape.
- Supplying and laying turfs in rows with staggered joints, trimming edges and marrying in with adjacent levels.
- For working turfs to any levels, ground contour or plan shape.

GENERAL INFORMATION/ REQUIREMENTS

115 SEEDED AND TURFED AREAS
- Growth and development: Healthy, vigorous grass sward, free from the visible effects of pests, weeds and disease.
- Appearance: A closely knit, continuous ground cover of even density, height and colour.

120 CLIMATIC CONDITIONS
- General: Carry out the work while soil and weather conditions are suitable.

145 WATERING
- Quantity: Wet full depth of topsoil.
- Application: Even and without displacing seed, seedlings or soil.
- Frequency: As necessary to ensure the establishment and continued thriving of all seeding/ turfing.

146 WATERING
- Quantity: Wet full depth of topsoil.
- Application: Even and without displacing seed, seedlings or soil.
- Frequency: As instructed by CA.

150 WATER RESTRICTIONS
- Timing: If water supply is or is likely to be restricted by emergency legislation do not carry out seeding/ turfing until instructed. If seeding/ turfing has been carried out, obtain instructions on watering.

160 NOTICE
- Give notice before:
    - Setting out.
    - Applying herbicide.
    - Applying fertilizer.
    - Preparing seed bed.
    - Seeding or turfing.
    - Visiting site during maintenance period.
- Period of notice: 1 week or other period as agreed.

170 SETTING OUT
- Boundaries of seeding/ turfing areas: Mark clearly.

PREPARATION

205 PREPARATION MATERIALS
- General: Free from toxins, pathogens or other extraneous substances harmful to plant, animal or human life.
- Certification: Submit certificate giving supply source, content analysis, confirmation of suitability for purpose and confirmation of absence of harmful substances:
    - Give notice before ordering or using.

210 HERBICIDE
- Type: Suitable for suppressing perennial weeds.

231 PEAT
- Peat or products containing peat: Do not use.

250 CULTIVATION
- Compacted topsoil: Break up to full depth.
- Soil ameliorant/ Conditioner/ Fertilizer: As directed.

© AAS 2008- 2009

- Tilth: Reduce top 100 mm of topsoil to a tilth suitable for blade grading, particle size 10 mm (maximum).
- Material brought to the surface: Remove stones and clay balls larger than 50 mm in any dimension, roots, tufts of grass, rubbish and debris.

260 GRADING
- Topsoil condition: Reasonably dry and workable.
- Contours: Smooth and flowing, with falls for adequate drainage. Remove minor hollows and ridges.
- Finished levels after settlement: 25 mm above adjoining paving, kerbs, manholes etc.
- Give notice: If required levels cannot be achieved by movement of existing soil.

270 FERTILIZER
- Types: Apply both:
  - Superphosphate with a minimum of 18% water soluble phosphoric acid.
  - A sulfate of ammonia with a minimum of 20% nitrogen.
- Application: Before final cultivation and three to five days before seeding/ turfing.
- Coverage: Spread evenly, each type at 70 g/m², in transverse directions.

280 FINAL CULTIVATION
- Timing: After grading and fertilizing.
- Seed bed: Reduce to fine, firm tilth with good crumb structure.
  - Depth: 25 mm.
  - Surface preparation: Rake to a true, even surface, friable and lightly firmed but not over compacted.
  - Remove surface stones/ earth clods exceeding:
  General areas: 50mm.
  Fine lawn areas: 25 mm.
- Adjacent levels: Extend cultivation into existing adjacent grassed areas sufficient to ensure full marrying in of levels.

SEEDING

310 GRASS SEED
- Rate of application: 70g/m2.

319 QUALITY OF SEED
- Freshness: Produced for the current growing season.
- Certification: Blue label certified varieties to EC purity and germination regulations. When requested, submit an Official Seed Testing Station certificate of germination, purity and composition.
- Samples of mixtures: Submit when requested.

330 SOWING
- General: Establish good seed contact with the root zone to promote healthy, consistent growth.
- Method: To suit soil type, proposed usage of grassed area, location and weather conditions during and after sowing.

340 PRE-EMERGENT HERBICIDE
- General: Where soil has not been allowed to lie fallow apply a suitable pre-emergent herbicide immediately after sowing.

350 TURF EDGING TO SEEDED AREAS
- Turfs: To BS 3969, with no perennial ryegrass and of a similar seed composition to the seeded area.
- Timing: Before sowing.
- Preparation: Rake back a 750 mm wide margin around prepared seed beds.
  - Level of seed bed: Married in with turf.
- Placement: Single row laid end to end and trimmed to a line.
- Watering: On completion.

352 EDGES TO SEEDED AREAS
- Location: Planting beds and around newly planted trees.
- Timing: After seeded areas are well established.
- Edges: Cut to clean straight lines or smooth curves. Soil drawn back to permit edging.
- Arisings: Remove.

TURFING

410 TURF
- Standard: To BS 3969, free from undesirable grasses and weeds.
  - Grade: As Specified.
- Source: Submit proposals.
- Herbicide treatment: Apply not less than four weeks and not more than three months before lifting.

420 DELIVERY AND STORAGE
- Timing: Lay turf with minimum possible delay after lifting. If delay occurs, lay turf out on topsoil and keep moist.
- Frosty weather or waterlogged ground: Do not lift turf.
- Delivery: Arrange to avoid need for excessive stacking.
- Stacking height (maximum): 1 m.
- Dried out or deteriorated turf: Do not use.

423 INSPECTION OF TURF
- Give notice: Before lifting turf for relaying.

430 TURFING GENERALLY
- Timing of laying:
  - Spring and summer: Within 18 hours of delivery.
  - Autumn and winter: Within 24 hours of delivery.
- Weather conditions: Do not lay turf when persistent cold or drying winds are likely to occur or soil is frost bound, waterlogged or excessively dry.
- Working access: Planks laid on previously laid turf. Do not walk on prepared bed or newly laid turf.
- Jointing: Laid with broken joints, well butted up. Do not stretch turf.
- Edges: Whole turfs, trimmed to a true line.
- Adjusting levels: Remove high spots and fill hollows with fine soil.
- Consolidating: Lightly and evenly firm as laying proceeds to ensure full contact with substrate. Do not use rollers.
- Dressing, brushed well in to completely fill all joints.
- Watering: Thoroughly water completed turf immediately after laying. Check that water has penetrated to the soil below.

MAINTENANCE

610 FAILURES OF SEEDING/ TURFING
- General: Grassed areas that have failed to thrive (unless due to theft or malicious damage), during the first 2 months, will be regarded as defects due to materials or workmanship not in accordance with the Contract. Make good by recultivation and reseeding/ returfing.
- Timing of making good: Submit proposals.

Q40 FENCING

NOTE: rates allow for ready mixed concrete delivered to a point ne 50m horizontally to placing point.

GENERALLY

100 DEFINITIONS
- "Posts" and "Struts" include other supports.

103 METHODS OF MEASUREMENT
- Where existing posts are surrounded with concrete the surround is deemed to be not exceeding 0.10 m3. Any additional concrete both in breaking out and reinstatement will be measured and valued as additional work.
- Ironmongery partly fixed to hardwood and partly to softwood is measured as fixed to hardwood.

105 PRICES ALSO TO INCLUDE
- All cutting including cutting to profile necessary to follow contour of ground.
- Bolting is to include galvanized or other suitable and matching bolts.
- Posts, struts and other supports are to include pockets, mortices and holes for wires, bolts, strainers and the like.
- Welding is to include preparing and grinding smooth.

MATERIALS

110 CHAIN LINK FENCING
- Standard: To BS 1722-1, type As Specified in schedules.
- Height: As Specified in schedules.
- Mesh and wire: As Specified in schedules.
- Posts and struts: As Specified in schedules.
  - Treatment: As Specified in schedules.
- Centres of posts (maximum):
  - Straining posts: 69 m in straight runs and at all ends, corners, changes of direction and acute variations in level.
  - Intermediate posts: 3 m.
- Method of setting posts and struts:
  - Straining posts: As Specified in schedules.
  - Struts: As Specified in schedules.
  - Intermediate posts: As Specified in schedules.

150 GENERAL PATTERN STRAINED WIRE FENCING
- Standard: To BS 1722-3.
- Height: 925mm.
- Wire: 4.0mm.
- Posts and struts: As Specified in schedules.
  - Treatment: As Specified in schedules.
- Maximum centres of posts:
  - Straining posts: 150 m in straight runs and at all ends, corners, changes of direction and acute variations in level.
  - Intermediate posts: 3.5 m.
- Method of setting posts and struts:
  - Straining posts: As Specified in schedules.
  - Struts: As Specified in schedules.
  - Intermediate posts: As Specified in schedules.

210 WOODEN POST AND RAIL FENCING
- Standard: To BS 1722-7, type Ranch.
- Height: As Specified in schedules.
- Wood: Softwood.
  - Preservative treatment: As Specified by CA.
- Maximum centres of posts: 3000mm.
- Method of setting posts: Concrete.

250 PREFABRICATED WOOD PANEL FENCING
- Standard: To BS 1722-11.
- Type of infill: Larch Lap.
  - Treatment: As Specified by CA.
- Height: As Specified in schedules
- Posts: Concrete.
- Method of setting posts: In concrete in 300 mm square or diameter x 600 mm deep holes filled to not less than half the depth.

310 CLOSE BOARDED FENCING
- Standard: To BS 1722-5, type Featheredge.
- Height: As Specified in schedules.
- Wood: Softwood.
  - Treatment: As Specified by CA.
- Boards/ rails: As Specified in schedules.
- Posts: As Specified in schedules.
- Centres of posts (maximum): 2.4 m, or 3 m if a gravel board is specified.
- Method of setting posts: Set in concrete.

320 CLEFT CHESTNUT PALE FENCING
- Standard: To BS 1722-4,
- Height: 1200mm.
- Posts and struts: Chestnut.
- Centres of posts (maximum):
  - Straining posts: 70 m in straight runs and at all ends, corners, changes of direction and acute variations in level.
  - Intermediate posts: 2250mm.
- Method of setting posts and struts:
  - Straining posts: Driven.
  - Struts: driven.
  - Intermediate posts: Driven.

330 WOODEN PALISADE FENCING
- Standard: To BS 1722-5.
- Height: As Specified in schedules.
- Wood: Softwood.
  - Treatment: As Specified by CA.
- Palisades and rails: As Specified in schedules.
- Posts: As Specified in schedules.
- Centres of posts (maximum): 3 m.
- Method of setting posts: Concrete.

570 GATES
- Manufacturer: Contractors choice as approved.
- Sizes: As Specified in schedules.
- Posts: As specified by CA.
- Finish as delivered: As specified by CA.
- Fittings: As specified by CA.
- Method of fixing: As specified by CA.

610 INSTALLATION
- Set out and erect:
  - In straight lines or smoothly flowing curves as shown on drawings.
  - With tops of posts following profile of the ground.
  - With posts set rigid, plumb and to specified depth, or greater where necessary to ensure adequate support.
  - With correct fasteners and all components securely fixed.

615 COMPETENCE
- Operatives: Contractors must employ competent operatives.
- Qualifications: Submit certification in accordance with Sector Scheme 2A of training and experience in categories: as Appropriate.

620 SETTING POSTS IN CONCRETE
- Mix: To BS 5328, or to BS 8500-1 and -2 and BS EN 206-1, Designated mix not less than GEN1 or Standard mix not less than ST2.
- Alternative mix for small quantities: 50 kg Portland cement to 150 kg fine aggregate to 250 kg 20 mm nominal maximum size coarse aggregate, medium workability.
- Admixtures: Do not use.

- Holes: Excavate neatly and with vertical sides.
- Filling: Position post/ strut and fill hole with concrete to not less than the specified depth, well rammed as filling proceeds and consolidated.
- Backfilling of holes not completely filled with concrete: Excavated material, well rammed and consolidated.

630 EXPOSED CONCRETE FOUNDATIONS
- Filling: Compact until air bubbles cease to appear on the upper surface.
- Finishing: Weathered to shed water and trowelled smooth.

640 SETTING POSTS IN EARTH
- Holes: Excavated neatly, with vertical sides and as small as practicable to allow refilling.
- Filling: Position posts/ struts and replace excavated material, well rammed as filling proceeds.

650 DRIVEN POSTS
- Damage to heads: Minimize.
  - Repair: Neatly finish post tops after installation.

660 NAILED WOOD RAILS
- Length (minimum): Two bays, with joints in adjacent rails staggered.
- Fixing: Nail each length of rail to each post with two 100 mm galvanized nails.
- Rails with split ends: Replace.

665 CLEFT WOOD RAILS
- Length (maximum): 3.05 m.
- Fixing: Rail end section shaped to adequately fill the post mortice. Nail each rail to each prick post with two galvanized nails.
- Rails with split ends: Replace.

666 ARRIS RAILS
- Fixing:
  - Rail end section: Shaped to adequately fill the post mortice or recess.
  - Recessed posts: Rails bolted to each post.
  - Top rails: Fixed at both ends using galvanised nails.
- Rails with split ends: Replace.

670 SITE CUTTING OF WOOD
- General: Kept to a minimum.
- Below or near ground level: Cutting prohibited.
- Treatment of surfaces exposed by minor cutting and drilling: Two flood coats of solution recommended for the purpose by main treatment solution manufacturer.

680 DAMAGE TO GALVANIZED SURFACES
- Treatment of minor damage (including on fasteners and fittings): Low melting point zinc alloy repair rods or powders made for this purpose, or at least two coats of zinc-rich paint to BS 4652.
- Thickness: Apply sufficient material to provide a zinc coating at least equal in thickness to the original layer.

Q50 SITE/ STREET FURNITURE/ EQUIPMENT

GATES AND BARRIERS

130 GATES
- Manufacturer: As Specified by CA.
- Size: As Specified in schedules.
- Finish as delivered: As Specified by CA.
- Fittings/ Accessories: As Specified by CA.
- Method of setting posts: set in concrete.

190 BOLLARDS
- Manufacturer: As Specified by CA.
- Material/ Finish/ Colour: As Specified by CA.
- Height above ground: As Specified by CA.
- Method of fixing: Set in concrete.

SITE AND STREET FURNITURE

260 TREE GUARDS
- Manufacturer: Nichols & Clarke.
- Material/ Finish/ Colour: As Specified by CA.
- Method of fixing: As Specified by CA.

262 TREE GRILLES
- Manufacturer: Nichols & Clarke.
  - Product reference: As Specified in schedules.
- Size: As Specified in schedules.

INSTALLATION

510 CONCRETE FOUNDATIONS GENERALLY
- Standard: To BS 5328-1, -2, -3 and -4, or to BS 8500-1, -2 and BS EN 206-1.
- Mix: Designated mix not less than GEN 1 or Standard mix not less than ST2.
- Admixtures: Do not use.

515 SETTING COMPONENTS IN CONCRETE
- Foundation holes: Neat vertical sides; bottom covered with a 50 mm layer of concrete.
- Components: Accurately positioned and securely supported.
- Depth of foundations, bedding and haunching: Appropriate to provide adequate support and to receive overlying soft landscape or paving finishes.
- Concrete fill: Fully compacted as filling proceeds.
- Temporary support: Maintain for 48 hours (minimum) and prevent disturbance.
- Concrete foundations exposed to view: Compacted until air bubbles cease to appear on the upper surface, then weathered to shed water and trowelled smooth.

520 SETTING IN EARTH
- Holes: As small as practicable to allow refilling.
- Components being fixed: Accurately positioned and securely supported.
- Earth refilling: Well rammed as filling proceeds.

530 PRESERVATIVE TREATED TIMBER
- Surfaces exposed by minor cutting and drilling: Treated by immersion or with two flood coats of a solution recommended for the purpose by main treatment solution manufacturer.

540 BUILDING IN TO MASONRY WALLS
- Components being built in: Accurately positioned and securely supported. Set in mortar and pointed neatly to match adjacent walling.
- Temporary support: Maintain for 48 hours (minimum) and prevent disturbance.

545 ERECTION OF TIMBER AND PREFABRICATED STRUCTURES
- Checking: 5 days (minimum) before proposed erection date, check foundations, holding down bolts, etc.
- Inaccuracies or defects in prepared bases or supplied structures: Report immediately. Obtain instructions before proceeding.

550 DAMAGE TO GALVANIZED SURFACES
- Minor damage in areas up to 40 mm² (including on fixings and fittings): Make good using low melting point zinc alloy repair rods or powders made for this purpose, or at least two coats of zinc-rich paint to BS 4652. Apply sufficient material to provide a zinc coating at least equal in thickness to the original layer.

560 SITE PAINTING
- Timing: Prepare surfaces and apply finishes as soon as possible after fixing.

R: DISPOSAL SYSTEMS

R10 RAINWATER PIPEWORK/ GUTTERS

GENERALLY

100 DEFINITIONS
- "Connector". Any thimble, ferrule, cone, caulking bush, cap and lining or other fitting or adaptor whether equal or reducing or a combination of any of the above necessary to achieve the connection described.
- "Reducer". Deemed to exclude tapers but include diminishing pieces, reducing sets, socket reducers as applicable.

103 METHODS OF MEASUREMENT
- Fittings required to be inserted in existing pipework are deemed to match or be consistent with the existing pipework.
- Pipes are deemed to be straight unless otherwise stated.
- Where no reference is made to a background or to pipe supports, pipes are deemed fixed without supports.
- Gutters are deemed to be straight unless otherwise stated.

105 PRICES ALSO TO INCLUDE
- Pipes, pipework ancillaries and gutters to stated background to include fixing by whatever means are necessary to uphold the item including any shot firing, drilling and tapping necessary and the provision of fixing materials including matching screws where appropriate.
- Joints in the running length and for marking the position of holes, mortices, chases and the like.
- Couplers in the running length necessary for pipe jointing.
- Metric adaptors and conversion pieces between imperial and metric items.
- Additional reducers necessary to achieve correct sizes on reducing pipe fittings.
- Handing pipe sleeves over to others for fixing as necessary.
- Making final connections to equipment after the general installation work has been substantially completed.
- Preparation work (for painting) and priming of ferrous frameworks, brackets, hangers and the like.
- Tuition of Employer's operating staff.

TYPES OF PIPEWORK/ GUTTER

130 CAST IRON PIPEWORK FOR EXTERNAL USE
- Pipes, fittings and accessories:
  - Standard: To BS 460.
  - Manufacturer: Contactors choice or as specified in schedules.
  - Type: Eared.
  - Nominal sizes: 65 & 75mm.
  - Finish as supplied: As specified.
- Accessories: As Specified in schedules
- Jointing: Push fit.
- Fixing: Plugging & screwing

150 CAST IRON GUTTERS
- Gutters, fittings and accessories:
  - Standard: To BS 460.
  - Manufacturer: Contractors choice.
  - Profile: As Specified in schedules
  - Nominal sizes: As Specified in schedules
  - Finish as supplied: As specified.
- Accessories: As Specified in schedules
- Jointing: Bolted and mastic.
- Fixing: screwed to timber.

© AAS 2008– 2009

210 ALUMINIUM PIPEWORK FOR EXTERNAL USE
- Pipes, fittings and accessories:
  - Standard: Generally to BS 2997.
  - Manufacturer: Alumasc.
  - Type/ Grade: Manufacturers standard.
  - Sections/ Nominal sizes: As Specified in schedules.
  - Finish/ Colour: Powder coated, colour as Specified by CA.
- Accessories: As Specified in schedules.
- Jointing: Spigot & socket.
- Fixing: Plugging and screwing.

230 ALUMINIUM GUTTERS
- Gutters, fittings and accessories:
  - Standard: Generally to BS 2997.
  - Manufacturer: Alumasc.
  - Profile: As Specified in schedules
  - Type/ Grade: Manufacturers standard.
  - Nominal sizes: 102mm.
  - Finish/ Colour: Powder coated, colour as Specified by CA.
- Accessories: As Specified in schedules
- Jointing: Bolted and gasket
- Fixing: screwed to timber.

270 PVC-U PIPEWORK FOR EXTERNAL USE
- Pipes, fittings and accessories:
  - Manufacturer: Geberit Terrain
  Product reference: As Specified in schedules
  - Sections/ Nominal sizes: 69 mm.
  - Colour: Grey or As Specified by CA
- Accessories: As Specified in schedules
- Jointing: push fit
- Fixing: Plugging and screwing.

290 PVC-U GUTTERS
- Gutters, fittings and accessories:
  - Manufacturer: Geberit Terrain
  Product reference: As Specified in schedules
  - Profile: As Specified in schedules
  - Nominal sizes: As Specified in schedules.
  - Colour: Grey or As Specified by CA
- Accessories: As Specified in schedules.
- Fixing: screwed to timber.

310 FIBRE CEMENT PIPEWORK FOR EXTERNAL USE
- Pipes, fittings and accessories:
  - Manufacturer: Eternit.
  - Shape: Round.
  - Sizes: As Specified in schedules.
  - Finish/ Colour: Standard painted.
- Accessories: As Specified in schedules.
- Fixing: Plugging and screwing.

320 FIBRE CEMENT GUTTERS
- Gutters, fittings and accessories:
  - Manufacturer: Eternit.
  - Profile: As Specified in schedules
  - Sizes: As Specified in schedules
  - Finish/ Colour: Standard painted.
- Accessories: As Specified in schedules.
- Jointing: Bolted & gasket
- Fixing: screwed to timber.

INSTALLATION

410 INSTALLATION GENERALLY
- Discharge of rainwater: Complete, and without leakage or noise nuisance.
- Components: Obtain from the same manufacturer for each type of pipework/ guttering.
- Electrolytic corrosion: Avoid contact between dissimilar metals where corrosion may occur.
- Plastics and galvanized steel pipes: Do not bend.
- Protection:
  - Fit purpose made temporary caps to prevent ingress of debris.
  - Fit access covers, cleaning eyes and blanking plates as the work proceeds.
- Fixings/ Fasteners: As section Z20.

420 FIXING GUTTERS
- Joints: Watertight.
- Allowance for thermal and building movement: Provide clearances recommended by manufacturer and maintain as fixing proceeds.
- Roofing underlay: Dressed into gutter.

425 SETTING OUT EAVES GUTTERS - TO FALLS
- Setting out: To true line and even gradient to prevent ponding or backfall. Position high points of gutters as close as practical to the roof and low points not more than 50 mm below the roof.
- Outlets: Aligned with connections to below ground drainage.

426 SETTING OUT EAVES GUTTERS - LEVEL
- Setting out: Level and as close as practical to roof.
- Outlets: Aligned with connections to below ground drainage.

450 RAINWATER OUTLETS
- Fixing: Securely, before connecting pipework.
- Junctions between outlets and pipework: To accommodate movement in structure and pipework.

460 FIXING PIPEWORK
- Pipework: Fix securely at specified centres plumb and/ or true to line.
- Branches and low gradient sections: Fix with uniform and adequate falls to drain efficiently.
- Externally socketed pipes/ fittings: Fix with sockets facing upstream.
- Additional supports: Provide as necessary to support junctions and changes in direction.
- Vertical pipes:
  - Provide a loadbearing support at least at every storey level.
  - Tighten fixings as work proceeds so that every storey is self supporting.
  - Wedge joints in unsealed metal pipes to prevent rattling.
- Wall and floor penetrations: Isolate from structure.
  - Pipe sleeves: As Section P31.
- Allowance for thermal and building movement: Provide and maintain clearance as fixing and jointing proceeds.
- Expansion joint pipe sockets: Fix rigidly to building. Elsewhere, use brackets/ fixings that allow pipes to slide.

465 JOINTING PIPEWORK/ GUTTERS
- General: Joint using materials, fittings and techniques which will make effective and durable connections.
- Jointing differing pipework/ gutter systems: Use adaptors recommended by manufacturers.
- Cut ends of pipes/ gutters: Clean and square with burrs and swarf removed. Chamfer pipe ends before inserting into ring seal sockets.
- Jointing or mating surfaces: Clean, and where necessary lubricated, immediately before assembly.
- Junctions: Form using fittings intended for the purpose.
- Jointing material: Strike off flush and remove surplus. Do not allow to project into bore of pipes, fittings and appliances.
- Surplus flux, solvent, cement: Remove.

510 ELECTRICAL CONTINUITY - PIPEWORK
- Joints in metal pipes with flexible couplings: Use clips (or suitable standard pipe couplings) supplied for earth bonding by pipework manufacturer to ensure electrical continuity.

570 GUTTER TEST
- Preparation: Temporarily block all outlets.
- Testing: Fill gutters to overflow level and after 5 minutes closely inspect for leakage.

R11 FOUL DRAINAGE ABOVE GROUND

TYPES OF PIPEWORK

110 PVC-U PIPEWORK
- Pipes, fittings and accessories:
  - Standard: to BS EN 1329-1 or BS 4514, Kitemark certified.
- Manufacturer: Geberit terrain
  - Product reference: As Specified in schedules.
- Nominal sizes: DN As Specified in schedules
- Colour: As Specified by CA
- Accessories: As Specified in schedules
- Jointing: Solvent weld
- Fixing: As Specified in schedules

120 PLASTICS PIPEWORK
- Pipes, fittings and accessories:
  - Material/ Standard: BS EN 5255, Kitemark certified.
- Manufacturer: Marley.
  - Product reference: As Specified in schedules.
- Nominal sizes: DN As Specified in schedules.
- Colour: As Specified by CA.
- Accessories: As Specified in schedules.
- Jointing: Ring seal
- Fixing: As Specified in schedules

140 CAST IRON PIPEWORK
- Pipes and fittings:
  - Standard: To BS 416-1 with sockets.
- Manufacturer: Hargreaves or Timesaver as Specified in schedules
  - Product reference: As Specified in schedules.
- Nominal sizes: DN 100mm
- Finish: Black
- Accessories: As Specified in schedules
- Jointing: Caulked lead
- Fixing: Plugging

160 COPPER PIPEWORK
- Pipes: Copper tube:
  - Standard: To BS EN 1057, Kitemark certified.
  - Temper: Half hard R250.
  - Nominal wall thickness: 1.2 mm.
- Sizes: As Specified in schedules.
- Accessories: As Specified in schedules.
- Jointing generally: Integral lead free solder ring capillary fittings:
  - Standard: To BS EN 1254-1, Kitemark certified.
- Connections to appliances and equipment: Either: Compression fittings:
  - Standard: To BS EN 1254-2, Kitemark certified, or
  Fittings with threaded ends:
  - Standard: To BS EN 1254-4. Kitemark certified.
- Fixing: As Specified in schedules.

INSTALLATION

520 INSTALLATION GENERALLY
- Standard:
  - BS EN 12056-1, clauses 3 - 6.
  - BS EN 12056-2, clauses 3 - 6, National Annexes NA - NG, System III.
  - BS EN 12056-5, clauses 4 - 6, 8, 9 and 11.
- Drainage from appliances: Quick, quiet and complete, self-cleansing in normal use, without blockage, crossflow, backfall, leakage, odours, noise nuisance or risk to health.
- Pressure fluctuations in pipework (maximum): ±38 mm water gauge.
- Water seal retained in traps (minimum): 25 mm.
- Components: From the same manufacturer for each type of pipework.
- Access: Provide access fittings in convenient locations to permit cleaning and testing of pipework.
- Electrolytic corrosion: Avoid contact between dissimilar metals where corrosion may occur.
- Plastics and galvanized steel pipes: Do not bend.
- Concealed or inaccessible surfaces: Decorate before starting work specified in this section.

540 PIPE ROUTES
- General: The shortest practical, with as few bends as possible:
  - Bends in wet portion of soil stacks: Not permitted.
  - Routes not shown on drawings: Submit proposals before commencing work.

550 FIXING PIPEWORK
- Pipework: Fix securely at specified centres plumb and/ or true to line. Fix every length of discharge stack pipe at or close below socket collar or coupling.
- Branches and low gradient sections: Fix with uniform and adequate falls to drain efficiently.
- Externally socketed pipes/ fittings: Fix with sockets facing upstream.
- Additional supports: Provide as necessary to support junctions and changes in direction.
- Vertical pipes: Provide a load bearing support not less than every storey level. Tighten fixings as work proceeds so that every storey is self supporting.
- Wall and floor penetrations: Isolate from structure. Sleeve pipes as in section P31.
- Thermal and building movement: Provide and maintain clearance as fixing and jointing proceeds.
- Expansion joint sockets: Fix rigidly to the building.
- Fixings: To allow the pipe to slide.
  - Finish: Plated, sherardized, galvanized or other nonferrous.
  - Compatibility: Suitable for the purpose, material being fixed and background.

560 JOINTING PIPEWORK
- Jointing differing pipework systems: Use adaptors recommended by manufacturers.
- Cut ends of pipes: Clean and square with burrs and swarf removed. Chamfer pipe ends before inserting into ring seal sockets.
- Jointing or mating surfaces: Clean, and where necessary lubricate, immediately before assembly.
- Junctions: Form using fittings intended for the purpose.
- Jointing material: Do not allow to project into bore of pipes, fittings and appliances.
- Surplus flux/ solvent/ cement/ sealant: Remove from joints.

710 ELECTRICAL CONTINUITY
- Joints in metal pipes with flexible couplings: Use clips supplied for earth bonding by pipework manufacturer to ensure electrical continuity.

730 IDENTIFICATION OF INTERNAL FOUL DRAINAGE PIPEWORK
- Markings: To BS 1710.
- Type: Integral lettering on pipe wall, self-adhesive bands or identification clips.
- Location: Junctions, both sides of slab, bulkhead and wall penetrations, and elsewhere as agreed.

745 DISCHARGE AND VENTILATION STACKS
- Terminations: Perforated cover or cage that does not restrict airflow.
  - Material: Galvanised metal.

820 PIPEWORK TEST - AIRTIGHTNESS
- Preparation:
  - Open ends of pipework: Temporarily seal with plugs.
  - Test apparatus: Connect a 'U' tube water gauge and air pump to pipework via a plug or through trap of an appliance.
- Testing: Pump air into pipework until gauge registers 38 mm.
- Required performance: Allow a period for temperature stabilisation, after which the pressure of 38 mm is to be maintained without loss for not less than three minutes.

850 PREHANDOVER CHECKS
- Temporary caps: Remove.
- Permanent blanking caps, access covers, rodding eyes, floor gratings and the like: Secure complete with all fixings.

R12 DRAINAGE BELOW GROUND

GENERALLY

100 METHODS OF MEASUREMENT
- All Preambles in the following Work Sections apply equally to this section where appropriate:
  - D20 - Excavating and filling
  - E10 – Mixing/casting/curing in-situ concrete
  - E20 - Formwork for in-situ concrete
  - E30 - Reinforcement for insitu concrete
  - E50 - Precast concrete framed structures
  - F10 - Brick/block walling
  - M20 - Plastered/Rendered/Roughcast coatings
- Pipes are deemed laid in trenches unless stated otherwise.
- Granular fillings and coverings other than those using excavated materials, are measured separately.

102 CONTRARY TO SMM
- Notwithstanding clauses R12 & 13. 3.1, a separate item has not been included. The cost is deemed to be included in the rates for excavation.
- Benchings have been measured in m2 stating an average thickness (SMM R12.11.9).

104 PRICES ALSO TO INCLUDE
- Jointing fittings and accessories to pipes to include jointing as pipework unless otherwise stated.
- Rainwater shoes, gullies and traps to include setting and surrounding with concrete as specified including any formwork and extra excavation, disposal and earthwork support required unless otherwise stated.
- Excavations for pipes without beds are to include handholes.
- Concrete and fillings are to include finishing to falls, cross falls, slopes and cross slopes.
- Traps, gullies and other fittings and accessories are to include extra excavation and disposal.
- Traps, gullies and other fittings and accessories are to include items of metalwork not sold separately (e.g. mud buckets and strainers).
- Pitch fibre pipes and fittings and accessories are to include cutting and tapering of ends and straight and 5-degree couplings.

106 EXISTING DRAINS
- Setting out: Before starting work, check invert levels and positions of existing drains, sewers, inspection chambers and manholes against drawings. Report discrepancies.
- Protection: Protect existing drains to be retained and maintain normal operation.

108 IN SITU CONCRETE FOR USE IN DRAINAGE BELOW GROUND
- Standard: To BS 5328-1, -2, -3 and -4, or to BS 8500-1, -2 and BS EN 206-1.
- Mix: C15 .
- Equivalent or better mix: Submit proposals.
- Different mixes may be used for different parts of the drainage work.

TYPES OF PIPELINE

121 CLAY PIPELINES
- Pipes, bends and junctions: Vitrified clay to BS EN 295-1, with flexible joints, Kitemark certified.
    - Manufacturer: Hepworth.
  Product reference: As Specified in schedules.
    - Strength: Manufacturers standard.
    - Sizes: DN As Specified in schedules.
    - Jointing: Flexible couplings .
- Bedding class: As Specified by CA.

151 PLAIN WALL PLASTICS PIPELINES
- Pipes, bends and junctions: PVC-U to BS EN 1401-1, class SN4, with flexible joints, Kitemark certified.
    - Manufacturer: Geberit Terrain.
  Product reference: As Specified in schedules.
    - Sizes: As Specified in schedules.

EXCAVATING/ BACKFILLING

202 PUBLIC ROADS AND PAVINGS
- Excavating and backfilling of trenches: To Stationery Office 'Specification for the reinstatement of openings in highways'.

205 EXCAVATED MATERIAL
- Turf, topsoil, hardcore, etc: Set aside for use in reinstatement.

210 LOWER PART OF TRENCH - GENERAL
- Trench from bottom up to 300 mm above crown of pipe: With vertical sides and of a width as small as practicable but not less than external diameter of pipe plus 300 mm.

230 TYPE OF SUBSOIL
- General: Where type of subsoil at level of crown of pipe differs from that stated for the type of pipeline, obtain instructions before proceeding.

240 FORMATION FOR BEDS
- Timing: Excavate to formation immediately before laying beds or pipes.
- Mud, rock projections, boulders and hard spots: Remove. Replace with consolidated bedding material.
- Local soft spots: Harden by tamping in bedding material.
- Inspection of excavated formations: Give notice.

270 BACKFILLING TO PIPELINES
- Backfilling from top of surround or protective cushion: Material excavated from trench, compacted in layers 300 mm (maximum) thick.
- Heavy compactors: Do not use before there is 600 mm of material over pipes.

280 BACKFILLING UNDER ROADS AND PAVINGS
- Backfilling from top of surround or protective cushion up to formation level: Granular Sub-base Material Type 1 to HA Specification for Highway Works, Clause 803, laid and compacted in 150 mm layers.

295 WARNING MARKER TAPES
- Type: Heavy gauge polyethylene.
- Installation: During backfilling, lay continuously over pipelines.
- Depth: 300-400 mm.

BEDDING/ JOINTING

310 INSTALLATION GENERALLY
- Pipes and fittings: From same manufacturer for each pipeline.
- Jointing: Use recommended lubricants. Leave recommended gaps at ends of spigots to allow for movement.
- Jointing differing pipes and fittings: With adaptors recommended by pipe manufacturer.
- Laying pipes: To true line and regular gradient on even bed for full length of barrel with sockets (if any) facing up the gradient.
    - Protect from damage and ingress of debris. Seal all exposed ends during construction.
- Timing: Minimize time between laying and testing. Backfill after successful initial testing.

340 CLASS D NATURAL BED
- Trench: Excavate slightly shallower than final levels.
    - Trimming: By hand to accurate gradients, replacing overdig with compacted spoil.
- Pipes: Resting uniformly on their barrels, adjusted to line and gradient. Do not use hard packings under pipes.
- Backfilling:
    - Timing: Laid after successful initial testing.
    - Material: Protective cushion of selected fill, free from vegetable matter, rubbish, frozen soil and material retained on a 40 mm sieve.
    - Depth: 150 mm (250 mm for adoptable sewers) above crown of pipe.
    - Compaction: By hand in 100 mm layers.

350 CLASS F GRANULAR BED
- Granular material to BS 882:
  Pipe size (DN)   Nominal single size (mm)
  100 & 150   10
  225 & 300   10 or 20
- Bedding: Granular material, compacted over full width of trench:
    - Thickness (minimum): 50 mm for sleeve jointed pipes, 100 mm for socket jointed pipes. Where trench bottom is uneven due to hard spots or other reason, increase thickness by 100 mm.
- Pipes: Digging slightly into bed, resting uniformly on their barrels and adjusted to line and gradient.
- Backfilling:
    - Timing: Laid after successful initial testing.
    - Material: Protective cushion of selected fill, free from vegetable matter, rubbish, frozen soil and material retained on a 40 mm sieve.
    - Depth: 150 mm (250 mm for adoptable sewers) above crown of pipe.
    - Compaction: By hand in 100 mm layers.

360 CLASS N AS-DUG MATERIAL BED
- Material: One of the following:
    - As-dug material with a compaction fraction of not more than 0.3.
    - All-in aggregate to BS 882, nominal size 10 mm.
    - Fine aggregate to BS 882.
- Bedding: Granular material, compacted over full width of trench:
    - Thickness (minimum): 50 mm for sleeve jointed pipes, 100 mm for socket jointed pipes. Where trench bottom is uneven due to hard spots or other reason, increase thickness by 100 mm.
- Pipes: Digging slightly into bed, resting uniformly on their barrels and adjusted to line and gradient.
- Backfilling:
    - Timing: Laid after successful initial testing.
    - Material: Protective cushion of selected fill, free from vegetable matter, rubbish, frozen soil and material retained on a 40 mm sieve.
    - Depth: 150 mm (250 mm for adoptable sewers) above crown of pipe.

- Compaction: By hand in 100 mm layers.

380 CLASS O FULL DEPTH GRANULAR SUPPORT
- Granular material to BS 882:
  Pipe size (DN)   Nominal single size (mm)
  100 & 150   10
  225 & 300   10 or 20
- Bedding: Granular material, compacted over full width of trench.
  - Thickness (minimum): 100 mm.
- Pipes: Digging slightly into bed, resting uniformly on their barrels and adjusted to line and gradient.
- Granular support:
  - Timing: Laid after successful initial testing.
  - Depth: To slightly above crown of pipe.
  - Compaction: By hand.
- Backfilling:
  - Material/ Depth: Either:
  Protective cushion of selected fill, free from vegetable matter, rubbish, frozen soil and material retained on a 40 mm sieve, to 300 mm above crown of pipe, or:
  Additional granular material, to 100 mm above crown of pipe.
  - Compaction: By hand in 100 mm layers.

390 CLASS P FULL DEPTH GRANULAR SUPPORT
- Granular material to BS 882:
  Pipe size (DN)   Nominal single   Graded
  size (mm)   size (mm)
  100 & 150   10   Not permitted
  225 & 300   10 or 20   20 to 5
- Bedding: Granular material, compacted over full width of trench.
  - Thickness (minimum): 100 mm.
- Pipes: Digging slightly into bed, resting uniformly on their barrels and adjusted to line and gradient.
- Granular support:
  - Timing: Laid after successful initial testing.
  - Depth: To slightly above crown of pipe.
  - Compaction: By hand.
- Backfilling:
  - Material/ Depth: Either:
  Protective cushion of selected fill, free from vegetable matter, rubbish, frozen soil and material retained on a 40 mm sieve, to 300 mm above crown of pipe, or:
  Additional granular material, to 100 mm above crown of pipe.
  - Compaction: By hand in 100 mm layers.

430 CLASS W GRANULAR SURROUND
- Timing: Excavate trench after hardcore has been laid and compacted.
- Granular material to BS 882:
  Pipe size (DN)   Nominal single size (mm)
  100 & 150   10
  225 & 300   10 or 20
- Granular bedding, compacted over full width of trench:
  - Thickness (minimum): 100 mm.
- Pipes: Digging slightly into bed, resting uniformly on their barrels and adjusted to line and gradient.
- Granular surround:
  - Timing: Laid after successful initial testing.
  - Depth: To 100 mm above crown of pipe.
  - Compaction: By hand.
- Backfilling:
  - Material: Hardcore or granular material.
  - Depth: Up to slab formation.
  - Compaction: In 300 mm (maximum) thick layers.

451 CLASS Y CONCRETE SURROUND FOR SHALLOW PIPES UNDER BUILDINGS
- Locations: Where crown of pipe is less than 300 mm below underside of slab.
- Timing: Excavate trench after hardcore has been laid and compacted.
- Concrete blinding (over full width of trench): Allow to set before laying pipes.
  - Thickness: 25 mm.
- Pipes:
  - Temporary support: Folding wedges of compressible board. Prevent flotation.
  - Height above blinding (minimum): 100 mm.
- Surround: Encase pipe in concrete of same mix as slab and cast integrally with the slab.
- Extent of surround: To within 150 mm of next nearest flexible joint.

461 CLASS Z CONCRETE SURROUND
- Concrete blinding (over full width of trench): Allow to set before laying pipes.
  - Thickness: 25 mm.
- Temporary pipe support: Folding wedges of compressible board. Prevent flotation.
- Surround, to full width of trench:
  - Timing: After successful initial testing.
  - Depth: To 150 mm above crown of pipe or as shown on drawings.
  - Vertical construction joints: At face of flexible pipe joints using 18 mm thick compressible board precut to profile of pipe.
- Socketed pipes: Fill gaps between spigots and sockets with resilient material to prevent entry of concrete.

470 PIPE RUNS NEAR FOUNDATIONS
- Class Z concrete surround: Provide in locations where bottom of trench is lower than bottom of foundation. Measurements are horizontal clear distances between nearest edges of foundations and pipe trenches.
- Trenches less than one metre from foundations: Top of concrete surround not lower than bottom of foundation.
- Trenches more than one metre from foundations: Top of concrete surround not lower than D mm below bottom of foundation, where D mm is horizontal distance of trench from foundation, less 150 mm.

512 PIPELINES PASSING THROUGH STRUCTURES
- Pipelines that must be cast in or fixed to structures (including manholes, catchpits and inspection chambers): Provide short length (rocker) pipes near each external face, with flexible joint at each end:

Pipe size (DN)	Distance to first joint from structure (mm)	Short length (mm)
100 & 150	150	600
225	225	600

- Pipelines that need not be cast in or fixed to structures (e.g. walls to footings): Provide either:
  - Short length (rocker) pipes as specified above, or
  - Openings in the structures to give 50 mm minimum clearance around the pipeline. Closely fit a rigid sheet to each side of opening to prevent ingress of fill or vermin.

520 BENDS AT BASE OF SOIL STACKS
- Type: As specified in schedules.
- Stabilizing bends: Bed in concrete without impairing flexibility of couplings.

570 FLEXIBLE COUPLINGS
- Standard: To BS EN 295-4, WIS 04-41-01, or Agrément certified.
- Manufacturer: Contractors choice.
- Ends of pipes to be joined: Cleanly cut and square.
- Outer surfaces of pipes to be joined: Clean and smooth. Where necessary, e.g. on concrete or iron pipes, smooth out mould lines and/ or apply a cement grout over the sealing area.

TERMINAL/ ACCESS FITTINGS

690 INSTALLATION OF FITTINGS
- Setting out: Square with and tightly jointed to adjacent construction.
- Bedding and surrounds: Concrete, 150 mm thick.
- Permissible deviation in level of gully gratings: +0 to -10 mm.
- Exposed openings in fittings: Fit purpose made temporary caps. Protect from site traffic.

MANHOLES/ CHAMBERS/ SOAKAWAYS/ TANKS

710 BRICK MANHOLES/ INSPECTION CHAMBERS
- Bases: Plain concrete, thickness As Specified by CA.
- Brickwork: Engineering class B, frogs facing upwards.
- Steps: Cast iron to BS 1247. Bed in joints to all chambers over 900 mm deep at 300 mm vertical centres staggered 300 mm horizontally, with lowest step not more than 300 mm above benching and top step not more than 450 mm below top of cover.
- Channels, branches and benching: As Specified by CA.
- Cover slabs: As Specified by CA.
    - Thickness: As Specified by CA.
    - Openings: To suit required access covers.
    - Reinforcement: Steel fabric to BS 4483, reference As Specified in schedules.
- Access covers and seating: As Specified in schedules.

730 CONCRETE MANHOLES/ INSPECTION CHAMBERS
- Manholes/ Inspection chambers:
    - Standard: To BS5911 Part 1.
    - Manufacture: All components from same manufacturer.
    - Manufacturer: RMC.
  Product reference: Various.
    - Chamber sections: As Specified in schedules.
    - Cover slabs: As Specified in schedules.
    - Joints: as recommended by chamber manufacturer.
  Joint surfaces: Prime if required.
      Inner joint surface: Trim surplus jointing material extruded into chamber and point neatly.
    - Steps: Required in chambers over 900mm deep.

760 CONVENTIONAL CHANNELS, BRANCHES AND BENCHING
- Main channel: Bed solid in 1:3 cement:sand mortar.
    - Branches: Connect to channel, preferably at half pipe level, so that discharge flows smoothly in direction of main flow.
    - Branches greater than nominal size 150 mm: Connect with the soffit level with that of the main drain.
    - Connecting angle more than 45° to direction of flow: Use three-quarter section channel bends.
- Plastics channels: Use clips or ensure adequate mechanical key when bedding on to mortar.
- Benching: Form in concrete, to rise vertically from top of main channel to a level not lower than soffit of outlet pipe, then slope upwards at 10% to walls.

762 PREFORMED PLASTICS CHANNELS, BRANCHES AND BENCHING
- Manufacturer: Hepworth.
    - Product reference: As Specified in schedules.
    - Sizes and integral branches: To suit each manhole.
- Temporary caps: Remove as necessary.
- Bedding: 1:3 cement:sand mortar.
- Benching: Form in concrete, with 10% fall from manhole walls to component rim.

811 CAST IRON ACCESS COVERS AND SEATING
- Covers: Grey iron or ductile iron to BS EN 124.
  - Manufacturer: Contractors choice .
  - Types: As Specified in schedules.
- Seating: Engineering bricks to BS 3921, Class B, laid in 1:3 cement:sand mortar, or precast concrete cover frame units, Type 1 or Type 2 to suit cover shape.
- Bedding and haunching frame: Solidly in 1:3 cement:sand mortar over its whole area, centrally over opening, top level and square with joints in surrounding finishes. Cut back top of haunching to 30 mm below top of surface material.

821 STEEL ACCESS COVERS AND SEATING
- Covers: Steel to BS EN 124, where applicable.
  - Manufacturer: Contractors choice .
  - Types: As Specified in schedules.
- Seating: Engineering bricks to BS 3921, Class B, laid in 1:3 cement:sand mortar or precast concrete cover frame units, Type 1 or Type 2 to suit cover shape.
- Bedding and haunching frame: Solidly in 1:3 cement:sand mortar over its whole base area, centrally over opening, top level and square with joints in surrounding finishes. Cut back top of haunching to 30 mm below top of surface material.

861 CONNECTIONS TO SEWERS
- General: Connect new pipework to existing adopted sewers to the requirements of the Adopting Authority or its agent.

CLEANING/ TESTING/ INSPECTION/ COMPLETION

900 REMOVAL OF DEBRIS AND CLEANING
- Preparation: Before cleaning, final testing, CCTV inspection if specified and immediately before handover, lift covers to manholes, inspection chambers and access points. Remove wrappings, mortar droppings and any other debris.
- Cleaning: Thoroughly flush with water to remove silt and check for blockages. Rod pipelines between access points if there is any indication that they may be obstructed.
- Washings and detritus: Do not discharge into sewers or watercourses.
- Covers: Securely replace after cleaning and testing.

910 TESTING/ INSPECTION
- Dates for inspection: Give notice.
- Attendance: Provide water, assistance and apparatus as required.
- Remedial work: Repair leaks.

920 WATER/ AIR TESTING OF GRAVITY DRAINS AND PRIVATE SEWERS UP TO DN 300
- Initial testing: To ensure that pipelines are sound and properly installed, air test short lengths to BS 8301, clause 25.6.3 immediately after completion of bedding/ surround.
- Final testing: For final checking and statutory authority approval, water test to BS 8301, clause 25.6.2 all lengths of pipeline from terminals and connections to manholes/ chambers and between manholes/ chambers.

925 WATER/ AIR TESTING OF GRAVITY DRAINS AND PRIVATE SEWERS UP TO DN 300
- Air testing preparation and method: To BS 8301, clauses 25.6.3.1 and 25.6.3.2.
- Initial testing: To ensure that pipelines are sound and properly installed, air test short lengths to BS EN 1610, clauses 13.1 and 13.2 immediately after completion of bedding/ surround.
- Final testing: For final checking and statutory authority approval, water test to BS EN 1610, clause 13.1 and 13.3 all lengths of pipeline from terminals and connections to manholes/ chambers and between manholes/ chambers.

940 WATER TESTING OF MANHOLES/ INSPECTION CHAMBERS
- Timing: Before backfilling.
- Standard:
  - Exfiltration: To BS EN 1610, clause 13.3.
  - Infiltration: No identifiable flow of water penetrating the chamber.

S: PIPED SUPPLY SYSTEMS

S12 HOT AND COLD WATER SUPPLY SYSTEMS

GENERAL

100 DEFINITIONS
- "Connector". Any thimble, ferrule, cone, caulking bush, cap and lining or other fitting or adaptor whether equal or reducing or a combination of any of the above necessary to achieve the connection described.
- "Pipes". Includes "pipes" and "tubes".

105 METHODS OF MEASUREMENT
- Types of joints to pipework and ductwork have only been stated in the description of equipment and ancillaries where these are not implicit.
- The largest outlet size only has been given in fitting sectional insulation coverings around reducing pipe fittings.
- Where, in the opinion of the Contract Administrator, it is necessary to drain down a system, tank, cylinder etc. in order to repair or renew pipes and fittings, the cost of this operation and subsequent refilling, items C8350/200-240 and C8510/050 refer, shall be added to the relative items of repair or renewal.
- Fittings required to be inserted in existing pipework are deemed to match or be consistent with the existing pipework.
- Pipes are deemed to be straight unless otherwise stated.
- Where no reference is made to a background or to pipe supports, pipes are deemed fixed without supports.

110 PRICES ALSO TO INCLUDE
- Testing and stamping pipe fittings etc where required by the water undertakers bye-laws as described in the specification.
- Joints in the running length and for marking the position of holes, mortices, chases and the like.
- Couplers in the running length necessary for pipe jointing.
- Metric adaptors and conversion pieces between imperial and metric items.
- Additional reducers necessary to achieve correct sizes on reducing pipe fittings.
- Spacers and special support units necessary to maintain uninterrupted the run of -insulation on pipework and ducting.
- Handing pipe sleeves over to others for fixing as necessary.
- Using preformed sections for fitting sectional insulation around fittings and ancillaries.
- Making final connections to equipment after the general installation work has been substantially completed.
- Preparation work (for painting) and priming of ferrous frameworks, brackets, hangers and the like
- Tuition of Employer's operating staff

PRODUCTS

310 DEZINCIFICATION
- Fittings, pipelines, equipment located below ground or in concealed or inaccessible locations: Resistant to dezincification, e.g. gunmetal.

380 DIRECT INSULATED COMBINATION UNITS
- Standard: To BS 3198, Kitemark certified.
- Manufacturer: Contractors choice.
- Hot water capacity: As specified in schedules.

430 INDIRECT INSULATED COMBINATION UNITS,
- Standard: To BS 3198, Kitemark certified.
- Manufacturer: Contractors choice.
- Hot water capacity: As specified in schedules.

510 COPPER PIPELINES FOR GENERAL USE
- Standard: To BS EN 1057, Kitemark certified.
- Temper: Half hard R250.
- Wall thickness (nominal):
  - OD 6, 8, 10 and 12 mm: 0.6 mm.
  - OD 15 mm: 0.7 mm.
  - OD 22 and 28 mm: 0.9 mm.
  - OD 35 and 42 mm: 1.2 mm.
- Jointing generally: Integral lead free solder ring capillary fittings to BS EN 1254-1, Kitemark certified.
- Connections to appliances and equipment: Select from:
  - Compression fittings: To BS EN 1254-2, Kitemark certified.
  - Fittings with threaded ends: To BS EN 1254-4.
- Supports: As specified in schedules.

515 COPPER PIPELINES FOR UNDERGROUND USE
- Standard: To BS EN 1057, Kitemark certified.
- Manufacturer: Contractors choice.
- Temper: Half Hard R250.
- Finish: Seamless polyethylene to BS 3412.
  - Colour: As specified by CA.
- Wall thickness (nominal):
  - OD 6, 8, 10 and 12 mm: 0.8 mm.
  - OD 15 mm: 1.0 mm.
  - OD 22 and 28 mm: 1.2 mm.
  - OD 35 and 42 mm: 1.5 mm.
- Jointing generally: Integral lead free solder ring capillary fittings to BS EN 1254-1, Kitemark certified.
- Connections to appliances and equipment: Select from
  - Compression fittings: To BS EN 1254-2, Kitemark certified.
  - Fittings with threaded ends: To BS EN 1254-4.
- Supports: As specified in schedules.

520 POLYETHYLENE PIPELINES FOR UNDERGROUND USE
- Standard: To BS 6572, Kitemark certified.
- Jointing: Compression fittings or electro fusion as specified.
- Colour: Blue.

540 STEEL PIPELINES
- Standard: To BS 1387.
- Grade: Heavy duty.
- Concealed fittings $\leq$125 mm: Welded to BS EN 10253.
- Above ground fittings: Screwed to BS EN 10242.
- Finish: Galvanised.
- Sizes: As specified in schedules.

570 INSULATION TO PIPELINES
- Material: Preformed flexible closed cell or mineral fibre split tube.
- Thermal conductivity (maximum): 0.04 W/m·K.
- Thickness:
  - Hot water pipelines: Equal to the outside diameter of the pipe up to a maximum of 40 mm.
  - Internal cold water pipelines: 25 mm.
  - Roof space cold water pipelines: 32 mm.
  - External cold water pipelines: 38 mm.
- Fire performance: Class 1 spread of flame when tested to BS 476-7.

620 VALVES GENERALLY
- Types: Approved for the purpose by local water supply undertaker and of appropriate pressure and/or temperature ratings.
- Control of valves: Fit with handwheels for isolation and lockshields for isolation and regulation of circuits or equipment.

630 DOUBLE CHECK VALVE ASSEMBLIES
- Standard: Copper alloy check valves to BS 6282-1 with test cock to BS 2879 between.

640 DRAINING TAPS
- Standard: Copper alloy to BS 2879, Type 1, hose connection pattern, Kitemark certified.

660 GATE VALVES
- Standard: To BS 5154, Series B, Kitemark certified.

670 STOP VALVES AND DRAW-OFF TAPS, ABOVE GROUND
- Standard: Copper alloy to BS 1010-2, Kitemark certified.

675 STOP VALVES, UNDERGROUND
- Standard: DZR copper alloy CZ 132 to BS 5433.

EXECUTION

710 INSTALLATION GENERALLY
- Installation: To BS 6700.
- Performance: Free from leaks and the audible effects of expansion, vibration and water hammer.
- Fixing of equipment, components and accessories: Fix securely, parallel or perpendicular to the structure of the building.
- Preparation: Immediately before installing tanks and cisterns on a floor or platform, clear the surface completely of debris and projections.
- Corrosion resistance: In locations where moisture is present or may occur, provide corrosion resistant fittings/ fixings and avoid contact between dissimilar metals by use of suitable washers, gaskets, etc.

720 INSTALLING CISTERNS
- Outlet positions: Connect lowest outlets at least 30 mm above bottom of cistern.
- Access: Fix cistern with a minimum clear space of 350 mm above, or 225 mm if the cistern does not exceed 450 mm in any dimension.

725 INSTALLING WARNING/ OVERFLOW PIPES TO CISTERNS
- Difference (minimum) between normal water level and overflow level:
    - Cold water storage cisterns: The greater of 32 mm or the bore of warning pipe.
    - Feed and expansion cisterns: Sufficient to allow 20% increase in the volume of water in the tank, plus 25 mm.
- Vertical distance (minimum) of water supply inlet above overflow level: Bore of warning pipe.
- Fall (minimum): 1 in 10.
- Installation: Support to prevent sagging. Terminate pipes separately in prominent positions with turned down ends. Turn down within the cistern. Terminate 50 mm below normal water level.
- Insulation: Insulate within the building where the pipe is in an uninsulated space and subject to freezing.

727 INSTALLING VENT PIPES OVER CISTERNS
- Route: Install with no restrictions or valves and rising continuously from system connection to discharge over cistern.
- Internal diameter (minimum): 20 mm.

740 INSTALLING GAS FIRED HOT WATER BOILERS/ CIRCULATORS
- Standards: In accordance with BS 6798 and BS 5546.

790 PIPELINES INSTALLATION
- Appearance: Install pipes straight, and parallel or perpendicular to walls, floors, ceilings, and other building elements.
- Pipelines finish: Smooth, consistent bore, clean, free from defects, e.g. external scratching, toolmarks, distortion, wrinkling, and cracks.
- Concealment: Generally conceal pipelines within floor, ceiling and/ or roof voids.

- Access: Locate runs to facilitate installation of equipment, accessories and insulation and allow access for maintenance.
- Arrangement of hot and cold pipelines: Run hot pipelines above cold where routed together horizontally. Do not run cold water pipelines near to heating pipelines or through heated spaces.
- Electrical equipment: Install pipelines clear of electrical equipment. Do not run pipelines through electrical enclosures or above switch gear distribution boards or the like.
- Insulation allowance: Provide space around pipelines to fit insulation without compression.

800 PIPELINES FIXING
- Fixing: Secure and neat.
- Joints, bends and offsets: Minimize.
- Pipeline support: Prevent strain, e.g. from the operation of taps or valves.
- Drains and vents: Fix pipelines to falls. Fit draining taps at low points and vents at high points.
- Thermal expansion and contraction: Allow for thermal movement of pipelines. Isolate from structure. Prevent noise or abrasion of pipelines caused by movement. Sleeve pipelines passing through walls, floors or other building elements.
- Dirt, insects or rodents: Prevent ingress.

810 SUPPORTS FOR COPPER AND STEEL PIPELINES
- Spacing: Fix securely and true to line at the following maximum centres:
  - 15 and 22 mm pipe OD: 1200 mm horizontal, 1800 mm vertical.
  - 28 and 35 mm pipe OD: 1800 mm horizontal, 2400 mm vertical.
  - 42 and 54 mm pipe OD: 2400 mm horizontal, 3000 mm vertical.
- Additional supports: Locate within 150 mm of connections, junctions and changes of direction.

815 SUPPORTS FOR EXPOSED THERMOPLASTICS PIPELINES
- Spacing: Fix securely and true to line at the following maximum centres:
  - Up to 16 mm pipe OD: 300 mm horizontal, 500 mm vertical.
  - 17-25 mm pipe OD: 500 mm horizontal, 800 mm vertical.
  - 26-32 mm pipe OD: 800 mm horizontal, 1000 mm vertical.
- Additional supports: Locate within 150 mm of connections, junctions and changes of direction.

820 BENDS IN THERMOPLASTICS PIPELINES
- Bends: Do not use 90° elbow fittings instead of 90° bends.
- Large radius bends: Support at maximum centres.
- 90° bends: Fix pipe clips either side of bend.
- Small radius bends: Fully support 90° bends with cold form bend fixtures.

830 PIPELINE SPACING
- Clearance (minimum) to face of wall-fixed pipes or pipe insulation:
  - From floor: 150 mm.
  - From ceiling: 50 mm.
  - From wall: 15 mm.
  - Between pipes: 25 mm.
  - From electrical conduit, cables, etc: 150 mm.

840 JOINTS IN COPPER AND STEEL PIPELINES
- Preparation: Cut pipes square. Remove burrs.
- Joints: Neat, clean and fully sealed. Install pipe ends into joint fittings to full depth.
- Bends: Do not use formed bends on exposed pipework, except for small offsets. Form changes of direction with radius fittings.
- Adaptors for connecting dissimilar materials: Purpose designed.
- Substrate and plastics pipes and fittings: Do not damage, e.g. by heat when forming soldered joints.
- Flux residue: Clean off.

841 CAPILLARY JOINTS IN PLASTICS COATED PIPELINES
- Plastics coating: Do not damage, e.g. by direct or indirect heat. Wrap completed joint (when cool) with PVC tape of matching colour, half lapped.

845 JOINTS IN THERMOPLASTICS PIPELINES
- Fittings and accessories for joints: Purpose designed.
- Preparation: Cut pipes square. Remove burrs.
- Joints: Neat, clean and fully sealed. Install pipe ends into joint fittings to full depth.
- Compression fittings: Do not overtighten.

850 PIPELINES ENTERING BUILDINGS
- Depth: Lay pipes at least 750 mm below finished ground level.
- Pipelines rising into building within 750 mm of the external face of the external wall or passing through a ventilated void below floor level: Insulate from finished floor level to 600 mm beyond external face of building.
- Ends of pipeducts: Seal both ends to a depth of at least 150 mm.

855 EXTERNAL SUPPLY PIPELINES
- Pipelines exposed to air or less than 750 mm below finished ground level: Insulate.

860 INSTALLING INSULATION TO PIPELINES
- Cold water pipelines: Insulate in unheated spaces. Insulate potable cold water pipelines.
- Hot water pipelines: Insulate, except for short lengths in prominent positions next to appliances.
- Appearance: Fix securely and neatly. Make continuous over fittings and at supports. Leave no gaps. Locate split on 'blind' side of pipeline.
- Timing: Fit insulation after testing.

865 INSTALLING INSULATION TO CISTERNS
- General: Fix securely to sides and top of cisterns. Leave no gaps.
- Access cover: Allow removal of cover with minimum disturbance to insulation.
- Underside of cistern: Insulate where exposed in unheated spaces.

870 INSTALLING VALVES
- Isolation and regulation valves: Provide on equipment and subcircuits.
- Access: Locate where valves can be readily operated and maintained and next to equipment which is to be isolated.
- Connection to pipework: Fit with joints to suit the pipe material.

COMPLETION

910 FLUSHING AND FILLING
- Standard: To BS 6700.

920 SYSTEM DISINFECTION
- Disinfection: To BS 6700.

930 TESTING
- Standard: To BS 6700.
- Notice (minimum): 3 days.
- Preparation: Secure and clean pipework and equipment. Fit cistern and tank covers.
- Leak testing: Start boiler and run the system until all parts are at normal operating temperatures and then allow to cool to cold condition for a period of 3 h.
- Pressure testing: At both hot and cold conditions joints, fittings and components must be free from leaks and signs of physical distress when tested for at least 1 h as follows:
    - Systems fed directly from the mains, and systems downstream of a booster pump: Apply a test pressure equal to 1.5 times the maximum pressure to which the installation or relevant part is designed to be subjected in operation.
    - Systems fed from storage: Apply a test pressure equal to the pressure produced when the storage cistern is filled to its normal maximum operating level.
    - Inaccessible or buried pipelines: Carry out hydraulic pressure test to twice the working pressure.

940 COMMISSIONING
- Standard: To BS 6700.
- Equipment: Check and adjust operation of equipment, controls and safety devices.
- Outlets: Check operation of outlets for satisfactory rate of flow and temperature.

950 TESTING SERVICE PIPELINES
- Test method: Disconnect from the mains, fill with potable water, exclude air, and apply at least twice the working pressure for 1 h.
- Test criterion: No leakage.

960 DOCUMENTATION
- Manufacturers' operating and maintenance instructions: Submit for equipment and controls.
- System operating and maintenance instructions: Submit for the system as a whole giving optimum settings for controls.
- Record drawings: Submit drawings showing the location of circuits and operating controls.

970 OPERATING TOOLS
- Tools: Supply tools for operation, maintenance and cleaning purposes.
- Valve keys: Supply keys for valves and vents.

980 LABELS
- Valve labels: Provide labels on isolating and regulating valves on primary circuits, stating their function.

V: ELECTRICAL SUPPLY/POWER/LIGHTING SYSTEMS

V90 ELECTRICAL INSTALLATION

GENERAL

100 METHODS OF MEASUREMENT
- Conduits, cable trunking, cable tray, cable ladders and racks and busbar trunking are deemed straight unless otherwise stated.
- Conduits are deemed fixed to surfaces unless otherwise stated.

105 PRICES ALSO TO INCLUDE
- Manufacturers' standard couplers for jointing trunking.
- Short lengths of fixed lid where trunking passes through walls, floors etc.
- Bolted lapped joints on trays where required.
- Temporary supports, etc for items described as embedded in floor screeds or concrete.
- Phase marking, circuit lists and labels on equipment and control gear.
- Making final connections to equipment after the general installation work has been substantially completed.
- Preparation work (for painting) and priming of ferrous frameworks, brackets, hangers and the like.
- Tuition of Employer's operating staff.

110 PRICES DO NOT INCLUDE
- These rates do not include for the NIC periodic inspections, as these are deemed to be part of a maintenance inspection programme. However, the Contractor is expected to report to the Client any failures to comply, which are apparent without testing, of the existing equipment and installation on which no work has been ordered. These reports should be in writing but would be limited to simple observations of the existing system made by the tradesmen in the area where work has been carried out.

115 INSPECTION AND TESTING GENERALLY
- The post installation testing to ensure compliance with IEE regulations, Electricity at Work Regulations, and NIC guidelines is limited to the work ordered and executed. The Electrician is to satisfy himself that the circuit is safe, i.e. carry out continuity and polarity checks, which are included in the rates. Further testing such as earth loop impedence, insulation resistance or under sized earthing are not included in the rates. The reimbursement for these costs are to be included within the Contractors percentage adjustment.

PRODUCTS

310 PRODUCTS GENERALLY
- Standard: To BS 7671.
- CE Marking: Required.
- Proposals: Submit drawings, technical information and manufacturer's literature.

320 DISTRIBUTION BOARDS AND CONSUMER UNITS
- Standards: To BS 5486-12, -13 and BS EN 60439-3. ASTA certified.
- Manufacturer: Wylex.
  - Product reference: As specified in schedules.
- Main control: As specified in schedules.
  - Rating: To suit maximum demand.
- Number of ways: As specified in schedules.
  - Circuit protection: Miniature circuit-breakers.
  - Standard: To BS EN 60898.
- Additional circuit protection: As specified by CA.
  - Standard: To BS EN 61008-1 or BS EN 61009-1.

340 CONDUIT, TRUNKING AND DUCTING
- Standard: To BS 50086-1.
- Type: Suitable for location and use.

342 STEEL CONDUIT AND FITTINGS
- Standards: To BS 4568-1 and BS EN 50086.
- Manufacturer: Contractors choice.
- Sizes: 20mm.
- Jointing: As specified in schedules.
- Fittings: As specified in schedules.
- Finish: Black.

410 CABLES
- Standard: BASEC certified.

420 PROTECTIVE CONDUCTORS
- Type: Cable conductors with yellow/ green sheath.

430 ELECTRICAL ACCESSORIES
- Standard: To BS EN 60669-1.
- Manufacturer: As specified in schedules.
- Finish: As specified in schedules.

510 LUMINAIRES
- Manufacturer: Thorn.
  - Product reference: As specified in schedules.
- Mounting: As specified in schedules.
- Lamp: As specified in schedules.
  - Power rating: As specified in schedules.

515 LUMINAIRE SUPPORTING COUPLERS
- Standard: To BS 6972.

580 EARTHING AND BONDING
- Earth electrodes: In accordance with BS 7430.
- Earth clamps: To BS 951.

585 EARTH BARS
- Separate earth bar: Required.
- Size: Determine.
- Material: Copper.

EXECUTION

610 EXECUTION GENERALLY
- Standard: To BS 7671.

630 CONNECTION TO INCOMING SUPPLY
- Main switchboard/ distribution board: Connect to main incoming metering equipment.

650 SWITCHGEAR INSTALLATION
- Clearance in front of switchgear (minimum): 1 m.
- Labelling: Permanently label each way, identifying circuit function, rating and cable size.

700 CABLES LAID DIRECTLY IN THE GROUND
- Cable bedding: 75 mm of sand.
- Backfilling: 75 mm of sand over cables, then as-dug material.
- Multiple cables in same trench: Set 150 mm apart.
- Cables below roads and hardstandings: Duct, derate if longer than 10 m.

710 CABLES ENTERING BUILDINGS FROM BELOW GROUND
- Pipeducts: Seal at both ends.
- Proposals: Submit drawings.

740 CONDUIT AND FITTINGS
- Fixing: Fix securely. Fix boxes independently of conduit.
- Location: Position vertically and horizontally in line with equipment served and parallel with building lines. Locate where accessible.
- Jointing:
  - Number of joints: Minimize.
  - Lengths of conduit: Maximize.
  - Cut ends: Remove burrs and plug during building works.
  - Movement joints in structure: Manufactured expansion coupling.
  - Threaded steel conduits: Tightly screw to ensure electrical continuity, with no thread showing.
  - Conduit connections to boxes and items of equipment, other than those with threaded entries: Earthing coupling/ male brass bush and protective conductor.
- Changes of direction: Site machine-formed bends, junction boxes and proprietary components. Do not use elbows or tees. Alternatively, use conduit boxes.
- Connections to boxes, trunking, equipment and accessories: Screwed couplings, adaptors, connectors and glands, with rubber bushes at open ends.
- Mounting and support: As specified in schedules.

748 DRAINAGE OF CONDUIT
- Drainage outlets: Locate at lowest points in conduit installed externally, and where condensation may occur.

750 INSTALLING TRUNKING/ DUCTING/ CABLE MANAGEMENT SYSTEMS
- Positioning: Accurate with respect to equipment served and parallel with other services, and where relevant, floor level and other building lines.
- Access: Provide space encompassing cable trunking to permit access for installing and maintaining cables.
- Jointing:
  - Number of joints: Minimize.
  - Lengths of conduit: Maximize.
  - Steel systems: Mechanical couplings. Do not weld. Fit a copper link at each joint to ensure electrical continuity.
- Movement: Fix securely. Restrain floor mounted systems during screeding.
- Junctions and changes of direction: Proprietary jointing units.
- Cable entries: Fit grommets, bushes or liners.
- Protection: Fit temporary blanking plates. Prevent ingress of screed and other extraneous materials.
- Service outlet units: Fit when cables are installed.

800 CABLE ROUTES
- Cables generally: Conceal wherever possible.
  - Concealed cable runs to wall switches and outlets: Align vertically with the accessory.
- Exposed cable runs: Submit proposals.
  - Orientation: Straight, vertical and/ or horizontal and parallel to walls.
- Distance from other services running parallel: 150 mm minimum.
  - Heating pipes: Position cables below.

810 INSTALLING CABLES
- General: Install cables neatly and securely. Protect against accidental damage, adverse environmental conditions, mechanical stress and deleterious substances.
- Timing: Do not start internal cabling until building enclosure provides permanently dry conditions.
- Jointing: At equipment and terminal fittings only.
- Cables passing through walls: Sleeve with conduit bushed at both ends.
- Cables surrounded or covered by insulation: Derate.

811 CABLES IN PLASTER
- Protection: Cover with galvanized steel channel nailed to substrate.

812 **CABLES IN VERTICAL TRUNKING/ DUCTS**
- Support: Pin racks or cleats at each floor level or at 5 m vertical centres, whichever is less.
- Heat barrier centres (maximum): 5 m.
- Heat barriers: Required except where fire resisting barriers are not provided.

813 **CABLES IN ACCESSIBLE ROOF SPACES**
- Cables running across ceiling joists: Fix to timber battens which are nailed to joists.

820 **ARMOURED CABLE**
- Temperature: Do not start installation if cable or ambient temperature is below 0°C, or has been below 0°C during the previous 24 h.
- Galvanized steel guards: Fit where cables are vulnerable to mechanical damage.
- Earthing: Bond armour to equipment and main earthing system.
- Connections to apparatus: Moisture proof, sealed glands and PVC shrouds.

825 **PVC SHEATHED CABLE**
- Temperature: Do not install cables if ambient temperature is below 5°C.

840 **ELECTRICAL ACCESSORIES AND EQUIPMENT**
- Location: Coordinate with other wall or ceiling mounted equipment.
- Positioning: Accurately and square to vertical and horizontal axes.
- Alignment: Align adjacent accessories on the same vertical or horizontal axis.
- Mounting heights (finished floor level to underside of equipment/ accessory): As specified by CA.

845 **FINAL CONNECTIONS**
- Size: Determine.
- Cable: Heat resisting white flex.
- Length: Allow for equipment removal and maintenance.

850 **MULTIGANG SWITCHES**
- General: Connect switches so that there is a logical relationship with luminaire positions. Fit blanks to unused switch spaces.
- Segregation: Internally segregate each phase with phase barriers and warning plates.

860 **INSTALLING LUMINAIRES**
- Supports: Adequate for weight of luminaire.
- Locations: Submit proposals.

865 **INSTALLING EMERGENCY LUMINAIRES**
- Permanent electrical supplies: Derive from adjacent local lighting circuit.
- Charge indicator: Position in a conspicuous location.

870 **INSTALLING EXTERNAL LUMINAIRES**
- Locations: Submit proposals.
- Seals: Check for particle ingress and clean.

880 **INSTALLING EARTH BARS**
- Location: At incoming electrical service position.
- Mounting: Wall mounted on insulated supports.

890 **LABELLING**
- Identification and notices:
    - Standards: To BS 5499-5 and BS 5378-2.
    - Equipment: Label when a voltage exceeding 230 V is present.
- Distribution boards and consumer units: Card circuit chart within a reusable clear plastic cover. Fit to the inside of each unit. Include typed information identifying the outgoing circuit references, their device rating, cable type, size, circuit location and details. Label each outgoing way corresponding to the circuit chart.
- Sub-main cables: Label at both ends with proprietary cable marker sleeves.

895 ENGRAVING
- Metal and plastic accessories: Engrave, indicating their purpose.
- Emergency lighting test key switches: Describe their function.
- Multigang light switches: Describe the luminaire arrangement.

COMPLETION

910 FINAL FIX
- Accessory faceplates, luminaires and other equipment: Fit after completion of building painting.

915 CLEANING
- Electrical equipment: Clean immediately before handover.
- Equipment not supplied but installed and electrically connected: Clean immediately before handover.

920 INSPECTION AND TESTING
- Standard: To BS 7671.
- Notice before commencing tests (minimum): 24 hours.
- Labels and signs: Fix securely before system is tested.
- Inspection and completion certificates: Submit.
    - Number of copies: 2.

940 INSPECTION AND TESTING OF EXTERNAL LIGHTING SYSTEM
- Switching: Check correct operation of photoelectric control units, time switches and other switching devices over at least one switching cycle.
- Orientation: Adjust luminaires to achieve optimal performance.

990 DOCUMENTATION
- Timing: Submit at practical completion.
- Contents:
    - Full technical description of each system installed.
    - Manufacturer's operating and maintenance instructions for fittings and apparatus.
    - Manufacturer's guarantees and warranties.
    - As-installed drawings showing circuits and their ratings and locations of fittings and apparatus.
    - List of normal consumable items.

W: COMMUNICATIONS/SECURITY/CONTROL SYSTEMS

W10 RADIO/TV/CCTV

PRODUCTS

310 TELEVISION ANTENNAE
- Standards: To BS 5640-1, -2 and in accordance with BS 6330.
- Size and arrangement: Submit proposals.

315 TELEVISION CABLES
- Standard: In accordance with BS 6330.

320 TELEVISION OUTLETS
- Standards: To BS 5733 and BS EN 60669-1.
- Approvals: Kitemark certified.
- Faceplate configuration: Single coaxial outlet.
- Connection: Push fit.
- Finish: Match electrical accessories.

EXECUTION

600 INSTALLING CABLES GENERALLY
- General: Install cables neatly and securely. Conceal wherever possible. Protect against accidental damage, adverse environmental conditions, mechanical stress and deleterious substances.
    - Concealed cable runs to outlets: Align vertically with the accessory.
- Exposed cable runs: Submit proposals.
    - Orientation: Straight, vertical and/ or horizontal and parallel to walls.
- Distance from other services running parallel: 150 mm minimum.
    - Heating pipes: Position cables below.
- Timing: Do not start internal cabling until building enclosure provides permanently dry conditions.
- Jointing: At equipment and terminal fittings only.
- Cables passing through walls: Sleeve with conduit bushed at both ends.
- Cables running across ceiling joists: Fix to timber battens which are nailed to joists.
- Length of final connection: Sufficient for equipment removal and maintenance.

610 INSTALLING OUTLETS AND EQUIPMENT GENERALLY
- Location: Coordinate with other wall or ceiling mounted equipment.
- Positioning: Accurate and square to vertical and horizontal axes.
- Alignment: Align adjacent accessories on the same vertical or horizontal axis.
- Mounting heights (finished floor level to underside of equipment/ accessory): As specified by CA.

620 INSTALLING TELEVISION ANTENNAE
- Standard: In accordance with BS 6330.
- Signal field strength: Survey at the intended location for reception. Submit results.
- Location: Submit proposals.
- Fixing: Submit proposals.

630 INSTALLING TELEVISION CABLING AND OUTLETS
- Standards: To BS 7671 and in accordance with BS 6330.
- Cable route: Submit proposals.
- Fixing: Proprietary clips.
- External wall entry: Slope downwards towards external wall, form drip loop or provide proprietary rain cover.

COMPLETION

900 TELEVISION DISTRIBUTION SYSTEMS TESTING AND COMMISSIONING
- Standards: To BS EN 50083-7 and in accordance with BS 6330.
- Head end input: Reduce signal strength by 3 dB and demonstrate via inspection no impairment in signal quality at each outlet.

- Amplifier output: Reduce by 3 dB and demonstrate via inspection no impairment in signal quality at each outlet.

990 DOCUMENTATION
- Timing: Submit at completion.
- Contents:
    - Full technical description of each system installed.
    - Manufacturers' operating and maintenance instructions for fittings and apparatus.
    - Manufacturers' guarantees and warranties.
    - As-installed drawings showing circuits and their ratings and locations of fittings and apparatus.
    - List of normal consumable items.

W40 ACCESS CONTROL

PRODUCTS

365 INTERCOM ENTRANCE PANELS
- Entrance panel call buttons: As specified in schedules.
- Microphone: Integral.
- Speaker: Integral.
- Camera: Integral within video intercom entrance panels.
- Finish: Brushed stainless steel.
- Backlight: Required.

370 INTERCOM REMOTE HANDSETS
- Buzzer: Integral.
- Monitor: Integral within video remote handsets.
- Door release button: Required.
- Mounting: Wall mounted.
- Finish: Manufacturers standard.

375 ACCESS CONTROL CABLING
- Standard: To BS EN 50133-1.
- Conductor size and number of cores: Submit proposals.
- Sheath: PVC.

EXECUTION

600 INSTALLING CABLES GENERALLY
- General: Install cables neatly and securely. Conceal wherever possible. Protect against accidental damage, adverse environmental conditions, mechanical stress and deleterious substances.
    - Concealed cable runs to outlets: Align vertically with the accessory.
- Exposed cable runs: Submit proposals.
    - Orientation: Straight, vertical and/ or horizontal and parallel to walls.
- Distance from other services running parallel: 150 mm minimum.
    - Heating pipes: Position cables below.
- Timing: Do not start internal cabling until building enclosure provides permanently dry conditions.
- Jointing: At equipment and terminal fittings only.
- Cables passing through walls: Sleeve with conduit bushed at both ends.
- Cables running across ceiling joists: Fix to timber battens which are nailed to joists.
- Length of final connection: Sufficient for equipment removal and maintenance.

610 INSTALLING OUTLETS AND EQUIPMENT GENERALLY
- Location: Coordinate with other wall or ceiling mounted equipment.
- Positioning: Accurate and square to vertical and horizontal axes.
- Alignment: Align adjacent accessories on the same vertical or horizontal axis.
- Mounting heights (finished floor level to underside of equipment/ accessory): As specified by CA.

670 INSTALLING ACCESS CONTROL SYSTEMS
- Standard: To BS 7671 and BS EN 50133-1.

- Location of the access controller: As instructed.
- Mounting heights: Submit proposals.
- Reader mounting: As instructed.

680 INSTALLING ACCESS CONTROL CABLES
- Appearance: As instructed.
- Route: Submit proposals.
- Jointing: At equipment terminals only.
- Device wiring: Individual radial circuit from control panel.
- Containment: As instructed.

700 INSTALLING INTERCOM SYSTEMS
- Entrance panel:
  - Mounting: As instructed.
  - Location: As instructed.
  - Height (finished floor level to underside of equipment): As instructed.
- Handset locations: As instructed.
- Labelling: Identify call buttons.

COMPLETION

930 ACCESS CONTROL SYSTEM TESTING AND COMMISSIONING
- Standard: To BS EN 50133-1.
- Tokens to be supplied: As instructed.
- System programming:
  - Set up tokens with holder information.
  - Set up access permissions.
  - Set up access times.
- Access points: Verify the correct operation of reader, and of release/ closure mechanism.

990 DOCUMENTATION
- Timing: Submit at completion.
- Contents:
  - Full technical description of each system installed.
  - Manufacturers' operating and maintenance instructions for fittings and apparatus.
  - Manufacturers' guarantees and warranties.
  - As-installed drawings showing circuits and their ratings and locations of fittings and apparatus.
  - List of normal consumable items.

Z: BUILDING FABRIC REFERENCE SPECIFICATION

Z12 PRESERVATIVE/ FIRE RETARDANT TREATMENT

110 TREATMENT APPLICATION
- Timing: After cutting and machining timber, and before assembling components.
- Processor: Licensed by manufacturer of specified treatment solution.
- Certification: For each batch of timber provide a certificate of assurance that treatment has been carried out as specified.

120 COMMODITY SPECIFICATIONS
- Standard: Current edition of the British Wood Preserving and Damp-proofing Association (BWPDA) Manual.

130 PRESERVATIVE TREATMENT SOLUTION STRENGTHS/ TREATMENT CYCLES
- General: Select to achieve specified service life and to suit timber treatability.

160 ORGANIC SOLVENT PRESERVATIVE TREATMENT
- Solution:
    - a. 5.0% W/W pentachlorophenol
    - b. 1.75% W/W pentachlorophenol and 0.40% W/W tributylin oxide
    - c. 1.0% W/W zinc present as zinc naphthenate and 5.0% W/W pentachlorophenol
    - d. 1.0% W/W tributylin oxide
- Pentachlorophenol to confirm to BS 5707, zinc naphthenate and tributylin to BS 5666.
- The solutions must also contain a water repellent, colour or chemical tracer which are compatible with the paint system and adhesive to be used, and contact insecticide of not less than 0.50% W/W gamma-hexachlorocyclo-hexane or dieldrin.
    - Manufacturer: Contractors choice approved by client .
    - Application: Double vacuum + low pressure impregnation.
- Moisture content of timber at time of treatment: As specified for the timber/ component at time of fixing. After treatment, timber to be surface dry before use.

170 CREOSOTE PRESERVATIVE TREATMENT
- Solution:
    - Manufacturer: Contractors choice .
    - Application: High pressure impregnation, or immersion.
- Moisture content of timber at time of treatment: Not more than 28%. After treatment, allow timber to dry before using.

Z20 FIXINGS/ ADHESIVES

110 FIXINGS GENERALLY
- Integrity of supported components: Types, sizes and quantities of fasteners/ packings and spacings of fixings selected to retain supported components without distortion or loss of support.
- Components/ Substrates/ Fasteners of dissimilar metals: Fixed with isolating washers/ sleeves to avoid bimetallic corrosion.
- General usage: To recommendations of fastener manufacturers and/ or manufacturers of components, products or materials fixed and fixed to.
- Appearance: As approved samples.

120 FASTENER DURABILITY
- Fasteners in external construction: Of corrosion resistant material or with a corrosion resistant finish.

130 FASTENER DURABILITY
- Fasteners in external construction:
    - Fasteners not directly exposed to weather: Of corrosion resistant material or with a corrosion resistant finish.
    - Fasteners directly exposed to weather: Of corrosion resistant material.

140 FIXINGS THROUGH FINISHES
- Penetration of fasteners/ plugs into substrate: To achieve a secure fixing.

150 PACKINGS
- Function: To take up tolerances and prevent distortion of materials/ components.
- Materials: Noncompressible, noncorrodible, rot proof.
- Locations: Not within zones to be filled with sealant.

160 CRAMP FIXINGS
- Cramp positions: Maximum 150 mm from each end of frame sections and at 600 mm maximum centres.
- Fasteners: Cramps fixed to frames with screws of same material as cramps.
- Cramp fixings in masonry work: Fully bedded in mortar.

170 NAILED TIMBER FIXINGS
- Nails:
  - Steel: To BS 1202-1 or BS EN 10230-1.
  - Copper: To BS 1202-2.
  - Aluminium: To BS 1202-3.
- Penetration: Fully driven in without splitting or crushing timber.
- Surfaces visible in completed work: Nail heads punched below surfaces.
- Nailed timber joints: Two nails per joint (minimum), opposed skew driven.

180 FIXINGS TO MASONRY
- Fasteners:
  - Light duty: Plugs and screws.
  - Heavy duty: Expansion anchors or chemical anchors.

210 PLUGS
- Type: Proprietary types suited to background, loads to be supported and conditions expected in use.

220 SCREW FIXINGS
- Screws: To BS 1210.
- Finished level of countersunk screw heads:
  - Exposed: Flush with timber surface.
  - Concealed (holes filled/ stopped): Sunk minimum 2 mm below surface.
- Washers and screw cups: Of same material as screw.

230 PELLETED COUNTERSUNK FIXINGS
- Finished level of countersunk screw heads: Minimum 6 mm below timber surface.
- Pellets: Cut from matching timber, grain matched and glued in to full depth of hole.
- Finished level of pellets: Flush with surface.

240 PLUGGED COUNTERSUNK FIXINGS
- Finished level of countersunk screw heads: Minimum 6 mm below timber surface.
- Plugs: Glued in to full depth of hole.
- Finished level of plugs: Projecting above surface.

250 POWDER ACTUATED FIXING SYSTEMS
- Powder actuated fixing tools: To BS 4078-2 and Kitemark certified. Operatives trained and certified as competent by tool manufacturer.
- Types of fastener, accessories and consumables: As recommended by tool manufacturer.
- Protective coating to exposed fasteners used externally or in other locations subject to dampness: Zinc rich primer to fastener heads.

510 ADHESIVES
- Storage/ Usage: In accordance with manufacturer's and statutory requirements.
- Surfaces: Clean. Regularity and texture adjusted to suit bonding and gap filling characteristics of adhesive.
- Finished adhesive joints: Fully bonded. Free of surplus adhesive.

Z21 MORTARS

CEMENT GAUGED MORTARS

110 CEMENT GAUGED MORTAR MIXES
- Specification: Proportions and additional requirements for mortar materials are specified elsewhere.

120 SAND FOR SITE MADE CEMENT GAUGED MASONRY MORTARS
- Standard: To BS EN 13139.
- Grading: 0/2 (FP or MP).
  - Fines content where the proportion of sand in a mortar mix is specified as a range (e.g. 1:1: 5-6):
  Lower proportion of sand: Use category 3 fines.
  Higher proportion of sand: Use category 2 fines.
- Sand for facework mortar: Maintain consistent colour and texture. Obtain from one source.

131 READY-MIXED LIME/ SAND FOR CEMENT GAUGED MASONRY MORTARS
- Standard: To BS 4721 or BS EN 998-2.
- Lime: Nonhydraulic to BS EN 459-1.
  - Type: CL 90S.
- Pigments for coloured mortars: To BS EN 12878.

135 SITE MADE LIME/ SAND FOR CEMENT GAUGED MASONRY MORTARS
- Permitted use: Where a special colour is not required and in lieu of factory made ready-mixed material.
- Lime: Nonhydraulic to BS EN 459-1.
  - Type: CL 90S.
- Mixing: Thoroughly mix lime with sand, in the dry state. Add water and mix again. Allow to stand, without drying out, for at least 16 hours before using.

160 CEMENTS FOR MORTARS
- Cement: To BS EN 197-1 and CE marked.
  - Types: Portland cement, CEM I.
  Portland slag cement, CEM II-S.
  Portland fly ash cement, CEM II-V or W.
  - Strength class: 32.5, 42.5 or 52.5.
- White cement: To BS EN 197-1 and CE marked.
  - Type: Portland cement, CEM I.
  - Strength class: 52.5.
- Sulphate resisting cement: To BS 4027 and Kitemarked.
  - Strength class: 42.5.
- Masonry cement: To BS 5224 and Kitemarked.
  - Class: MC 12.5 (with air entraining agent).

180 ADMIXTURES FOR SITE MADE CEMENT GAUGED MORTARS
- Air entraining (plasticizing) admixtures: To BS 4887-1 and compatible with other mortar constituents.
- Other admixtures: Submit proposals.
- Prohibited admixtures: Calcium chloride and any admixture containing calcium chloride.

190 RETARDED READY TO USE CEMENT GAUGED MASONRY MORTARS
- Standard: To BS 4721 or BS EN 998-2.
- Lime for cement:lime:sand mortars: Nonhydraulic to BS EN 459-1.
  - Type: CL 90S.
- Pigments for coloured mortars: To BS EN 12878.
- Time and temperature limitations: Use within limits prescribed by mortar manufacturer.
  - Retempering: Restore workability with water only within prescribed time limits.

200 **STORAGE OF CEMENT GAUGED MORTAR MATERIALS**
- Sands and aggregates: Keep different types/ grades in separate stockpiles on hard, clean, free-draining bases.
- Factory made ready-mixed lime:sand/ ready to use retarded mortars: Keep in covered containers to prevent drying out or wetting.
- Bagged cement/ hydrated lime: Store off the ground in dry conditions.

210 **MAKING CEMENT GAUGED MORTARS**
- Batching: By volume. Use clean and accurate gauge boxes or buckets.
  - Mix proportions: Based on dry sand. Allow for bulking of damp sand.
- Mixing: Mix materials thoroughly to uniform consistency, free from lumps.
  - Mortars containing air entraining admixtures: Mix mechanically. Do not overmix.
- Working time (maximum): Two hours at normal temperatures
- Contamination: Prevent intermixing with other materials.

LIME/ SAND MORTARS

310 **LIME/ SAND MORTAR MIXES**
- Specification: Proportions and additional requirements for mortar materials are specified elsewhere.

320 **SAND FOR LIME/ SAND MASONRY MORTARS**
- Type: Sharp, well graded.
  - Quality, sampling and testing: To BS EN 13139.
  - Grading/ Source: As specified elsewhere in relevant mortar mix items.

335 **READY PREPARED LIME PUTTY**
- Manufacturer: contractors choice .

350 **STORAGE OF LIME/ SAND MORTAR MATERIALS**
- Sands and aggregates: Keep different types/ grades in separate stockpiles on hard, clean, free-draining bases.
- Ready prepared nonhydraulic lime putty: Prevent drying out and protect from frost.
- Nonhydraulic lime:sand mortar: Store on clean bases or in clean containers that allow free drainage. Prevent drying out or wetting and protect from frost.
- Bagged hydrated hydraulic lime: Store off the ground in dry conditions.

360 **MAKING LIME/ SAND MORTARS GENERALLY**
- Batching: By volume. Use clean and accurate gauge boxes or buckets.
- Mixing: Mix materials thoroughly to uniform consistency, free from lumps.
- Contamination: Prevent intermixing with other materials, including cement.

Z22 SEALANTS

120 **SUITABILITY OF JOINTS**
- Presealing checks:
  - Joint dimensions: Within limits specified for the sealant.
  - Substrate quality: Surfaces regular, undamaged and stable.
- Joints not fit to receive sealant: Submit proposals for rectification.

130 **PREPARING JOINTS**
- Surfaces to which sealant must adhere:
  - Remove temporary coatings, tapes, loosely adhering material, dust, oil, grease, surface water and contaminants that may affect bond.
  - Clean using materials and methods recommended by sealant manufacturer.
- Vulnerable surfaces adjacent to joints: Mask to prevent staining or smearing with primer or sealant.
- Primer, backing strip, bond breaker: Types recommended by sealant manufacturer.
  - Backing strip and/ or bond breaker installation: Insert into joint to correct depth, without stretching or twisting, leaving no gaps.
- Protection: Keep joints clean and protect from damage until sealant is applied.

160 APPLYING SEALANTS
- Substrate: Dry (unless recommended otherwise) and unaffected by frost, ice or snow.
- Environmental conditions: Mix and apply primers and sealants within temperature and humidity ranges recommended by manufacturers. Do not dry or raise temperature of joints by heating.
- Sealant application: Unless specified otherwise, fill joints completely and neatly, ensuring firm adhesion to substrates.
- Sealant profiles:
    - Butt and lap joints: Slightly concave.
    - Fillet joints: Flat or slightly convex.
- Protection: Protect finished joints from contamination or damage until sealant has cured.

SECTION H

COMPILER'S NOTES

The *Access Audit Schedule* is intended for use as a stand alone document. Therefore the preliminaries and preambles from the *National Schedule* have been reproduced in the *Access Audit Schedule* for ease of use.

Although this document has been compiled using the *National Schedule,* it has been necessary to introduce numerous additional items, which although coded the same as the *National Schedule*, are not included in the *National Schedule* of the same year. Having said this, item codes used in the *Access Audit Schedule* which are found in the *National Schedule* are the same and the user is entitled to use the *National Schedule* to supplement the *Access Audit schedule* if they so wish. Due to the nature of the work, certain items from the *Mechanical* and *Electrical Schedules* have also been used.

USERS ARE REMINDED THAT THE NATIONAL SCHEDULE IS NOT A PRICING BOOK; FURTHER, THEY ARE ADVISED TO TAKE A SELECTION OF RATES IN THE NATIONAL SCHEDULE AND TEST THEM AGAINST THEIR OWN DATA ON A REGULAR BASIS.

Effective date of schedule

The *Access Audit Schedule* is published annually in August of each year to include the latest promulgated wage rates and prices for materials and plant applicable at that time.

Users are advised that the rates in this, the first issue of the *Access Audit Schedule* are deemed to be effective as from 1 August 2007.

MTC Contract

Users are to note that the 2006 version of the MTC contract (MTC 2006) has now been published. A number of changes have been introduced and consequently sections C-F have been revised to accommodate these. Where MTC 98 contracts are being used, users should refer to the prelims in the 2006 – 2007 version of the schedules

BASIS OF PRICES

Labour Rates

The labour rates have been calculated in accordance with the principles given in the Code of Estimating Procedure published by The Chartered Institute of Building, modified as necessary to suit the following criteria :

		BUILDING OPERATIVES	PLUMBERS	ELECTRICIANS
a	Basis	Current wages and allowances as agreed by CIJC	Current wages and allowances as agreed by JIB for Plumbing & Mechanical Engineering Services	Current wages and allowances as agreed by JIB for Electrical Contracting Industry
b	Labour rate	Basic rate (Guaranteed minimum earnings) + 16% site determined bonus	Basic rate + 30% bonus	Basic rate + 30% bonus
c	Additional payments	Skills allowance and tool money as appropriate	Responsibility allowance included for advanced plumber tool money as appropriate	
d	Gross hours worked	39 hour week	39 hour week	37½ hour week
e	Resultant nonproductive overtime	75 hours per annum	75 hours per annum	Nil
f	Sickness	3 days per annum without pay	see 'Holidays with pay'	see 'Holidays with pay'
g	Inclement weather	2% of total hours	2% of total hours	2% of total hours
h	Public holidays per annum	8 days	8 days	8 days

		BUILDING OPERATIVES	PLUMBERS	ELECTRICIANS
i	Annual holidays with pay	21 days duration Payment as annual holidays agreement	21 days duration Published payments for HWP and sickness benefits	21 days duration Published payments for HWP and sickness benefits
j	National Insurance	£0.00 (on earnings up to and including £97.01 per week.) 12.8% on earnings above £97.01 per week.	£0.00 (on earnings up to and including £97.01 per week.) 12.8% on earnings above £97.01 per week.	£0.00 (on earnings up to and including £97.01 per week.) 12.8% on earnings above £97.01 per week.
k	Industry pension scheme	Nil	6.50%	Nil
l	CITB levy	0.50% on earnings	Nil	Nil
m	Severance pay	2%	2%	2%
n	Employer's liability and third party insurance	2%	2%	2%
o	Trade supervision	6%	6%	6%

Calculation of listed labour rates

The labour rates listed under BASIC PRICES are calculated as follows. This information is given to assist the USER in making any adjustments that are deemed necessary to suit the particular circumstances that are applicable.

Building operatives

Hours worked in full year	1802	1802
Less inclement weather 2%	36	
Productive hours	1766	
Non-productive overtime		75
Public holidays (8 x 7.8)		62.5
Hours for which payment is made		1940

	General Labourer £	Craftsman (Bricklayer) £
Basic rate per week (Rate as per employers offer March 2006)	285.09	379.08
Additional payments for skills - see notes	N/A	N/A
Cost per week	285.09	379.08
Cost per hour (divide by 39)	7.31	9.72
Add site determined bonus 16%	1.17	1.56
Cost per hour	£ 8.48	£ 11.27

© AAS 2008 - 2009

Building Operatives

Build-up of all-in labour rates per annum

				General Labourer £	Craftsman (Bricklayer) £
Earnings hrs	1940	x £ 8.48		16,450.42	-
		x £ 11.27		-	21,873.89
Payment for sickness (7 days with pay)*				141.53	141.53
Sub-total (A)				16,591.95	22,015.42
National Insurance : secondary contribution (Employer) (rates as from April 2005)					
0% on first £97.01 per week				-	-
12.8% on remainder		12.8% on £11,547.43		1,478.07	
		12.8% on £16,970.90			2,172.27
CITB Levy - 0.50% on (A)				82.96	110.08
Annual holiday template Payment 164 hours		x £8.48		1,390.65	
		x £11.27			1,849.13
Total				19,543.64	26,146.90
Allowance for severance pay		2%		390.87	522.94
Employer's liability		2%		390.87	522.94
Trade supervision		6%		1172.62	1,568.81
Total cost per annum				£ 21,498.00	£ 28,761.59
Rate per hour (Divide by productive hours 1766)				£ 12.17	£ 16.29

Notes :

The foregoing calculations apply to General Labourers and Bricklayers. Rates for other Tradesmen (e.g. Banksmen, Timbermen, Roofers, etc) are built up in a similar way by incorporating the appropriate allowances and extra payments with respect to the following:

(i) Continuous extra skill or responsibility, in accordance with Working Rule 1.2.2

(ii) As from 22 June 1995 the London addition has been abolished for craft operatives and adult labourers

(iii) As from June 1997 the tools and clothing allowance is now included in the Basic rate

It should also be noted that as from April 2002 the CITB Levy is now calculated as a percentage (0.50%) of payroll

Plumbers

The productive hours worked by plumbers have been taken as for the Building Operatives.

Hours worked in full year	1802	1802
Less inclement weather 2%	36	
Productive hours	1766	
Non-productive overtime		75
Public holidays (8 x 7.8 hours)		62.5
Hours for which payment is made		1940

	Advanced Plumber £	Trained Plumber £
Basic rate per hour	11.60	9.94
Additional payments for responsibility	0.44	N/A
Sub-total	12.04	9.94
Add bonus 30%	3.612	2.98
Total Cost per hour	£ 15.652	£ 12.92

Plumbers

Build-up of all-in labour rates per annum

				Advanced Plumber £	Trained Plumber £
Earnings	1940 hrs	x £ 15.652		30,364.88	-
		X £ 12.92		-	25,068.68
Sub-total (A)				30,364.88	25,068.68
National Insurance : secondary contribution (Employer) (rates as from April 2006)					
0% on first £97.10 per week				-	-
12.8% on remainder		12.8% on £25,322.36		3,241.01	
		12.8% on £20,024.16			2,563.09
Industry pension 6.5% of (A)				1,973.72	1,629.46
Annual holiday credit and sickness benefit stamp					
	52 weeks	x £35.10		1,825.20	
		x £31.21			1,622.92
Sub-total				37,404.80	30,884.16
Allowance for severance pay			2%)	748.10	617.68
Employer's liability			2%)	748.10	617.68
Trade supervision			6%	2,244.29	1,853.05
Total cost per annum			£	41,145.28	£ 33,972.57
Rate per hour (Divide by productive hours 1766)			£	23.30	£ 19.24

Electricians

The hours worked by Electricians are based on a standard working week without any overtime, as follows.

Hours worked in full year	1732.50	1732.50
Less inclement weather 2%	34.65	
Productive hours	1697.85	
Public holidays (8 x 7.5 hours)		60.00
Hours for which payment is made		1792.50

	Electrician £	Electrician's Labourer £
Basic rate per hour	12.58	9.30
Add bonus 30%	3.774	2.79
Total Cost per hour	£ 16.354	£ 12.090

Electricians

Build-up of all-in labour rates per annum

			Electrician £	Electrician's Labourer £
Earnings	1792.5 hrs	x £ 16.354	29,314.55	-
		x £ 12.09	-	21,671.33
National Insurance : secondary contribution (Employer) (rates as from April 2001)				
0% on first £97.01 per week			-	-
12.8% on remainder		12.8% on £24,270.03	3,106.56	
		12.8% on £16,626.81		2,128.23
Industry Pension 6.5%			1,905.45	1,408.64
Combined benefit stamp + 'Top up' 52 weeks		x £40.35	2,098.20	-
		x £37.26	-	1,937.52
Total			36,424.75	27,145.71
Allowance for severance pay		2%	728.50	542.91
Employer's liability		2%	728.50	542.91
Trade supervision		6%	2,185.49	1,628.74
Total cost per annum			£ 40,067.23	£ 29,860.28
Rate per hour (Divide by productive hours 1698)			£ 23.60	£ 17.59

Materials Costs

Appendix B lists the firms who have supplied the prices of materials used in the calculation of the unit rates. It is emphasised that the prices thus obtained are only indicative and the User should satisfy himself as to their validity.

Allowances have been made for waste and no account has been taken of discounts available under annual purchasing agreements.

Plant Costs

Plant costs have been obtained wherever possible from HSS 2007 directory of hire rates.

QUANTITIES, RATES, EXTENSIONS AND TOTALS

All quantities, rates, extensions and totals are to be kept to two places of decimals.

DESCRIPTIONS

The National Schedule is firmly based on the Standard Method of Measurement of Building Works, 7th Edition Revised 1998 (SMM7R). This implies that any disputes over measurement could be directly related to the Method and any rulings on the Method would apply to the Schedule.

Certain minor departures from SMM7 have perforce been necessary, for example, in order to keep the same item numbering E10 is described as In situ concrete whereas SMM7R has now split this into E05, E10 and E11, other departures are detailed under Measurement and Pricing Preambles and in appendix A

In addition, the National Schedule has been compiled using British Standard spellings, for example "lintel", and British Standard nomenclatures, for example of timber.

In order to keep the National Schedule simple and to enable items of a complex nature to be sensibly measured, the composite items, particularly those that encompass more than one section, have been dissected into their components. This avoids the proliferation of items that would otherwise be caused by minor variations in the parts of a composite item. It also means that a more logical aggregate rate, based on the component parts of a description, can be applied.

COMPUTER-BASED LIBRARY

The National Schedule is fully computer-based. This means that items of work are produced by using the computer to generate descriptions from a custom-compiled Library placed in its memory. This Library is divided into Sections corresponding to SMM7. Each of these Sections, e.g. In situ concrete/Large precast concrete, is designated by its SMM7 work section letter. A detailed description of the logic and function of the Library is given in Appendix A.

The benefits from using a Library are:

    the codes are identifiable in the Schedule:

    coding is permanent, compact and does not change with amendments or updates to the Schedule;

    several possibilities for amending the Schedule are available : these are described in Appendix A.

COMPILER'S NOTES - APPENDIX A

COMPUTER-BASED LIBRARY

This library is divided into Work Groups corresponding to SMM7. Each of these is designated by its SMM7 Work Groups letter. For example, the Work Group of **In-situ concrete/Large precast concrete** bears the Letter E. Work Groups are further broken down into Work Sections to which the relevant SMM7 work section number is appended. By this method a unique code is allocated to each sub-section. The Work Section **In-situ concrete** for example, is labelled E10.

Each Work Section contains two levels of descriptive text : Combined Headings (CHs) and Work Heads (WHs). CHs are numbered from 01-99 and WHs from 001-999. For every full description, consisting of a Combined Heading and Work Head, there is therefore a unique fixed-length code in the form A0000/000. In practice each separate part of a full description is identified in the Schedule by its own part-code.

EXAMPLE OF LIBRARY

The following examples show how the Library works :

Work Groups are defined by a single letter, e.g.

D Groundwork
E In-situ concrete/Large precast concrete
F Masonry
G Structural/Carcassing metal/timber

and so on (see Contents for full list)

Each Work Group is made up of Work Sections and each Work Section is defined by a number within the Group letter. For example, Work Group E (In situ concrete) consists of the following Work Sections: :

E10 In-situ concrete
E20 Formwork for in-situ concrete
E30 Reinforcement for in-situ concrete

and so on (see Contents for full list)

Each Work Section consists of Combined Headings and Work Heads (item descriptions), e.g. Work Section E10 (In-situ Concrete) contains Combined Headings :

10 Plain prescribed mix ST4
25 Plain prescribed mix ST5

and so on

followed by Work Heads

10   Foundations ne 150 mm thick poured on or against earth or unblinded hardcore

15   Foundations 150 mm to 450 mm thick poured on or against earth or unblinded hardcore

20   Foundations over 450 mm thick poured on or against earth or unblinded hardcore

30   Isolated foundations ne 150 mm thick poured on or against earth or unblinded hardcore

35   Isolated foundations 150 mm to 450 mm thick poured on or against earth or unblinded hardcore

and so on

© AAS 2008– 2009

These codes and texts appear in the Schedule as follows :

    E :   **In situ concrete/Large precast concrete** (Work Group title)

    E10:   **In situ concrete** (Work Section E10)

    E1010 Plain described mix C20 (Combined Heading 10)

        010   Foundations ne 150 mm thick poured on or against earth or unblinded hardcore (Work Head 010)

    E1025 Plain described mix C25 (Combined Heading 25)
        010   Foundations ne 150 mm thick poured on or against earth or unblinded hardcore (Work Head 010)

It will be noticed that identical descriptions, for example, "Foundations ne 150 mm thick poured on or against earth or unblinded hardcore" above (Work Heads) are produced by the same code irrespective of the headings under which they lie.

The Library contains a number of headings and items that, for compactness, it has been decided not to call up for use in the Schedule. These provide scope for future amendments to the Schedule.
Several possibilities for future amendment of the Schedule are available :-

- additional coding can produce an item not currently printed in the Schedule by a combination of CHs and WHs already in the Library. Such an item could be said to be in the form

    existing Work Section/existing CH/existing WH

- by amending the **Library**, a simple process, other combinations can be used to produce any heading or item desired in the alternative forms

    existing Work Section/existing CH/<u>new</u> WH
    existing Work Section/<u>new</u> CH/existing WH
    existing Work Section/<u>new</u> CH/<u>new</u> WH

- a <u>new</u> Work Section could use any of the four foregoing combinations;

in every case a unique code will be produced so that the required new item goes into its proper place in the Schedule and keeps its number in perpetuity.

Users, introducing their own descriptions into the Schedule for their own use, are advised not to use the omitted numbers in the published Schedule since these "missing" numbers may well be taken up by the Co-Authors in the future. This could lead to an obvious conflict but can easily be avoided if users introduce their own references in the form of suffix letters,

    e.g.   E1010/010A to define a new item (unique to the user organisation) which requires to be entered into the Schedule for that organisation immediately following the standard published item coded E1010/010

Further information on this subject can be obtained from the Technical Secretary.

COMPILER'S NOTES - APPENDIX B

SOURCES OF INFORMATION

Acknowledgements are extended to the following firms who have supplied information or advice in the compilation of the Access Audit Schedule:-

HSS Hire shops 25 Willow Lane, Mitcham, Surrey, CR4 4TS	020 8260 5005
Gradus Accessories Park Green Macclesfield Cheshire SK11 7LZ	01625 428922
Kee Klamps Ltd 1 Bolton Road, Reading, Berkshire RG2 0NH	01734 311022
Nicholls & Clarke Ltd Unit 10, Nelson Trading Estate, Morden Road, Wimbledon, London, SW19 3BL	020 8540 1141
Travis Perkins Lodge way House, Lodge Way Harlestone Road, Northampton NN5 7UG	0870 333 2111
Configure The Old School House Forshaw Heath Lane Earlswood Solihull B94 5LH	0870 066 1880

© AAS 2008– 2009

APPENDIX C

Schedule of waste allowances (not exhaustive)

Generally all items of furniture or fittings etc, which are enumerated have no wastage allowance. Items which are machined, cut etc or the full pack quantity is not used e.g. paint, adhesives etc have wastage factors.

Timber	10%	
Sheet Material	10%	
Pipes, gutters etc	5%	
Insulation	5%	
Wood filler and the like	5%	
Roofing tiles, slates etc.	5%	
Nails, screws, bolts etc	5%	
Fencing materials	5%	
Timber posts	10%	(Where measured in ml)
Aggregates	5%	
Reinforcement	5%	
Plasterboards	5%	
Plasters	10%	
Angle beads etc	5%	
Bricks/blocks	5%	
DPC/Mortars/concrete	10%	
Joint Fillers/mastics etc	5%	
Paints	5%	
Wallpapers	15%	
Ceramic/quarry tiles	5%	
Flooring	5%	
Roofing felts	10%	
Asphalt/liquid roofing products	5%	
Glass	10%	
Window sections	10%	(Where measure in ml)

PART 2

Global Schedule of Rates

The Global Schedule of Rates computer system can be used on Measured Term Contracts to which any of the NSR Schedule of Rates apply or any other Schedule of Rates. Addendum Schedules are easily added.

For a Works Order or Estimate, the Schedule is either searched on screen and description selected or the required code number is typed in, followed by full dimensions or quantity. Non Schedule codes, such as Star Rates, Net Rates, Dayworks, Invoices, etc., are easily added to a measure. When the completed Order is displayed, dimensions are squared, the resultant quantity is multiplied by the rate and a total Order value generated. Relevant percentages are automatically applied including the Contractor's percentage.

Many reports are available enabling tight control of the contract. These include;
Orders issued but not completed.
Orders issued, completed or billed within a given period.
Orders processed for a selected Building.
Orders completed either within or outside a defined period.
Interim Payments.
How often a particular Nsr code or section of codes has been used on a Contract.
Order printout using, instead of total rate, plant, labour or material rates.

Order details can also be exported to Microsoft Excel and Word or Emailed to another office.

The software will run on any of the Full versions of Windows.

For further information;

Email: john@barcellos.co.uk

Telephone: 0116 233 5559   Fax: 0116 233 5560

Address: Barcellos Limited, Sandbach House, 8 Salisbury Road, Leicester, LE1 7QR

Website: www.barcellos.co.uk

General Index

Part 2

Page

Schedule of Description and Prices

Y15	Car Parking	Y/1
Y17	Footpaths	Y/11
Y18	Street Furniture	Y/19
Y19	Signage and Wayfinding (Externally)	Y/23
Y20	External Ramps and Steps	Y/29
Y21	Finishes and Surface Treatments (Externally)	Y/37
Y22	Windows, Doors and Doorways	Y/41
Y24	Internal Ramps and Steps	Y/63
Y25	Fixtures and Fittings	Y/69
Y26	Finishes and Surface Treatments (Internally)	Y/73
Y27	Decorating (Internally)	Y/85
Y28	Decorating (Externally)	Y/89
Y29	Signage and Wayfinding (Internally)	Y/91
Y30	Sanitary Conveniences	Y/95
Y32	General Building Works	Y/115
Y33	Mechanical and Electrical Alterations	Y/121
Y34	Lifts and Lifting Devices	Y/135

W T PARKER LIMITED

ELECTRICAL & MECHANICAL TERM CONTRACTORS TO DEFENCE ESTATES AND THE NHS

**W.T. Parker Ltd
24-28 Moor Street
Burton on Trent
Staffordshire
DE14 3SX
Tel: 01283 542661
Fax: 01283 536189
www.wtparker.co.uk**

			Mat. £	Lab. £	Plant £	Total £
	Y15 : CAR PARKING					
Y1510	**TARMACADAM**					
010	**Break out existing hardsurface and renew in tarmacadam n.e 5m2**					
	Coated macadam or asphalt paving 50 to 75 mm thick (C9020/335 - 1. m2), Compacting ground (Q0535/300 - 1. m2), Excavated material off site (Q1025/100 - 0.075 m3), Pavings to falls and crossfalls and to slopes ne 15 degrees from horizontal to blinded hardcore, in areas 2 to 5 m2 (Q2210/030 - 1. m2), Pavings 15 mm thick to falls and crossfalls and to slopes ne 15 degrees from horizontal to blinded hardcore, in areas 2 to 5 m2 (Q2220/120 - 1. m2)	m2	15.14	61.98	14.09	91.21
012	**Break out existing hardsurface and renew in tarmacadam exc 5m2**					
	Coated macadam or asphalt paving 50 to 75 mm thick (C9020/335 - 1 m2), Compacting ground (Q0535/300 - 1 m2), Excavated material off site (Q1025/100 - 0.075 m3), Pavings to falls and crossfalls and to slopes ne 15 degrees from horizontal to blinded hardcore, in areas 10 to 25 m2 (Q2210/020 - 1 m2), Pavings 15 mm thick to falls and crossfalls and to slopes ne 15 degrees from horizontal to blinded hardcore, in areas 10 to 25 m2 (Q2220/110 - 1 m2)	m2	12.98	47.20	12.64	72.82
015	**Resurface tarmac car park with 30mm wearing course**					
	Coated macadam or asphalt paving wearing course ne 25 mm thick (C9020/325 - 1 m2), Pavings 25 mm thick to falls and crossfalls and to slopes ne 15 degrees from horizontal to blinded hardcore, in areas 10 to 25 m2 (Q2220/140 - 1.2 m2)	m2	8.27	21.44	4.06	33.77
020	**Create new tarmacadam parking bay in existing landscaped area; including International Access Symbol stencilled in bay and non reflective International Access Symbol sign, screw fixed to wall**					
	Clearing site vegetation (D2010/250 - 18 m2), Concrete kerbs 155 x 255 mm complete with bed, haunching and the like (D2015/225 - 4 m), Yellow permanent paint line marking to tarmac/concrete/block paved surface Disabled access symbol (406mm) (M6175/135 - 1 nr), 1.5mm rigid plastic sign size 297 x 210mm ; screw fixed to wall (N1510/140 - 1 nr), To reduce levels 0.25 m max depth (Q0515/025 - 5.7 m3), Excavated material off site (Q0530/150 - 5.7 m3), Filling to excavations ne 250 mm average thick (Q0532/200 - 4.32 m3), Compacting ground (Q0540/300 - 18 m2), Compacting filling (Q0540/302 - 18 m2), Trenches ne 300 mm wide 0.25 m max depth (Q1015/010 - 2.04 m3), Excavated material off site (Q1027/100 - 2.04 m3), Kerbs 125 x 255 mm half battered type HB2, foundation 450 x 150 mm, haunching one side (Q1050/200 - 13 m), Kerbs 125 x 255 mm half battered type HB2, foundation 450 x 150 mm haunching one side curved (Q1050/205 - 1 m), Extra over for dropped kerbs (Q1050/207 - 4 nr), Pavings to falls and crossfalls and to slopes ne 15 degrees from horizontal to blinded hardcore, in areas 10 to 25 m2 (Q2210/020 - 18 m2), Pavings 15 mm thick to falls and crossfalls and to slopes ne 15 degrees from horizontal to blinded hardcore, in areas 10 to 25 m2 (Q2220/110 - 18 m2)	Item	714.07	956.42	591.65	2262.14

© NSR 01 Aug 2008 - 31 Jul 2009

Y1

	Y15 : CAR PARKING		Mat. £	Lab. £	Plant £	Total £
Y1510						
023	**Create new tarmacadam parking bay in existing highway; including International Access Symbol stencilled in bay and non reflective International Access Symbol sign, screw fixed to wall**					
	Clearing site vegetation (D2010/250 - 18 m2), Concrete kerbs 155 x 255 mm complete with bed, haunching and the like (D2015/225 - 7 m), Extra over excavation irrespective of depth for breaking out coated macadam or asphalt paving ne 75 mm thick (D2020/815 - 12 m2), Yellow permanent paint line marking to tarmac/concrete/block paved surface Disabled access symbol (406mm) (M6175/135 - 1 nr), 1.5mm rigid plastic sign size 297 x 210mm ; screw fixed to wall (N1510/140 - 1 nr), To reduce levels 0.25 m max depth (Q0515/025 - 5.7 m3), Excavated material off site (Q0530/150 - 5.7 m3), Filling to excavations ne 250 mm average thick (Q0532/200 - 4.32 m3), Compacting ground (Q0540/300 - 18 m2), Compacting filling (Q0540/302 - 18 m2), Trenches ne 300 mm wide 0.25 m max depth (Q1015/010 - 2.04 m3), Excavated material off site (Q1027/100 - 2.04 m3), Kerbs 125 x 255 mm half battered type HB2, foundation 450 x 150 mm, haunching one side (Q1050/200 - 13 m), Kerbs 125 x 255 mm half battered type HB2, foundation 450 x 150 mm haunching one side curved (Q1050/205 - 1 m), Extra over for dropped kerbs (Q1050/207 - 4 nr), Pavings to falls and crossfalls and to slopes ne 15 degrees from horizontal to blinded hardcore, in areas 10 to 25 m2 (Q2210/020 - 18 m2), Pavings 15 mm thick to falls and crossfalls and to slopes ne 15 degrees from horizontal to blinded hardcore, in areas 10 to 25 m2 (Q2220/110 - 18 m2)	Item	714.07	1084.37	602.84	2401.28
025	**extra over parking bay for post mounted aluminium or plastic sign, in lieu of wall mounted sign, setting post in concrete**					
	Standard warning sign on post, post set in concrete mix C15, excavating and removing excavated material from site (Q5010/055 - 1 nr)	nr	68.92	21.98		90.90
Y1520	**CONCRETE**					
030	**Break out existing hardsurface and renew in reinforced concrete 150mm thick in areas n.e. 5m2**					
	Slabs ne 150 mm thick (E1041/355 - 0.15 m3), Square mesh ref A142 2.22 kg/m2 (E3030/210 - 1 m2), To reduce levels 0.25 m max depth (Q0510/025 - 0.15 m3), Extra over excavation irrespective of depth for breaking out concrete (Q0510/065 - 0.15 m3), Compacting ground (Q0535/300 - 1 m2), Excavated material off site (Q1025/100 - 0.15 m3), Tamping by mechanical means to falls (Q2140/205 - 1 m2)	m2	19.42	49.50	10.90	79.82
032	**Break out existing hardsurface and renew in reinforced concrete 150mm thick in areas exc. 5m2**					
	Beds 150 mm to 450 mm thick (E1040/335 - 0.15 m3), Square mesh ref A142 2.22 kg/m2 (E3030/210 - 1 m2), Extra over excavation irrespective of depth for breaking out concrete (Q0510/065 - 0.15 m3), To reduce levels 0.25 m max depth (Q0515/025 - 0.15 m3), Compacting ground (Q0535/300 - 1 m2), Excavated material off site (Q1025/100 - 0.15 m3), Tamping by mechanical means to falls (Q2140/205 - 1 m2)	m2	18.16	42.82	11.15	72.13

© NSR 01 Aug 2008 - 31 Jul 2009   Y2

	Y15 : CAR PARKING		Mat. £	Lab. £	Plant £	Total £
Y1520						
035	**Create new concrete parking bay in existing landscaped area; including International Access Symbol stencilled in bay and non reflective International Access Symbol sign, screw fixed to wall**					
	Clearing site vegetation (D2010/250 - 18 m2), Concrete kerbs 155 x 255 mm complete with bed, haunching and the like (D2015/225 - 4 m), Yellow permanent paint line marking to tarmac/concrete/block paved surface Disabled access symbol (406mm) (M6175/135 - 1 nr), 1.5mm rigid plastic sign size 297 x 210mm ; screw fixed to wall (N1510/140 - 1 nr), To reduce levels 0.25 m max depth (Q0515/025 - 5.7 m3), Excavated material off site (Q0530/150 - 5.7 m3), Filling to excavations ne 250 mm average thick (Q0532/200 - 4.32 m3), Compacting ground (Q0540/300 - 18 m2), Compacting filling (Q0540/302 - 18 m2), Trenches ne 300 mm wide 0.25 m max depth (Q1015/010 - 2.04 m3), Excavated material off site (Q1027/100 - 2.04 m3), Kerbs 125 x 255 mm half battered type HB2, foundation 450 x 150 mm, haunching one side (Q1050/200 - 13 m), Kerbs 125 x 255 mm half battered type HB2, foundation 450 x 150 mm haunching one side curved (Q1050/205 - 1 m), Extra over for dropped kerbs (Q1050/207 - 4 nr), Beds ne 150 mm thick (Q2110/010 - 3.9 m3), Tamping by mechanical means (Q2140/200 - 26 m2)	Item	915.48	669.32	565.91	2150.71
037	**Create new concrete parking bay in existing highway; including International Access Symbol stencilled in bay and non reflective International Access Symbol sign, screw fixed to wall**					
	Clearing site vegetation (D2010/250 - 18 m2), Concrete kerbs 155 x 255 mm complete with bed, haunching and the like (D2015/225 - 7 m), Extra over excavation irrespective of depth for breaking out concrete paving ne 100 mm thick (D2020/800 - 26 m2), Extra for each additional 25 mm thickness (D2020/801 - 52 m2), Yellow permanent paint line marking to tarmac/concrete/block paved surface Disabled access symbol (406mm) (M6175/135 - 1 nr), 1.5mm rigid plastic sign size 297 x 210mm ; screw fixed to wall (N1510/140 - 1 nr), To reduce levels 0.25 m max depth (Q0515/025 - 5.7 m3), Excavated material off site (Q0530/150 - 5.7 m3), Filling to excavations ne 250 mm average thick (Q0532/200 - 4.32 m3), Compacting ground (Q0540/300 - 18 m2), Compacting filling (Q0540/302 - 18 m2), Trenches ne 300 mm wide 0.25 m max depth (Q1015/010 - 2.04 m3), Excavated material off site (Q1027/100 - 2.04 m3), Kerbs 125 x 255 mm half battered type HB2, foundation 450 x 150 mm, haunching one side (Q1050/200 - 13 m), Kerbs 125 x 255 mm half battered type HB2, foundation 450 x 150 mm haunching one side curved (Q1050/205 - 1 m), Extra over for dropped kerbs (Q1050/207 - 4 nr), Beds ne 150 mm thick (Q2110/010 - 3.9 m3), Tamping by mechanical means (Q2140/200 - 26 m2)	Item	915.48	943.87	577.10	2436.45
038	**extra over parking bay for post mounted aluminium or plastic sign, in lieu of wall mounted sign, setting post in concrete**					
	Standard warning sign on post, post set in concrete mix C15, excavating and removing excavated material from site (Q5010/055 - 1 nr)	nr	68.92	21.98		90.90

© NSR 01 Aug 2008 - 31 Jul 2009   Y3

	Y15 : CAR PARKING		Mat. £	Lab. £	Plant £	Total £
Y1530	**BLOCK PAVIOURS**					
040	**Break out existing hardsurface and renew in concrete block paviours in areas n.e. 5m2**					
	To reduce levels 0.25 m max depth (Q0510/025 - 0.15 m3), Extra over excavation irrespective of depth for breaking out concrete (Q0510/065 - 0.15 m3), Compacting ground (Q0535/300 - 1 m2), Excavated material off site (Q1025/100 - 0.15 m3), Filling to make up levels ne 250 mm av thick, compacting (Q2010/010 - 0.075 m3), Pavings to falls and crossfalls and to slopes ne 15 degrees from horizontal in areas ne 2 m2 (Q2540/025 - 1 m2)	m2	39.05	66.56	8.99	114.60
042	**Break out existing hardsurface and renew in concrete block paviours in areas exc. 5m2**					
	To reduce levels 0.25 m max depth (Q0515/025 - 0.15 m3), Extra over excavation irrespective of depth for breaking out concrete (Q0515/065 - 0.15 m3), Compacting ground (Q0535/300 - 1 m2), Excavated material off site (Q1025/100 - 0.15 m3), Filling to make up levels ne 250 mm av thick, compacting (Q2010/010 - 0.075 m3), Pavings to falls and crossfalls and to slopes ne 15 degrees from horizontal (Q2540/020 - 1 m2)	m2	32.89	42.25	14.69	89.83
044	**Lift existing loose/broken block paving and relay in areas n.e. 5m2**					
	Brick block stone paving 70 mm thick, storing for re-use (C9020/320 - 1 m2), Pavings level or to falls in areas ne 2 m2 (C9047/377 - 1 m2)	m2	3.23	50.91	1.79	55.93
046	**Lift existing loose/broken block paving and relay in areas exc. 5m2**					
	Brick block stone paving 70 mm thick, storing for re-use (C9020/320 - 1 m2), Pavings level or to falls (C9047/375 - 1 m2)	m2	3.23	40.74	1.24	45.21
048	**Create new parking bay in existing landscaped area; including International Access Symbol stencilled in bay and non reflective International Access Symbol sign, screw fixed to wall**					
	Clearing site vegetation (D2010/250 - 18 m2), Concrete kerbs 155 x 255 mm complete with bed, haunching and the like (D2015/225 - 4 m), Yellow permanent paint line marking to tarmac/concrete/block paved surface Disabled access symbol (406mm) (M6175/135 - 1 nr), 1.5mm rigid plastic sign size 297 x 210mm ; screw fixed to wall (N1510/140 - 1 nr), To reduce levels 0.25 m max depth (Q0515/025 - 5.7 m3), Excavated material off site (Q0530/150 - 5.7 m3), Filling to excavations ne 250 mm average thick (Q0532/200 - 4.32 m3), Compacting ground (Q0540/300 - 18 m2), Compacting filling (Q0540/302 - 18 m2), Trenches ne 300 mm wide 0.25 m max depth (Q1015/010 - 2.04 m3), Excavated material off site (Q1027/100 - 2.04 m3), Kerbs 125 x 255 mm half battered type HB2, foundation 450 x 150 mm, haunching one side (Q1050/200 - 13 m), Kerbs 125 x 255 mm half battered type HB2, foundation 450 x 150 mm haunching one side curved (Q1050/205 - 1 m), Extra over for dropped kerbs (Q1050/207 - 4 nr), Pavings level or to falls (Q2540/010 - 26 m2)	Item	1260.69	1008.18	570.07	2838.94

© NSR 01 Aug 2008 - 31 Jul 2009

Y4

	Y15 : CAR PARKING		Mat. £	Lab. £	Plant £	Total £
Y1530						
049	**Create new parking bay in existing highway; including International Access Symbol stencilled in bay and non reflective International Access Symbol sign, screw fixed to wall**					
	Clearing site vegetation (D2010/250 - 18 m2), Concrete kerbs 155 x 255 mm complete with bed, haunching and the like (D2015/225 - 7 m), Extra over excavation irrespective of depth for breaking out coated macadam or asphalt paving ne 75 mm thick (D2020/815 - 12 m2), To reduce levels 0.25 m max depth (Q0515/025 - 5.7 m3), Excavated material off site (Q0530/150 - 5.7 m3), Filling to excavations ne 250 mm average thick (Q0532/200 - 4.32 m3), Compacting ground (Q0540/300 - 18 m2), Compacting filling (Q0540/302 - 18 m2), Trenches ne 300 mm wide 0.25 m max depth (Q1015/010 - 2.04 m3), Excavated material off site (Q1027/100 - 2.04 m3), Kerbs 125 x 255 mm half battered type HB2, foundation 450 x 150 mm, haunching one side (Q1050/200 - 13 m), Kerbs 125 x 255 mm half battered type HB2, foundation 450 x 150 mm haunching one side curved (Q1050/205 - 1 m), Extra over for dropped kerbs (Q1050/207 - 4 nr), Pavings level or to falls (Q2540/010 - 26 m2)	Item	1219.89	1125.47	580.98	2926.34
050	**extra over parking bay for post mounted aluminium or plastic sign, in lieu of wall mounted sign, setting post in concrete**					
	Standard warning sign on post, post set in concrete mix C15, excavating and removing excavated material from site (Q5010/055 - 1 nr)	nr	68.92	21.98		90.90
Y1540	BRICK PAVIOURS					
058	**Break out existing hardsurface and renew in clay paviours in areas n.e. 5m2**					
	To reduce levels 0.25 m max depth (Q0510/025 - 0.075 m3), Extra over excavation irrespective of depth for breaking out concrete (Q0510/065 - 0.15 m3), To reduce levels 0.25 m max depth (Q0515/025 - 0.075 m3), Compacting ground (Q0535/300 - 1 m2), Excavated material off site (Q1025/100 - 0.15 m3), Filling to make up levels ne 250 mm av thick, compacting (Q2010/010 - 0.075 m3), Pavings to falls and crossfalls and to slopes ne 15 degrees from horizontal in areas ne 2 m2 (Q2545/025 - 1 m2)	m2	59.43	64.11	9.25	132.79
060	**Break out existing hardsurface and renew in clay paviours in areas exc 5m2**					
	Extra over excavation irrespective of depth for breaking out concrete (Q0510/065 - 0.15 m3), To reduce levels 0.25 m max depth (Q0515/025 - 0.15 m3), Compacting ground (Q0535/300 - 1 m2), Excavated material off site (Q1025/100 - 0.15 m3), Filling to make up levels ne 250 mm av thick, compacting (Q2010/010 - 0.075 m3), Pavings to falls and crossfalls and to slopes ne 15 degrees from horizontal (Q2545/020 - 1 m2)	m2	49.87	51.49	8.95	110.31
062	**Lift existing loose/broken clay brick paving and relay in areas n.e. 5m2**					
	Brick block stone paving 70 mm thick, storing for re-use (C9020/320 - 1 m2), Pavings level or to falls in areas ne 2 m2 (C9047/377 - 1 m2)	m2	3.23	50.91	1.79	55.93

© NSR 01 Aug 2008 - 31 Jul 2009   Y5

	Y15 : CAR PARKING		Mat. £	Lab. £	Plant £	Total £
Y1540						
063	**Lift existing loose/broken block paving and relay in areas exc. 5m2**					
	Brick block stone paving 70 mm thick, storing for re-use (C9020/320 - 1 m2), Pavings level or to falls (C9047/375 - 1 m2)	m2	3.23	40.74	1.24	45.21
065	**Create new parking bay in existing landscaped area; including International Access Symbol stencilled in bay and non reflective International Access Symbol sign, screw fixed to wall**					
	Clearing site vegetation (D2010/250 - 18 m2), Concrete kerbs 155 x 255 mm complete with bed, haunching and the like (D2015/225 - 4 m), Yellow permanent paint line marking to tarmac/concrete/block paved surface Disabled access symbol (406mm) (M6175/135 - 1 nr), 1.5mm rigid plastic sign size 297 x 210mm ; screw fixed to wall (N1510/140 - 1 nr), To reduce levels 0.25 m max depth (Q0515/025 - 5.7 m3), Excavated material off site (Q0530/150 - 5.7 m3), Filling to excavations ne 250 mm average thick (Q0532/200 - 4.32 m3), Compacting ground (Q0540/300 - 18 m2), Compacting filling (Q0540/302 - 18 m2), Trenches ne 300 mm wide 0.25 m max depth (Q1015/010 - 2.04 m3), Excavated material off site (Q1027/100 - 2.04 m3), Kerbs 125 x 255 mm half battered type HB2, foundation 450 x 150 mm, haunching one side (Q1050/200 - 13 m), Kerbs 125 x 255 mm half battered type HB2, foundation 450 x 150 mm haunching one side curved (Q1050/205 - 1 m), Extra over for dropped kerbs (Q1050/207 - 4 nr), Pavings level or to falls (Q2545/010 - 26 m2)	Item	1702.17	1008.18	570.07	3280.42
067	**Create new parking bay in existing highway; including International Access Symbol stencilled in bay and non reflective International Access Symbol sign, screw fixed to wall**					
	Clearing site vegetation (D2010/250 - 18 m2), Concrete kerbs 155 x 255 mm complete with bed, haunching and the like (D2015/225 - 7 m), Extra over excavation irrespective of depth for breaking out coated macadam or asphalt paving ne 75 mm thick (D2020/815 - 12 m2), Yellow permanent paint line marking to tarmac/concrete/block paved surface Disabled access symbol (406mm) (M6175/135 - 1 nr), 1.5mm rigid plastic sign size 297 x 210mm ; screw fixed to wall (N1510/140 - 1 nr), To reduce levels 0.25 m max depth (Q0515/025 - 5.7 m3), Excavated material off site (Q0530/150 - 5.7 m3), Filling to excavations ne 250 mm average thick (Q0532/200 - 4.32 m3), Compacting ground (Q0540/300 - 18 m2), Compacting filling (Q0540/302 - 18 m2), Trenches ne 300 mm wide 0.25 m max depth (Q1015/010 - 2.04 m3), Excavated material off site (Q1027/100 - 2.04 m3), Kerbs 125 x 255 mm half battered type HB2, foundation 450 x 150 mm, haunching one side (Q1050/200 - 13 m), Kerbs 125 x 255 mm half battered type HB2, foundation 450 x 150 mm haunching one side curved (Q1050/205 - 1 m), Extra over for dropped kerbs (Q1050/207 - 4 nr), Pavings level or to falls (Q2545/010 - 26 m2)	Item	1702.17	1136.13	581.26	3419.56
068	**extra over parking bay for post mounted aluminium or plastic sign, in lieu of wall mounted sign, setting post in concrete**					
	Standard warning sign on post, post set in concrete mix C15, excavating and removing excavated material from site (Q5010/055 - 1 nr)	nr	68.92	21.98		90.90

© NSR 01 Aug 2008 - 31 Jul 2009   Y6

	Y15 : CAR PARKING		Mat. £	Lab. £	Plant £	Total £
Y1550	**KERBS**					
070	**Break out existing kerbs and replace with concrete dropped/flush kerbs**					
	Concrete kerbs 155 x 255 mm complete with bed, haunching and the like (D2015/225 - 1 m), Compacting ground (Q0535/300 - 1 m2), Excavated material off site (Q1025/100 - 0.15 m3), Kerbs 125 x 150 mm half battered flush, foundation 450 x 150 mm, haunching one side (Q1050/209 - 1 m)	m	13.38	30.24	10.93	54.55
072	**Excavate and lay new 125 x 255mm concrete kerb (Plain)**					
	Compacting ground (Q0535/300 - 1 m2), Trenches ne 300 mm wide 1 m max depth (Q1010/012 - 0.067 m3), Excavated material off site (Q1025/100 - 0.15 m3), Kerbs 125 x 255 mm half battered type HB2, foundation 450 x 150 mm, haunching one side (Q1050/200 - 1 m)	m	13.76	17.11	7.20	38.07
073	**Excavate and lay new 125 x 150mm concrete kerb (flush)**					
	Compacting ground (Q0535/300 - 1 m2), Trenches ne 300 mm wide 0.25 m max depth (Q1010/010 - 0.075 m3), Excavated material off site (Q1025/100 - 0.075 m3), Kerbs 125 x 150 mm half battered flush, foundation 450 x 150 mm, haunching one side (Q1050/209 - 1 m)	m	13.38	16.01	3.60	32.99
075	**Excavate and lay new 125 x 255mm concrete kerb (Plain curved)**					
	Trenches ne 300 mm wide 1 m max depth (Q1010/012 - 0.068 m3), Excavated material off site (Q1025/100 - 0.068 m3), Compacting bottoms of excavations (Q1030/150 - 0.45 m2), Kerbs 125 x 255 mm half battered type HB2, foundation 450 x 150 mm haunching one side curved (Q1050/205 - 1 m)	m	19.57	18.40	3.26	41.23
080	**Excavate and lay new stone kerb (Plain)**					
	Compacting ground (Q0535/300 - 0.45 m2), Trenches ne 300 mm wide 1 m max depth (Q1010/012 - 0.068 m3), Excavated material off site (Q1025/100 - 0.068 m3), Kerbs, stone PC £30/ml, foundation 450 x 150 mm, haunching one side (Q1050/215 - 1 m)	m	37.21	14.47	3.26	54.94
Y1560	**ROAD MARKING**					
090	**Burn off/obliterate existing road markings**					
	Burn off/obliterate existing road markings (C9025/365 - 1 m)	m	3.34	4.31	0.17	7.82
100	**Provide permanent paint line markings to road surface up to 50mm wide**					
	Yellow permanent paint line marking to tarmac/concrete/block paved surface up to 50mm wide (M6175/120 - 1 m)	m	0.44	0.86	0.08	1.38
110	**Provide permanent paint line markings to road surface 50 - 100mm wide**					
	Yellow permanent paint line marking to tarmac/concrete/block paved surface 50 - 100mm wide (M6175/130 - 1 m)	m	0.67	1.44	0.14	2.25

© NSR 01 Aug 2008 - 31 Jul 2009   Y7

	Y15 : CAR PARKING		Mat. £	Lab. £	Plant £	Total £
Y1560						
115	**Provide temporary paint line markings to road surface up to 50mm wide**					
	Yellow temporary paint line marking to tarmac/concrete/block paved surface up to 50mm wide (M6175/150 - 1 m)	m	0.13	0.86	0.08	1.07
120	**Provide temporary paint line markings to road surface 50 - 100mm wide**					
	Yellow temporary paint line marking to tarmac/concrete/block paved surface 50 - 100mm wide (M6175/155 - 1 m)	m	0.26	1.44	0.14	1.84
130	**Provide thermoplastic line markings to road surface 50 - 100mm wide**					
	Yellow Thermoplastic line marking to tarmac/concrete/block paved surface 100mm wide (M6175/210 - 1 m)	m	4.18	1.73	0.05	5.96
140	**Provide disabled access symbol in permanent line markings to road surface (406mm)**					
	Yellow permanent paint line marking to tarmac/concrete/block paved surface Disabled access symbol (406mm) (M6175/135 - 1 nr)	nr	21.20	2.88	0.28	24.36
150	**Provide disabled access symbol in yellow thermoplastic to road surface (1300mm)**					
	1300mm Yellow thermoplastic disabled Road marking to tarmac/concrete/block paved surface (M6175/220 - 1 nr)	nr	96.25	5.76	0.17	102.18
155	**Create accessible bay on car park with disabled access symbol and hatched transfer area in yellow permanent paint to road surface**					
	Yellow permanent paint line marking to tarmac/concrete/block paved surface 50 - 100mm wide (M6175/130 - 20 m), Yellow permanent paint line marking to tarmac/concrete/block paved surface Disabled access symbol (406mm) (M6175/135 - 1 nr)	nr	34.60	31.68	3.08	69.36
160	**Create accessible bay on car park with 1300 disabled access symbol and hatched transfer area in yellow thermoplastic to road surface**					
	Yellow Thermoplastic line marking to tarmac/concrete/block paved surface 100mm wide (M6175/210 - 20 m), 1300mm Yellow thermoplastic disabled Road marking to tarmac/concrete/block paved surface (M6175/220 - 1 nr)	nr	179.85	40.36	1.17	221.38

© NSR 01 Aug 2008 - 31 Jul 2009   Y8

			Mat. £	Lab. £	Plant £	Total £
	Y17 : FOOTPATHS					
Y1720	**TARMACADAM**					
012	**Break out existing hardsurface and renew in tarmacadam n.e 5m2**					
	Coated macadam or asphalt paving 50 to 75 mm thick (C9020/335 - 1 m2), Compacting ground (Q0535/300 - 1 m2), Excavated material off site (Q1025/100 - 0.075 m3), Pavings to falls and crossfalls and to slopes ne 15 degrees from horizontal to blinded hardcore, in areas 2 to 5 m2 (Q2210/030 - 1 m2)	m2	10.32	45.34	12.46	68.12
016	**Break out existing hardsurface and renew in tarmacadam exc 5m2**					
	Coated macadam or asphalt paving 50 to 75 mm thick (C9020/335 - 1 m2), Compacting ground (Q0535/300 - 1 m2), Excavated material off site (Q1025/100 - 0.075 m3), Pavings to falls and crossfalls and to slopes ne 15 degrees from horizontal to blinded hardcore, in areas 10 to 25 m2 (Q2210/020 - 1 m2)	m2	8.85	36.10	11.55	56.50
030	**Create tarmacadam resting place 3600 x 1100mm, comprising 75mm thick macadam and concrete framed seat with arm rests and timber slats. (PC £250)**					
	Filling to excavations ne 250 mm av thick (D2070/035 - 0.594 m3), To reduce levels 0.25 m max depth (Q0515/025 - 0.891 m3), Compacting ground (Q0540/300 - 3.96 m2), Compacting filling (Q0540/302 - 3.96 m2), Excavated material off site (Q1027/100 - 0.891 m3), Edging 50 x 150 mm type EF, foundation 300 x 100 mm, haunching one side (Q1050/225 - 6 m), Pavings to falls and crossfalls and to slopes ne 15 degrees from horizontal to blinded hardcore, in areas 10 to 25 m2 (Q2210/020 - 3.96 m2), Pavings 15 mm thick to falls and crossfalls and to slopes ne 15 degrees from horizontal to blinded hardcore, in areas 10 to 25 m2 (Q2220/110 - 3.96 m2), Concrete framed seating with timber slats placed in position, PC £250 (Q5050/100 - 1 nr)	nr	364.85	185.99	73.70	624.54
032	**Extra over for metal framed seating with timber slats (PC £350) in lieu of concrete framed seat with arm rests and timber slats**					
	Concrete framed seating with timber slats placed in position, PC £250 (Q5050/100 - -1 nr), Metal framed seating with timber slats Bolted to ground, PC £350 (Q5050/110 - 1 nr)	nr	100.00	6.94		106.94
050	**Create new 2 metre wide tarmacadam footpath 75mm thick macadam and straight concrete edging, to edge of existing road**					
	Filling to excavations ne 250 mm av thick (D2070/035 - 0.3 m3), To reduce levels 0.25 m max depth (Q0515/025 - 0.45 m3), Compacting ground (Q0540/300 - 2 m2), Compacting filling (Q0540/302 - 2 m2), Excavated material off site (Q1027/100 - 0.45 m3), Edging 50 x 150 mm type EF, foundation 300 x 100 mm, haunching one side (Q1050/225 - 1 m), Pavings to falls and crossfalls and to slopes ne 15 degrees from horizontal to blinded hardcore, in areas 10 to 25 m2 (Q2210/020 - 2 m2), Pavings 15 mm thick to falls and crossfalls and to slopes ne 15 degrees from horizontal to blinded hardcore, in areas 10 to 25 m2 (Q2220/110 - 2 m2)	m	44.26	72.81	37.22	154.29

© NSR 01 Aug 2008 - 31 Jul 2009   Y9

	Y17 : FOOTPATHS		Mat. £	Lab. £	Plant £	Total £
Y1720						
052	**Extra over for curved edgings**					
	Edging 50 x 150 mm type EF, foundation 300 x 100 mm, haunching one side (Q1050/225 - -1 m), Edging 50 x 150 mm type EF, foundation 300 x 100 mm, haunching one side curved (Q1050/230 - 1 m)	m	1.75	2.13		3.88
065	**Widen existing tarmacadam footpath n.e 500mm using 75mm thick macadam and straight concrete edging**					
	Concrete kerbs 50 x 150 mm complete with bed, haunching and the like (D2015/215 - 1 m), Filling to excavations ne 250 mm av thick (D2070/035 - 0.075 m3), To reduce levels 0.25 m max depth (Q0515/025 - 0.11 m3), Compacting ground (Q0540/300 - 0.5 m2), Compacting filling (Q0540/302 - 0.5 m2), Excavated material off site (Q1027/100 - 0.11 m3), Edging 50 x 150 mm type EF, foundation 300 x 100 mm, haunching one side (Q1050/225 - 1 m), Pavings to falls and crossfalls and to slopes ne 15 degrees from horizontal to blinded hardcore, in areas 10 to 25 m2 (Q2210/020 - 0.5 m2), Pavings 25 mm thick to falls and crossfalls and to slopes ne 15 degrees from horizontal to blinded hardcore, in areas 10 to 25 m2 (Q2220/140 - 0.5 m2)	m	17.52	39.19	12.85	69.56
070	**Widen existing tarmacadam footpath exc 500mm using 75mm thick macadam and straight concrete edging**					
	Concrete kerbs 50 x 150 mm complete with bed, haunching and the like (D2015/215 - 1 m), Filling to excavations ne 250 mm av thick (D2070/035 - 0.15 m3), To reduce levels 0.25 m max depth (Q0515/025 - 0.22 m3), Compacting ground (Q0540/300 - 1 m2), Compacting filling (Q0540/302 - 1 m2), Excavated material off site (Q1027/100 - 0.22 m3), Edging 50 x 150 mm type EF, foundation 300 x 100 mm, haunching one side (Q1050/225 - 1 m), Pavings to falls and crossfalls and to slopes ne 15 degrees from horizontal to blinded hardcore, in areas 10 to 25 m2 (Q2210/020 - 1 m2), Pavings 25 mm thick to falls and crossfalls and to slopes ne 15 degrees from horizontal to blinded hardcore, in areas 10 to 25 m2 (Q2220/140 - 1 m2)	m2	28.28	56.48	22.07	106.83
Y1730	**CONCRETE**					
100	**Break out existing defective path and renew in concrete**					
	Take up and clear away defective concrete paving, reform base, lay concrete C20 100 mm (min) thickness with wood float finish to match existing including formwork and flush joints to existing (C9015/280 - 1 m2)	m2	17.66	40.28	10.42	68.36
110	**Create concrete resting place 3600 x 1100mm, comprising 150mm thick concrete and concrete framed seat with arm rests and timber slats. (PC £250)**					
	Filling to excavations ne 250 mm av thick (D2070/035 - 0.594 m3), Beds ne 150 mm thick (E1040/330 - 0.594 m3), Square mesh ref A142 2.22 kg/m2 (E3030/210 - 3.96 m2), To reduce levels 0.25 m max depth (Q0515/025 - 1.188 m3), Compacting ground (Q0540/300 - 3.96 m2), Compacting filling (Q0540/302 - 3.96 m2), Excavated material off site (Q1027/100 - 1.188 m3), Edging 50 x 150 mm type EF, foundation 300 x 100 mm, haunching one side (Q1050/225 - 6 m), Concrete framed seating with timber slats placed in position, PC £250 (Q5050/100 - 1 nr)	nr	385.38	104.47	84.03	573.88

© NSR 01 Aug 2008 - 31 Jul 2009   Y10

Y17 : FOOTPATHS			Mat. £	Lab. £	Plant £	Total £
Y1730						
115	**Extra over for metal framed seating with timber slats (PC £350) in lieu of concrete framed seat with arm rests and timber slats**					
	Concrete framed seating with timber slats placed in position, PC £250 (Q5050/100 - -1 nr), Metal framed seating with timber slats Bolted to ground, PC £350 (Q5050/110 - 1 nr)	nr	100.00	6.94		106.94
135	**Create new 2 metre wide concrete footpath 150mm thick reinforced concrete and straight concrete edging, to edge of existing road**					
	Filling to excavations ne 250 mm av thick (D2070/035 - 0.3 m3), Beds ne 150 mm thick (E1040/330 - 0.3 m3), Square mesh ref A142 2.22 kg/m2 (E3030/210 - 2 m2), To reduce levels 0.25 m max depth (Q0515/025 - 0.6 m3), Compacting ground (Q0540/300 - 2 m2), Compacting filling (Q0540/302 - 2 m2), Excavated material off site (Q1027/100 - 0.6 m3), Edging 50 x 150 mm type EF, foundation 300 x 100 mm, haunching one side (Q1050/225 - 1 m)	m	54.63	31.64	42.44	128.71
140	**Extra over for curved edgings**					
	Edging 50 x 150 mm type EF, foundation 300 x 100 mm, haunching one side (Q1050/225 - -1 m), Edging 50 x 150 mm type EF, foundation 300 x 100 mm, haunching one side curved (Q1050/230 - 1 m)	m	1.75	2.13		3.88
145	**Widen existing concrete footpath n.e 500mm using 150mm thick reinforced concrete on 150mm thick hardcore and straight concrete edging**					
	Concrete kerbs 50 x 150 mm complete with bed, haunching and the like (D2015/215 - 1 m), Excavated material off site (D2052/005 - 0.15 m3), Filling to excavations ne 250 mm av thick (D2070/035 - 0.075 m3), Beds ne 150 mm thick (E1040/330 - 0.075 m3), Square mesh ref A142 2.22 kg/m2 (E3030/210 - 0.5 m2), To reduce levels 0.25 m max depth (Q0515/025 - 0.15 m3), Compacting ground (Q0540/300 - 0.5 m2), Compacting filling (Q0540/302 - 0.5 m2)	m	11.96	21.83	14.24	48.03
150	**Widen existing concrete footpath exc 500mm using 150mm thick reinforced concrete on 150mm thick hardcore and straight concrete edging**					
	Concrete kerbs 50 x 150 mm complete with bed, haunching and the like (D2015/215 - 1 m), Excavated material off site (D2052/005 - 0.3 m3), Filling to excavations ne 250 mm av thick (D2070/035 - 0.15 m3), Beds ne 150 mm thick (E1040/330 - 0.15 m3), Square mesh ref A142 2.22 kg/m2 (E3030/210 - 1 m2), To reduce levels 0.25 m max depth (Q0515/025 - 0.3 m3), Compacting ground (Q0540/300 - 1 m2), Compacting filling (Q0540/302 - 1 m2)	m2	23.93	28.15	24.85	76.93

© NSR 01 Aug 2008 - 31 Jul 2009   Y11

	Y17 : FOOTPATHS		Mat. £	Lab. £	Plant £	Total £
Y1740	**PAVING SLABS**					
200	**Create flagged resting place 3600 x 1100mm, comprising 900 x 600mm concrete flags and concrete framed seat with arm rests and timber slats. (PC £250)**					
	Filling to excavations ne 250 mm av thick (D2070/035 - 0.594 m3), To reduce levels 0.25 m max depth (Q0515/025 - 0.792 m3), Compacting ground (Q0540/300 - 3.96 m2), Compacting filling (Q0540/302 - 3.96 m2), Excavated material off site (Q1027/100 - 0.792 m3), Edging 50 x 150 mm type EF, foundation 300 x 100 mm, haunching one side (Q1050/225 - 6 m), Pavings level or to falls (Q2512/010 - 3.96 m2), Concrete framed seating with timber slats placed in position, PC £250 (Q5050/100 - 1 nr)	nr	368.38	101.44	55.96	525.78
202	**Extra over for metal framed seating with timber slats (PC £350) in lieu of concrete framed seat with arm rests and timber slats**					
	Concrete framed seating with timber slats placed in position, PC £250 (Q5050/100 - -1 nr), Metal framed seating with timber slats Bolted to ground, PC £350 (Q5050/110 - 1 nr)	nr	100.00	6.94		106.94
208	**Create new 2 metre wide paved footpath with straight concrete edging, to edge of existing road**					
	Filling to excavations ne 250 mm av thick (D2070/035 - 0.3 m3), To reduce levels 0.25 m max depth (Q0515/025 - 0.4 m3), Compacting ground (Q0540/300 - 2 m2), Compacting filling (Q0540/302 - 2 m2), Excavated material off site (Q1027/100 - 0.4 m3), Edging 50 x 150 mm type EF, foundation 300 x 100 mm, haunching one side (Q1050/225 - 1 m), Pavings level or to falls (Q2510/010 - 2 m2)	m	50.78	33.81	28.27	112.86
210	**Extra over for curved edgings**					
	Edging 50 x 150 mm type EF, foundation 300 x 100 mm, haunching one side (Q1050/225 - -1 m), Edging 50 x 150 mm type EF, foundation 300 x 100 mm, haunching one side curved (Q1050/230 - 1 m)	m	1.75	2.13		3.88
212	**Remove and relay 600 x 600 x 50mm concrete slabs, bedding on sand**					
	Concrete slab paving 50 mm thick, storing for re-use (C9020/315 - 1 m2), Pavings to falls and crossfalls and to slopes ne 15 degrees from horizontal (C9030/380 - 1 m2)	m2	5.10	19.72	0.82	25.64
214	**Remove and relay single 600 x 600 x 50mm slab, bedding on sand**					
	Concrete slab paving 50 mm thick, storing for re-use (C9020/315 - 0.36 m2), Pavings to falls and crossfalls and to slopes ne 15 degrees from horizontal (C9030/380 - 0.36 m2)	nr	1.84	7.10	0.30	9.24
216	**Renew 600 x 600 x 50mm concrete slabs, bedding on sand**					
	Concrete slab paving 50 mm thick (C9020/314 - 1 m2), Pavings level or to falls (Q2520/010 - 1 m2)	m2	16.05	18.48	0.62	35.15
218	**Renew single 600 x 600 x 50mm slab, bedding on sand**					
	Concrete slab paving 50 mm thick (C9020/314 - 0.36 m2), Pavings level or to falls (Q2510/010 - 0.36 m2)	nr	5.85	6.32	0.22	12.39

© NSR 01 Aug 2008 - 31 Jul 2009   Y12

	Y17 : FOOTPATHS		Mat. £	Lab. £	Plant £	Total £
Y1740						
222	**Remove and relay 600 x 900 x 50mm concrete slabs, bedding on sand, in areas not exceeding 2 m2**					
	Concrete slab paving 50 mm thick, storing for re-use (C9020/315 - 0.5 m2), Pavings level or to falls in areas ne 2 m2 (C9035/377 - 1 m2)	m2	5.10	18.18	0.41	23.69
224	**Remove and relay single 600 x 900 x 50mm slab, bedding on sand**					
	Concrete slab paving 50 mm thick, storing for re-use (C9020/315 - 0.54 m2), Pavings level or to falls (C9035/375 - 0.54 m2)	nr	2.75	9.65	0.44	12.84
226	**Renew 600 x 900 x 50mm concrete slabs, bedding on sand**					
	Concrete slab paving 50 mm thick (C9020/314 - 1 m2), Pavings level or to falls (Q2512/010 - 1 m2)	m2	13.87	15.71	0.62	30.20
228	**Renew single 600 x 900 x 50mm concrete slabs, bedding on sand**					
	Concrete slab paving 50 mm thick (C9020/314 - 0.54 m2), Pavings level or to falls (Q2512/010 - 0.54 m2)	nr	7.49	8.48	0.33	16.30
Y1750	**BLOCK PAVING**					
300	**Create block paved resting place 3600 x 1100mm, comprising 200 x 100 x 65mm concrete paviours and concrete framed seat with arm rests and timber slats. (PC £250)**					
	Filling to excavations ne 250 mm av thick (D2070/035 - 0.594 m3), To reduce levels 0.25 m max depth (Q0515/025 - 0.792 m3), Compacting ground (Q0540/300 - 3.96 m2), Compacting filling (Q0540/302 - 3.96 m2), Excavated material off site (Q1027/100 - 0.792 m3), Edging 50 x 150 mm type EF, foundation 300 x 100 mm, haunching one side (Q1050/225 - 6 m), Pavings level or to falls (Q2540/010 - 3.96 m2), Concrete framed seating with timber slats placed in position, PC £250 (Q5050/100 - 1 nr)	nr	432.29	156.36	60.88	649.53
302	**Extra Over for clay paviours in lieu of concrete paviours**					
	Pavings level or to falls (Q2540/010 - -3.96 m2), Pavings level or to falls (Q2545/010 - 3.96 m2)	nr	67.24			67.24
304	**Extra over for metal framed seating with timber slats (PC £350) in lieu of concrete framed seat with arm rests and timber slats**					
	Concrete framed seating with timber slats placed in position, PC £250 (Q5050/100 - -1 nr), Metal framed seating with timber slats Bolted to ground, PC £350 (Q5050/110 - 1 nr)	nr	100.00	6.94		106.94
310	**Remove and relay 200 x 100 x 65mm Block paviours, bedding on sand**					
	Brick block stone paving 70 mm thick, storing for re-use (C9020/320 - 1 m2), Pavings level or to falls (C9047/375 - 1 m2)	m2	3.23	40.74	1.24	45.21

© NSR 01 Aug 2008 - 31 Jul 2009   Y13

	Y17 : FOOTPATHS		Mat. £	Lab. £	Plant £	Total £
Y1750						
315	**Create new 2 metre wide concrete block paved footpath and straight matching kerbs**					
	Filling to excavations ne 250 mm av thick (D2070/035 - 0.3 m3), To reduce levels 0.25 m max depth (Q0515/025 - 0.5 m3), Compacting ground (Q0540/300 - 2 m2), Compacting filling (Q0540/302 - 2 m2), Excavated material off site (Q1027/100 - 0.5 m3), Edging 200 x 125 mm type keykerb, foundation 300 x 100 mm, haunching one side (Q1050/250 - 1 m), Pavings level or to falls (Q2540/010 - 2 m2)	m	90.92	58.17	36.70	185.79
317	**Extra Over for clay paviours**					
	Pavings level or to falls (Q2540/010 - -2 m2), Pavings level or to falls (Q2545/010 - 2 m2)	m	33.96			33.96
350	**Widen existing concrete block paved footpath n.e 500mm on 150mm thick hardcore and matching edging**					
	Excavated material off site (D2052/005 - 0.13 m3), Filling to excavations ne 250 mm av thick (D2070/035 - 0.075 m3), To reduce levels 0.25 m max depth (Q0515/025 - 0.13 m3), Compacting ground (Q0540/300 - 0.5 m2), Compacting filling (Q0540/302 - 0.5 m2), Edging 200 x 125 mm type keykerb, foundation 300 x 100 mm, haunching one side (Q1050/250 - 1 m), Pavings level or to falls (Q2540/010 - 0.5 m2)	m	37.26	19.34	9.47	66.07
352	**Extra Over for clay paviours**					
	Pavings level or to falls (Q2540/010 - -0.5 m2), Pavings level or to falls (Q2545/010 - 0.5 m2)	m	8.49			8.49
355	**Widen existing concrete block paved footpath exc 500mm on 150mm thick hardcore and matching edging**					
	Excavated material off site (D2052/005 - 0.27 m3), Filling to excavations ne 250 mm av thick (D2070/035 - 0.15 m3), To reduce levels 0.25 m max depth (Q0515/025 - 0.27 m3), Compacting ground (Q0540/300 - 1 m2), Compacting filling (Q0540/302 - 1 m2), Edging 200 x 125 mm type keykerb, foundation 300 x 100 mm, haunching one side (Q1050/250 - 1 m), Pavings level or to falls (Q2540/010 - 1 m2)	m2	55.15	32.34	19.54	107.03
357	**Extra Over for clay paviours**					
	Pavings level or to falls (Q2540/010 - -1 m2), Pavings level or to falls (Q2545/010 - 1 m2)	m2	16.98			16.98

© NSR 01 Aug 2008 - 31 Jul 2009

			Mat. £	Lab. £	Plant £	Total £
	Y18 : STREET FURNITURE					
Y1810	**BOLLARDS**					
010	**Remove existing bollard and make good tarmacadam surface**					
	Removing steel bollard (C9010/220 - 1 nr), Cutting back edges of holes in pavings and replacing to match existing areas ne 0.1 m2 (C9025/359 - 1 nr)	nr	1.00	21.90		22.90
015	**Remove existing bollard, relocate and make good tarmacadam surface**					
	Cutting back edges of holes in pavings and replacing to match existing areas ne 0.1 m2 (C9025/359 - 1 nr), Removing bollard and refixing 914 mm high (C9089/940 - 1 nr)	nr	6.21	48.55		54.76
020	**Supply and install precast concrete bollard (PC £65.00) in concrete mix C15**					
	Concrete bollard (PC £65.00) set in concrete mix C15, excavating and removing excavated material from site (Q5030/260 - 1 nr)	nr	70.62	23.92		94.54
025	**Supply and install steel bollard (PC £100.00) in concrete mix C15**					
	Steel bollard (PC £100.00) set in concrete mix C15, excavating and removing excavated material from site (Q5010/060 - 1 nr)	nr	105.62	21.98		127.60
030	**Supply and install timber bollard (PC £45.00) in concrete mix C15**					
	Timber bollard (PC £45.00) set in concrete mix C15, excavating and removing excavated material from site (Q5040/270 - 1 nr)	nr	45.62	21.98		67.60
Y1820	**BENCHES**					
040	**Supply and fix concrete framed seating with timber slats (PC £250)**					
	Concrete framed seating with timber slats placed in position, PC £250 (Q5050/100 - 1 nr)	nr	250.00	16.18		266.18
045	**Supply and fix metal framed seating with timber slats bolted to ground (PC £350)**					
	Metal framed seating with timber slats Bolted to ground, PC £350 (Q5050/110 - 1 nr)	nr	350.00	23.12		373.12
050	**Supply and fix metal framed seating with metal slats bolted to ground (PC £550)**					
	Metal framed seating with metal slats Bolted to ground, PC £550 (Q5050/120 - 1 nr)	nr	550.00	23.12		573.12
055	**Supply and fix hardwood seating bolted to ground (PC £400)**					
	Solid hardwood seating, Bolted to ground, PC £400 (Q5050/130 - 1 nr)	nr	400.00	23.12		423.12

© NSR 01 Aug 2008 - 31 Jul 2009

Y15

	Y18 : STREET FURNITURE		Mat. £	Lab. £	Plant £	Total £
Y1830	**TREE PROTECTION**					
070	**Take up tree protection grid and remove to store (any type)**					
	Take up tree protection grid and remove to store (any type) (Q5060/050 - 1 nr)	nr		11.56		11.56
075	**Remove tree protection grid from store and refix (any type)**					
	Take tree protection grid from store and refix (any type) (Q5060/060 - 1 nr)	nr		32.36		32.36
080	**Provide and fit square 4 piece tree grid (750mm)**					
	Provide and fit square 4 piece tree grid (750mm) (Q5060/010 - 1 nr)	nr	333.78	32.36		366.14
085	**Provide and fit square 4 piece tree grid (1200mm)**					
	Provide and fit square 4 piece tree grid (1200mm) (Q5060/020 - 1 nr)	nr	538.10	34.67		572.77
090	**Provide and fit round 4 piece tree grid (1220mm)**					
	Provide and fit round 4 piece tree grid (1220mm dia) (Q5060/030 - 1 nr)	nr	548.21	34.67		582.88
095	**Provide and fit round tree guard (PC £150)**					
	supply and fix round tree guard (PC £150.00) (Q5060/040 - 1 nr)	nr	150.00	9.25		159.25
Y1850	**PLANTERS**					
145	**Remove existing planter to store**					
	Remove existing planter (C9010/230 - 1 nr)	nr		3.23		3.23
150	**Remove existing planter and relocate inc new fixing**					
	Remove existing planter (C9010/230 - 1 nr), Fix only hardwood planter inc new anchor kit (Q5070/070 - 1 nr)	nr	8.50	5.38		13.88
155	**Provide and fix hardwood planter, anchored to ground; 1000 x 500 x 700mm high**					
	Provide and fix hardwood planter anchored to ground 1000 x 500 x 700 high (Q5070/010 - 1 nr)	nr	520.50	2.15		522.65
160	**Provide and fix hardwood planter, anchored to ground; 1000 x 1000 x 700mm high**					
	Provide and fix hardwood planter anchored to ground 1000 x 1000 x 700 high (Q5070/020 - 1 nr)	nr	630.50	2.15		632.65
165	**Provide and fix hardwood planter, anchored to ground; 1000 x 1500 x 700mm high**					
	Provide and fix hardwood planter anchored to ground 1000 x 1500 x 700 high (Q5070/030 - 1 nr)	nr	833.50	2.15		835.65
170	**Provide and fix concrete hexagonal planter, (small 675mm dia)**					
	Provide and fit concrete hexagonal planter (Small) (Q5070/040 - 1 nr)	nr	125.00	6.46		131.46

© NSR 01 Aug 2008 - 31 Jul 2009

	Y18 : STREET FURNITURE		Mat. £	Lab. £	Plant £	Total £
Y1850						
175	**Provide and fix concrete hexagonal planter, (medium 900mm dia)**					
	Provide and fit concrete hexagonal planter (Medium) (Q5070/050 - 1 nr)	nr	285.00	8.62		293.62
180	**Provide and fix concrete hexagonal planter, (large 1450mm dia)**					
	Provide and fit concrete hexagonal planter (Large) (Q5070/060 - 1 nr)	nr	456.00	9.70		465.70
Y1860	**LITTER BINS**					
200	**Remove existing anchored litter bin and relocate inc new fixing anchors**					
	Remove existing planter (C9010/230 - 1 nr), Fix only hardwood planter inc new anchor kit (Q5070/070 - 1 nr)	nr	8.50	5.38		13.88
205	**Remove existing litter bin with concrete footing, relocate and make good.**					
	Remove existing litter bin (C9010/235 - 1 nr), Cutting back edges of holes in pavings and replacing to match existing areas ne 0.1 m2 (C9025/359 - 1 nr), Fix only below ground fixed litter bin, set in concrete mix C15, excavating and removing excavated material from site (Q5065/040 - 1 nr)	nr	6.62	32.03		38.65
210	**Provide and fix steel litter bin set in concrete**					
	Steel litter bin with galvanised liner set in concrete mix C15, excavating and removing excavated material from site (Q5065/010 - 1 nr)	nr	201.41	23.70		225.11
215	**Provide and fix precast concrete litter bin set in concrete**					
	Precast concrete litter bin with exposed aggregate finish set in concrete mix C15, excavating and removing excavated material from site (Q5065/020 - 1 nr)	nr	305.41	23.70		329.11
220	**Provide and fix thermoplastic (Victoriana) litter bin fixed with ground anchors**					
	Thermoplastic (Victoriana) round litter bin fixed with anchors (Q5065/030 - 1 nr)	nr	206.50	4.31		210.81
Y1870	**EXTERNAL HANDRAILS**					
250	**Provide and fix 50mm dia softwood handrail to wall on nylon coated elbow brackets, gloss painted to contrast with background**					
	Wood general surfaces in isolated surfaces ne 300 mm girth (M6135/580 - 1 m), Isolated handrails and grab rails 50 x 50 mm (P2010/190 - 1 m), Handrail Brackets (P2110/086 - 2 nr)	m	6.98	19.88		26.86

© NSR 01 Aug 2008 - 31 Jul 2009   Y17

	Y18 : STREET FURNITURE		Mat. £	Lab. £	Plant £	Total £
Y1870						
255	**Provide and fix 50mm dia softwood handrail on 75 x 75mm posts concreted into ground, gloss painted to contrast with background**					
	Wood general surfaces in isolated surfaces ne 300 mm girth (M6135/580 - 2 m), Isolated handrails and grab rails 50 x 50 mm (P2010/190 - 1 m), Independent gate posts 75 x 75 x 900 mm high, setting posts 450 mm below ground (Q4083/675 - 1 nr)	m	13.24	32.32		45.56
260	**Provide and fix 50mm dia hardwood handrail to wall on nylon coated elbow brackets, 3 coats lacquer**					
	Wood general surfaces in isolated surfaces ne 300 mm girth (M6155/580 - 1 m), Isolated handrails and grab rails 50 x 50 mm (P2020/190 - 1 m), Handrail Brackets (P2110/086 - 2 nr)	m	7.23	19.87		27.10
265	**Provide and fix 50mm dia hardwood handrail on 75 x 75mm posts concreted into ground, 3 coats lacquer**					
	Wood general surfaces in isolated surfaces ne 300 mm girth (M6155/580 - 2 m), Isolated handrails and grab rails 50 x 50 mm (P2020/190 - 1 m), Independent gate posts 75 x 75 x 900 mm high, setting posts 450 mm below ground (Q4084/675 - 1 nr)	m	23.53	31.15		54.68
270	**Galvanised mild steel handrail, Kee Klamps 42.4mm dia, with intermediate rail, PVC coated, bolted to concrete base**					
	Extra over for PVC coating to handrail (L3110/470 - 3 m), 42.4 mm diameter tube (7) (L3120/500 - 3 m), Tees (25-7) (L3120/550 - 1 nr), 90 degree crossovers (26-7) (L3120/560 - 1 nr), Base plates (62-7), plugging and screwing (L3120/590 - 1 nr)	m	103.04	21.38		124.42
275	**Extra over for end plate plugged and screwed to wall**					
	End plates (61-7), plugging and screwing (L3120/580 - 1 nr)	m	5.26	6.04		11.30
280	**Remove existing handrail and refix**					
	Timber handrails and brackets, resecuring (C5105/175 - 1 m), Plugging (isolated) (L2068/669 - 2 nr), Screwing (isolated) (L2068/674 - 2 nr)	m	0.33	6.32		6.65

© NSR 01 Aug 2008 - 31 Jul 2009   Y18

			Mat. £	Lab. £	Plant £	Total £
	Y19 : SIGNAGE AND WAYFINDING EXTERNALLY					
Y1910	**SIGNS**					
100	**152 x 305mm Disabled parking/access sign on 1.2m high steel enamel post set in concrete**					
	Non reflective steel sign size 152 x 305mm fixed to post (measured elsewhere) (N1520/100 - 1 nr), Steel enamel sign post 1.2m high, set in concrete mix C15, excavating and removing excavated material from site (Q5010/056 - 1 nr)	nr	120.32	23.00		143.32
105	**457 x 305mm Disabled parking/access sign on 1.2m high steel enamel post set in concrete**					
	Non reflective steel sign size 457 x 305mm fixed to post (measured elsewhere) (N1520/110 - 1 nr), Steel enamel sign post 1.2m high, set in concrete mix C15, excavating and removing excavated material from site (Q5010/056 - 1 nr)	nr	133.92	23.00		156.92
107	**152 x 305mm Disabled parking/access sign on 1.8m high galvanised steel post set in concrete**					
	Non reflective steel sign size 152 x 305mm fixed to post (measured elsewhere) (N1520/100 - 1 nr), Steel galvanised sign post 1.8m high, set in concrete mix C15, excavating and removing excavated material from site (Q5010/057 - 1 nr)	nr	140.82	23.00		163.82
109	**457 x 305mm Disabled parking/access sign on 1.8m high galvanised steel post set in concrete**					
	Non reflective steel sign size 457 x 305mm fixed to post (measured elsewhere) (N1520/110 - 1 nr), Steel galvanised sign post 1.8m high, set in concrete mix C15, excavating and removing excavated material from site (Q5010/057 - 1 nr)	nr	154.42	23.00		177.42
112	**152 x 305mm Disabled parking/access sign on 2.1m high galvanised steel post set in concrete**					
	Non reflective steel sign size 152 x 305mm fixed to post (measured elsewhere) (N1520/100 - 1 nr), Steel galvanised sign post 2.1m high, set in concrete mix C15, excavating and removing excavated material from site (Q5010/058 - 1 nr)	nr	135.12	24.08		159.20
115	**457 x 305mm Disabled parking/access sign on 2.1m high galvanised steel post set in concrete**					
	Non reflective steel sign size 457 x 305mm fixed to post (measured elsewhere) (N1520/110 - 1 nr), Steel galvanised sign post 2.1m high, set in concrete mix C15, excavating and removing excavated material from site (Q5010/058 - 1 nr)	nr	148.72	24.08		172.80
120	**152 x 305mm Disabled parking/access sign on aluminium post set in concrete**					
	Non reflective steel sign size 152 x 305mm fixed to post (measured elsewhere) (N1520/100 - 1 nr), Aluminium sign post 3m high, post set in concrete mix C15, excavating and removing excavated material from site (Q5010/059 - 1 nr)	nr	228.82	24.08		252.90

© NSR 01 Aug 2008 - 31 Jul 2009   Y19

	Y19 : SIGNAGE AND WAYFINDING EXTERNALLY		Mat. £	Lab. £	Plant £	Total £
Y1910						
125	457 x 305mm Disabled parking/access sign on aluminium post set in concrete					
	Non reflective steel sign size 457 x 305mm fixed to post (measured elsewhere) (N1520/110 - 1 nr), Aluminium sign post 3m high, post set in concrete mix C15, excavating and removing excavated material from site (Q5010/059 - 1 nr)	nr	242.42	24.08		266.50
130	Remove existing sign and post and make good tarmacadam surface					
	Removing steel clothes post (C9010/210 - 1 nr), Cutting back edges of holes in pavings and replacing to match existing areas ne 0.1 m2 (C9025/359 - 1 nr)	nr	1.00	20.18		21.18
Y1920	TACTILE WARNINGS					
400	Replace flags with blister paving slabs, 400 x 400mm, bedding on sand					
	Concrete slab paving 50 mm thick (C9020/314 - 0.16 m2), Pavings level or to falls in areas ne 2 m2 (Q2525/015 - 0.16 m2)	nr	10.30	3.77	0.10	14.17
402	Replace tarmacadam with blister paving slabs, 400 x 400mm, bedding on sand					
	Coated macadam or asphalt paving 50 to 75 mm thick (C9020/335 - 0.16 m2), Pavings level or to falls in areas ne 2 m2 (Q2525/015 - 0.16 m2)	nr	10.30	4.81	0.97	16.08
405	Replace block paving with blister paving slabs, 400 x 400mm, bedding on sand					
	Brick block stone paving 70 mm thick (C9020/319 - 0.16 m2), Pavings to falls and crossfalls and to slopes ne 15 degrees from horizontal in areas ne 2 m2 (Q2525/025 - 0.16 m2)	nr	10.30	5.15		15.45
410	Replace flags with blister paving slabs, 400 x 400mm, bedding on sand					
	Concrete slab paving 50 mm thick (C9020/314 - 1 m2), Pavings to falls and crossfalls and to slopes ne 15 degrees from horizontal (Q2525/020 - 1 m2)	m2	53.67	18.48	0.62	72.77
412	Replace tarmacadam with blister paving slabs, 400 x 400mm, bedding on sand					
	Coated macadam or asphalt paving 50 to 75 mm thick (C9020/335 - 1 m2), Pavings to falls and crossfalls and to slopes ne 15 degrees from horizontal (Q2525/020 - 1 m2)	m2	53.67	24.95	6.05	84.67
415	Replace block paving with blister paving slabs, 400 x 400mm, bedding on sand					
	Brick block stone paving 70 mm thick (C9020/319 - 1 m2), Pavings to falls and crossfalls and to slopes ne 15 degrees from horizontal (Q2525/020 - 1 m2)	m2	53.67	27.12		80.79

© NSR 01 Aug 2008 - 31 Jul 2009   Y20

	Y19 : SIGNAGE AND WAYFINDING EXTERNALLY		Mat. £	Lab. £	Plant £	Total £
Y1920						
418	**Replace flags with hazard warning paving slabs, 400 x 400mm, bedding on sand**					
	Concrete slab paving 50 mm thick (C9020/314 - 0.16 m2), Pavings to falls and crossfalls and to slopes ne 15 degrees from horizontal in areas ne 2 m2 (Q2527/025 - 0.16 m2)	nr	9.25	3.77	0.10	13.12
420	**Replace tarmacadam with hazard warning paving slabs, 400 x 400mm, bedding on sand**					
	Coated macadam or asphalt paving 50 to 75 mm thick (C9020/335 - 0.16 m2), Pavings to falls and crossfalls and to slopes ne 15 degrees from horizontal (Q2525/020 - 0.16 m2)	nr	8.59	3.99	0.97	13.55
422	**Replace block paving with hazard warning paving slabs, 400 x 400mm, bedding on sand**					
	Brick block stone paving 70 mm thick (C9020/319 - 0.16 m2), Pavings to falls and crossfalls and to slopes ne 15 degrees from horizontal in areas ne 2 m2 (Q2527/025 - 0.16 m2)	nr	9.25	5.15		14.40
424	**Replace flags with hazard warning paving slabs, 400 x 400mm, bedding on sand**					
	Concrete slab paving 50 mm thick (C9020/314 - 1 m2), Pavings to falls and crossfalls and to slopes ne 15 degrees from horizontal (Q2527/020 - 1 m2)	m2	48.17	18.48	0.62	67.27
426	**Replace tarmacadam with hazard warning paving slabs, 400 x 400mm, bedding on sand**					
	Coated macadam or asphalt paving 50 to 75 mm thick (C9020/335 - 1 m2), Pavings to falls and crossfalls and to slopes ne 15 degrees from horizontal (Q2527/020 - 1 m2)	m2	48.17	24.95	6.05	79.17
428	**Replace block paving with hazard warning paving slabs, 400 x 400mm, bedding on sand**					
	Brick block stone paving 70 mm thick (C9020/319 - 1 m2), Pavings to falls and crossfalls and to slopes ne 15 degrees from horizontal (Q2527/020 - 1 m2)	m2	48.17	27.12		75.29
430	**Replace flags with directional guidance paving slabs, 600 x 600mm, bedding on sand**					
	Concrete slab paving 50 mm thick (C9020/314 - 0.36 m2), Pavings level or to falls in areas ne 2 m2 (Q2529/015 - 0.36 m2)	nr	10.16	8.49	0.22	18.87
432	**Replace tarmacadam with directional guidance paving slabs, 600 x 600mm, bedding on sand**					
	Coated macadam or asphalt paving 50 to 75 mm thick (C9020/335 - 0.36 m2), Pavings level or to falls in areas ne 2 m2 (Q2529/015 - 0.36 m2)	nr	10.16	10.81	2.18	23.15
434	**Replace block paving with directional guidance paving slabs, 600 x 600mm, bedding on sand**					
	Brick block stone paving 70 mm thick (C9020/319 - 0.36 m2), Pavings level or to falls in areas ne 2 m2 (Q2529/015 - 0.36 m2)	nr	10.16	11.60		21.76

© NSR 01 Aug 2008 - 31 Jul 2009

Y21

	Y19 : SIGNAGE AND WAYFINDING EXTERNALLY		Mat. £	Lab. £	Plant £	Total £
Y1920						
436	**Replace flags with directional guidance paving slabs, 600 x 600mm, bedding on sand**					
	Concrete slab paving 50 mm thick (C9020/314 - 1 m2), Pavings level or to falls (Q2529/010 - 1 m2)	m2	23.49	18.48	0.62	42.59
438	**Replace tarmacadam with directional guidance paving slabs, 600 x 600mm, bedding on sand**					
	Coated macadam or asphalt paving 50 to 75 mm thick (C9020/335 - 1 m2), Pavings level or to falls (Q2529/010 - 1 m2)	m2	23.49	24.95	6.05	54.49
440	**Replace block paving with directional guidance paving slabs, 600 x 600mm, bedding on sand**					
	Brick block stone paving 70 mm thick (C9020/319 - 1 m2), Pavings level or to falls (Q2529/010 - 1 m2)	m2	23.49	27.12		50.61
444	**Create uncontrolled crossing point to existing paved footpath using 400 x 400mm blister paving slabs**					
	Concrete slab paving 50 mm thick (C9020/314 - 0.96 m2), Concrete kerbs 155 x 255 mm complete with bed, haunching and the like (D2015/225 - 1.2 m), Trenches ne 300 mm wide 0.25 m max depth (Q1010/010 - 0.11 m3), Excavated material off site (Q1025/100 - 0.11 m3), Extra over for transition kerbs (Q1050/208 - 2 nr), Kerbs 125 x 150 mm half battered flush, foundation 450 x 150 mm, haunching one side (Q1050/209 - 1.2 m), Pavings level or to falls (Q2525/010 - 0.96 m2)	nr	120.82	68.09	10.35	199.26
446	**Create uncontrolled crossing point to existing tarmac footpath using 400 x 400mm blister paving slabs**					
	Coated macadam or asphalt paving 50 to 75 mm thick (C9020/335 - 0.96 m2), Concrete kerbs 155 x 255 mm complete with bed, haunching and the like (D2015/225 - 1.2 m), Trenches ne 300 mm wide 0.25 m max depth (Q1010/010 - 0.11 m3), Excavated material off site (Q1025/100 - 0.11 m3), Extra over for transition kerbs (Q1050/208 - 2 nr), Kerbs 125 x 150 mm half battered flush, foundation 450 x 150 mm, haunching one side (Q1050/209 - 1.2 m), Pavings level or to falls (Q2525/010 - 0.96 m2)	nr	120.82	74.30	15.56	210.68
448	**Create uncontrolled crossing point to existing block paved footpath using 400 x 400mm blister paving slabs**					
	Brick block stone paving 70 mm thick (C9020/319 - 0.96 m2), Concrete kerbs 155 x 255 mm complete with bed, haunching and the like (D2015/225 - 1.2 m), Trenches ne 300 mm wide 0.25 m max depth (Q1010/010 - 0.11 m3), Excavated material off site (Q1025/100 - 0.11 m3), Extra over for transition kerbs (Q1050/208 - 2 nr), Kerbs 125 x 150 mm half battered flush, foundation 450 x 150 mm, haunching one side (Q1050/209 - 1.2 m), Pavings level or to falls (Q2525/010 - 0.96 m2)	nr	120.82	76.39	9.76	206.97

© NSR 01 Aug 2008 - 31 Jul 2009   Y22

	Y19 : SIGNAGE AND WAYFINDING EXTERNALLY		Mat. £	Lab. £	Plant £	Total £
Y1920						
450	**Create controlled crossing point to existing paved footpath using 400 x 400mm blister paving slabs**					
	Concrete slab paving 50 mm thick (C9020/314 - 3.84 m2), Concrete kerbs 155 x 255 mm complete with bed, haunching and the like (D2015/225 - 2 m), Trenches ne 300 mm wide 0.25 m max depth (Q1010/010 - 0.18 m3), Excavated material off site (Q1025/100 - 0.18 m3), Extra over for transition kerbs (Q1050/208 - 2 nr), Kerbs 125 x 150 mm half battered flush, foundation 450 x 150 mm, haunching one side (Q1050/209 - 2.4 m), Pavings level or to falls (Q2525/010 - 3.84 m2)	nr	291.44	149.56	18.48	459.48
455	**Create controlled crossing point to existing tarmac footpath using 400 x 400mm blister paving slabs**					
	Coated macadam or asphalt paving 50 to 75 mm thick (C9020/335 - 3.84 m2), Concrete kerbs 155 x 255 mm complete with bed, haunching and the like (D2015/225 - 2 m), Trenches ne 300 mm wide 0.25 m max depth (Q1010/010 - 0.18 m3), Excavated material off site (Q1025/100 - 0.18 m3), Extra over for transition kerbs (Q1050/208 - 2 nr), Kerbs 125 x 150 mm half battered flush, foundation 450 x 150 mm, haunching one side (Q1050/209 - 2.4 m), Pavings level or to falls (Q2525/010 - 3.84 m2)	nr	291.44	174.41	39.33	505.18
460	**Create controlled crossing point to existing block paved footpath using 400 x 400mm blister paving slabs**					
	Brick block stone paving 70 mm thick (C9020/319 - 3.84 m2), Concrete kerbs 155 x 255 mm complete with bed, haunching and the like (D2015/225 - 2 m), Trenches ne 300 mm wide 0.25 m max depth (Q1010/010 - 0.18 m3), Excavated material off site (Q1025/100 - 0.18 m3), Extra over for transition kerbs (Q1050/208 - 2 nr), Kerbs 125 x 150 mm half battered flush, foundation 450 x 150 mm, haunching one side (Q1050/209 - 2.4 m), Pavings level or to falls (Q2525/010 - 3.84 m2)	nr	291.44	182.74	16.10	490.28

© NSR 01 Aug 2008 - 31 Jul 2009   Y23

Y19 : SIGNAGE AND WAYFINDING EXTERNALLY		Mat. £	Lab. £	Plant £	Total £

			Mat. £	Lab. £	Plant £	Total £
	Y20 : EXTERNAL RAMPS AND STEPS					
Y2020	**ANCILLARY WORKS**					
010	**Excavate reduce level by hand n.e 250mm deep**					
	To reduce levels 0.25 m max depth (D2020/240 - 1 m3)	m3		33.83		33.83
012	**Excavate reduce level by machine n.e 250mm deep**					
	To reduce levels 0.25 m max depth (D2030/240 - 1 m3)	m3		1.16	3.41	4.57
014	**Extra over excavation for breaking out plain concrete 100mm thick**					
	Extra over excavation irrespective of depth for breaking out concrete paving ne 100 mm thick (D2020/800 - 1 m2)	m2		6.46		6.46
016	**Extra over excavation for breaking out plain concrete 150mm thick**					
	Extra over excavation irrespective of depth for breaking out concrete paving ne 100 mm thick (D2020/800 - 1 m2), Extra for each additional 25 mm thickness (D2020/801 - 2 m2)	m2		8.62		8.62
018	**Extra over excavation for breaking out plain concrete 200mm thick**					
	Extra over excavation irrespective of depth for breaking out concrete paving ne 100 mm thick (D2020/800 - 1 m2), Extra for each additional 25 mm thickness (D2020/801 - 4 m2)	m2		10.78		10.78
020	**Extra over excavation for breaking out reinforced concrete 100mm thick**					
	Extra over excavation irrespective of depth for breaking out reinforced concrete paving ne 100 mm thick (D2020/810 - 1 m2)	m2		11.85		11.85
022	**Extra over excavation for breaking out reinforced concrete 150mm thick**					
	Extra over excavation irrespective of depth for breaking out reinforced concrete paving ne 100 mm thick (D2020/810 - 1 m2), Extra for each additional 25 mm thickness (D2020/811 - 2 m2)	m2		15.29		15.29
024	**Extra over excavation for breaking out reinforced concrete 200mm thick**					
	Extra over excavation irrespective of depth for breaking out reinforced concrete paving ne 100 mm thick (D2020/810 - 1 m2), Extra for each additional 25 mm thickness (D2020/811 - 4 m2)	m2		18.73		18.73

© NSR 01 Aug 2008 - 31 Jul 2009

Y25

	Y20 : EXTERNAL RAMPS AND STEPS		Mat. £	Lab. £	Plant £	Total £
Y2020						
026	**Extra over excavation for breaking out tarmacadam 75mm thick**					
	Extra over excavation irrespective of depth for breaking out coated macadam or asphalt paving ne 75 mm thick (D2020/815 - 1 m2)	m2		6.46		6.46
028	**Extra over excavation for breaking out tarmacadam 100mm thick**					
	Extra over excavation irrespective of depth for breaking out coated macadam or asphalt paving ne 75 mm thick (D2020/815 - 1 m2), Extra for each additional 25 mm thickness (D2020/816 - 1 m2)	m2		7.54		7.54
030	**Take up concrete paving slabs 50mm thick**					
	Concrete slab paving 50 mm thick (D2015/112 - 1 m2)	m2		6.46	1.23	7.69
032	**Disposal of excavated material off site**					
	Excavated material off site (D2052/005 - 1 m3)	m3		2.09	56.14	58.23
034	**Disposal of excavated material on site**					
	Excavated material on site 50 m average distance and depositing in spoil heaps (D2050/015 - 1 m3)	m3		23.27		23.27
036	**Filling to excavation in hardcore n.e 250mm ave depth**					
	Filling to excavations ne 250 mm av thick (D2070/035 - 1 m3)	m3	38.44	4.42	12.95	55.81
038	**Compacting filling, blinding with sand**					
	Compacting filling, blinding with sand (D2080/080 - 1 m2)	m2	3.18	1.94		5.12
040	**Reinforced concrete designed mix C25 150 - 450mm thick**					
	Slabs 150 mm to 450 mm thick (E1040/375 - 1 m3)	m3	111.55	50.00	7.54	169.09
042	**Reinforced concrete designed mix C25 over 450mm thick**					
	Slabs over 450 mm thick (E1040/390 - 1 m3)	m3	111.55	45.04	6.79	163.38
044	**Plain concrete designed mix C20 150 - 450mm thick**					
	Beds 150 mm to 450 mm thick (E1010/105 - 1 m3)	m3	106.56	34.26		140.82
046	**Plain concrete designed mix C20 over 450mm thick**					
	Beds over 450 mm thick (E1010/110 - 1 m3)	m3	106.56	30.17		136.73
048	**Formwork for insitu concrete; 250 to 500mm high; plain smooth finish**					
	Sides of ground beams and edges of beds 250 mm to 500 mm high (E2025/092 - 1 m)	m	6.17	26.84		33.01

© NSR 01 Aug 2008 - 31 Jul 2009   Y26

	Y20 : EXTERNAL RAMPS AND STEPS		Mat. £	Lab. £	Plant £	Total £
Y2020						
050	**Formwork for insitu concrete; 500 to 1000mm high; plain smooth finish**					
	Sides of ground beams and edges of beds 500 mm to 1000 mm high (E2025/094 - 1 m)	m	12.63	46.42		59.05
052	**Formwork for insitu concrete; over 1000mm high; plain smooth finish**					
	Sides of ground beams and edges of beds over 1 m high (E2025/080 - 1 m2)	m2	12.63	43.90		56.53
054	**Formwork for insitu concrete; stairflights 2000mm wide; plain smooth finish**					
	Stairflights 2 m wide, waist 150 mm, risers vertical 150 mm high (E2025/444 - 1 m)	m	42.65	173.18	21.38	237.21
056	**Brickwork Walls 215mm thick in facing brick PC £375/1000**					
	Walls 215 mm thick, facework one side (F1034/104 - 1 m2)	m2	54.36	36.41	1.33	92.10
065	**Demolish existing 1.00 metre wide reinforced concrete steps and ramp down to 150mm below ground level; approx 450mm high; approx 3nr steps; ramp n.e. 7 metres**					
	To reduce levels 0.25 m max depth (D2030/240 - 2.5 m3), Extra over excavation irrespective of depth for breaking out reinforced concrete (D2030/760 - 2.5 m3), Excavated material off site (D2052/005 - 2.5 m3)	Item		243.32	260.45	503.77
067	**Demolish existing 1.00 metre wide reinforced concrete steps and ramp down to 150mm below ground level; approx 750mm high; approx 5nr steps; Ramp n.e. 12 metres**					
	To reduce levels 0.25 m max depth (D2030/240 - 6.15 m3), Extra over excavation irrespective of depth for breaking out reinforced concrete (D2030/760 - 6.15 m3), Excavated material off site (D2052/005 - 6.15 m3)	Item		598.58	640.71	1239.29
069	**Demolish existing 1.00 metre wide reinforced concrete steps and ramp down to 150mm below ground level; over 5nr steps; Ramp exc. 12 metres**					
	To reduce levels 0.25 m max depth (D2030/240 - 0.7 m3), Extra over excavation irrespective of depth for breaking out reinforced concrete (D2030/760 - 0.7 m3), Excavated material off site (D2052/005 - 0.7 m3)	m2		68.13	72.93	141.06
071	**Demolish existing 2.00 metre wide reinforced concrete steps and ramp down to 150mm below ground level; approx 450mm high; approx 3nr steps; ramp n.e. 7 metres**					
	To reduce levels 0.25 m max depth (D2030/240 - 5 m3), Extra over excavation irrespective of depth for breaking out reinforced concrete (D2030/760 - 5 m3), Excavated material off site (D2052/005 - 5 m3)	Item		486.65	520.90	1007.55

© NSR 01 Aug 2008 - 31 Jul 2009   Y27

	Y20 : EXTERNAL RAMPS AND STEPS		Mat. £	Lab. £	Plant £	Total £
Y2020						
073	**Demolish existing 2.00 metre wide reinforced concrete steps and ramp down to 150mm below ground level; approx 750mm high; approx 5nr steps; Ramp n.e. 12 metres**					
	To reduce levels 0.25 m max depth (D2030/240 - 12.3 m3), Extra over excavation irrespective of depth for breaking out reinforced concrete (D2030/760 - 12.3 m3), Excavated material off site (D2052/005 - 12.3 m3)	Item		1197.16	1281.41	2478.57
075	**Demolish existing 2.00 metre wide reinforced concrete steps and ramp down to 150mm below ground level; over 5nr steps; Ramp exc. 12 metres**					
	To reduce levels 0.25 m max depth (D2030/240 - 0.7 m3), Extra over excavation irrespective of depth for breaking out reinforced concrete (D2030/760 - 0.7 m3), Excavated material off site (D2052/005 - 0.7 m3)	m2		68.13	72.93	141.06
Y2030	**RAMPS**					
110	**Form reinforced concrete ramp to abut existing wall, n.e. 200mm high; 1800mm wide at gradient of 1:15; include for 1800mm long landings at top and bottom; including 100mm high kerbs to open side; (Handrails and surface finishes measured elsewhere)**					
	To reduce levels 0.25 m max depth (D2030/240 - 2.51 m3), Excavated material off site (D2052/005 - 2.51 m3), Filling to excavations ne 250 mm av thick (D2070/035 - 1.25 m3), Compacting filling, blinding with sand (D2082/080 - 12.54 m2), Slabs ne 150 mm thick (E1040/355 - 2.63 m3), Sides of ground beams and edges of beds ne 250 mm high (E2030/090 - 8 m), Square mesh ref A142 2.22 kg/m2 (E3030/210 - 12.54 m2), Trowelling to falls (E4110/072 - 12.54 m2)	Item	437.64	557.88	188.35	1183.87
115	**Form reinforced concrete ramp to abut existing wall, n.e. 350mm high; 1800mm wide at gradient of 1:15; include for 1800mm long landings at top and bottom; including 100mm high kerbs to open side; (Handrails and surface finishes measured elsewhere)**					
	To reduce levels 0.25 m max depth (D2030/240 - 4 m3), Excavated material off site (D2052/005 - 4 m3), Filling to excavations ne 250 mm av thick (D2070/035 - 2.4 m3), Compacting filling, blinding with sand (D2082/080 - 11.34 m2), Beds 150 mm to 450 mm thick (E1040/335 - 4.82 m3), Sides of foundations 250 mm to 500 mm high (E2030/062 - 9 m), Square mesh ref A142 2.22 kg/m2 (E3030/210 - 16 m2), Trowelling to falls (E4110/072 - 16 m2)	Item	771.58	775.01	300.61	1847.20
120	**Form reinforced concrete ramp to abut existing wall, n.e. 500mm high; 1800mm wide at gradient of 1:15; include for 1800mm long landings at top and bottom; including 100mm high kerbs to open side; (Handrails and surface finishes measured elsewhere)**					
	To reduce levels 0.25 m max depth (D2030/240 - 5 m3), Excavated material off site (D2052/005 - 5 m3), Filling to excavations ne 250 mm av thick (D2070/035 - 2.31 m3), Compacting filling, blinding with sand (D2082/080 - 20 m2), Beds 150 mm to 450 mm thick (E1040/335 - 7.64 m3), Sides of ground beams and edges of beds 250 mm to 500 mm high (E2025/092 - 10 m), Square mesh ref A142 2.22 kg/m2 (E3030/210 - 30 m2), Trowelling to falls (E4110/072 - 20 m2)	Item	1109.24	885.02	377.60	2371.86

© NSR 01 Aug 2008 - 31 Jul 2009   Y28

	Y20 : EXTERNAL RAMPS AND STEPS		Mat. £	Lab. £	Plant £	Total £
Y2030						
125	Form reinforced concrete ramp to abut existing wall, n.e. 750mm high; 1800mm wide at gradient of 1:15; include for 1800mm long landings at top and bottom and 1nr 1800mm long intermediate landing; including 100mm high kerbs to open side; (Handrails and surface finishes measured elsewhere)					
	To reduce levels 0.25 m max depth (D2030/240 - 7.91 m3), Excavated material off site (D2052/005 - 7.91 m3), Filling to excavations ne 250 mm av thick (D2070/035 - 4.75 m3), Compacting filling, blinding with sand (D2082/080 - 31.65 m2), Beds 150 mm to 450 mm thick (E1040/335 - 15.01 m3), Sides of ground beams and edges of beds ne 250 mm high (E2025/090 - 4 m), Sides of ground beams and edges of beds 250 mm to 500 mm high (E2025/092 - 3.5 m), Sides of ground beams and edges of beds 500 mm to 1000 mm high (E2025/094 - 7.5 m), Square mesh ref A142 2.22 kg/m2 (E3030/210 - 47.48 m2), Trowelling to falls (E4110/072 - 31.65 m2)	Item	2154.74	1625.17	629.60	4409.51
130	Form reinforced concrete ramp to abut existing wall, n.e. 1000mm high; 1800mm wide at gradient of 1:15; include for 1800mm long landings at top and bottom and 1nr 1800mm long intermediate landing; including 100mm high kerbs to open side; (Handrails and surface finishes measured elsewhere)					
	To reduce levels 0.25 m max depth (D2030/240 - 11.02 m3), Excavated material off site (D2052/005 - 11.02 m3), Filling to excavations ne 250 mm av thick (D2070/035 - 5.51 m3), Compacting filling, blinding with sand (D2082/080 - 36.72 m2), Beds 150 mm to 450 mm thick (E1040/335 - 21.65 m3), Sides of foundations 250 mm to 500 mm high (E2025/062 - 9 m), Sides of foundations 500 mm to 1000 mm high (E2025/064 - 11 m), Square mesh ref A142 2.22 kg/m2 (E3030/210 - 55 m2), Trowelling to falls (E4110/072 - 36.72 m2)	Item	3016.19	2207.86	866.32	6090.37
135	Form reinforced concrete ramp cast indepently of existing wall, n.e. 200mm high; 1800mm wide at gradient of 1:15; include for 1800mm long landings at top and bottom; including 100mm high kerbs to open side; (Handrails and surface finishes measured elsewhere)					
	To reduce levels 0.25 m max depth (D2030/240 - 2.64 m3), Filling to excavations ne 250 mm av thick (D2070/035 - 1.25 m3), Compacting filling, blinding with sand (D2082/080 - 13.2 m2), Slabs ne 150 mm thick (E1040/355 - 2.77 m3), Sides of ground beams and edges of beds ne 250 mm high (E2030/090 - 15 m), Square mesh ref A142 2.22 kg/m2 (E3030/210 - 13.2 m2), Trowelling to falls (E4110/072 - 13.2 m2)	Item	489.90	783.16	49.09	1322.15
140	Form reinforced concrete ramp cast indepently of existing wall, n.e. 350mm high; 1800mm wide at gradient of 1:15; include for 1800mm long landings at top and bottom; including 100mm high kerbs to open side; (Handrails and surface finishes measured elsewhere)					
	To reduce levels 0.25 m max depth (D2030/240 - 4.43 m3), Filling to excavations ne 250 mm av thick (D2070/035 - 2.66 m3), Compacting filling, blinding with sand (D2082/080 - 17.7 m2), Beds 150 mm to 450 mm thick (E1040/335 - 4.96 m3), Sides of foundations 250 mm to 500 mm high (E2030/062 - 12 m), Square mesh ref A142 2.22 kg/m2 (E3030/210 - 17.7 m2), Trowelling to falls (E4110/072 - 17.7 m2)	Item	847.42	924.36	82.63	1854.41

© NSR 01 Aug 2008 - 31 Jul 2009   Y29

	Y20 : EXTERNAL RAMPS AND STEPS		Mat. £	Lab. £	Plant £	Total £
Y2030						
145	**Form reinforced concrete ramp cast indepentently of existing wall, n.e. 500mm high; 1800mm wide at gradient of 1:15; include for 1800mm long landings at top and bottom; including 100mm high kerbs to open side; (Handrails and surface finishes measured elsewhere)**					
	To reduce levels 0.25 m max depth (D2030/240 - 5.55 m3), Filling to excavations ne 250 mm av thick (D2070/035 - 3.33 m3), Compacting filling, blinding with sand (D2082/080 - 22.2 m2), Beds 150 mm to 450 mm thick (E1040/335 - 8.01 m3), Sides of ground beams and edges of beds 250 mm to 500 mm high (E2025/092 - 18 m), Square mesh ref A142 2.22 kg/m2 (E3030/210 - 33.3 m2), Trowelling to falls (E4110/072 - 22.2 m2)	Item	1250.80	1140.09	114.58	2505.47
150	**Form reinforced concrete ramp cast indepentently of existing wall, n.e. 750mm high; 1800mm wide at gradient of 1:15; include for 1800mm long landings at top and bottom and 1nr 1800mm long intermediate landing; including 100mm high kerbs to open side; (Handrails and surface finishes measured elsewhere)**					
	To reduce levels 0.25 m max depth (D2030/240 - 4.7 m3), Excavated material off site (D2052/005 - 4.7 m3), Filling to excavations ne 250 mm av thick (D2070/035 - 5 m3), Compacting filling, blinding with sand (D2082/080 - 33.32 m2), Beds 150 mm to 450 mm thick (E1040/335 - 15.88 m3), Sides of ground beams and edges of beds ne 250 mm high (E2025/090 - 8 m), Sides of ground beams and edges of beds 250 mm to 500 mm high (E2025/092 - 7 m), Sides of ground beams and edges of beds 500 mm to 1000 mm high (E2025/094 - 15 m), Square mesh ref A142 2.22 kg/m2 (E3030/210 - 50 m2), Trowelling to falls (E4110/072 - 33.32 m2)	Item	2399.55	2199.63	447.28	5046.46
155	**Form reinforced concrete ramp cast indepentently of existing wall, n.e. 1000mm high; 1800mm wide at gradient of 1:15; include for 1800mm long landings at top and bottom and 1nr 1800mm long intermediate landing; including 100mm high kerbs to open sides; (Handrails and surface finishes measured elsewhere)**					
	To reduce levels 1 m max depth (D2030/242 - 12.24 m3), Excavated material off site (D2052/005 - 12.24 m3), Filling to excavations ne 250 mm av thick (D2070/035 - 6.12 m3), Compacting filling, blinding with sand (D2082/080 - 40.8 m2), Beds 150 mm to 450 mm thick (E1040/335 - 22.87 m3), Sides of foundations 250 mm to 500 mm high (E2025/062 - 18 m), Sides of foundations 500 mm to 1000 mm high (E2025/064 - 22 m), Square mesh ref A142 2.22 kg/m2 (E3030/210 - 60 m2), Trowelling to falls (E4110/072 - 40.8 m2)	Item	3389.77	3067.97	946.64	7404.38
170	**Extra over concrete ramps for facing in 215mm facing brick PC £375/1000**					
	Beds ne 150 mm thick (E1040/330 - -0.01 m3), Sides of ground beams and edges of beds over 1 m high (E2025/080 - -1 m2), Walls 215 mm thick, facework one side (F1034/104 - 1 m2)	m2	40.61	-7.99	1.25	33.87
175	**Extra over for brick on edge coping to edge of ramp**					
	Brick on edge copings 225 x 112 mm horizontal, set weathering (F1032/400 - 1 m)	m	6.96	10.29	0.38	17.63

© NSR 01 Aug 2008 - 31 Jul 2009   Y30

	Y20 : EXTERNAL RAMPS AND STEPS		Mat. £	Lab. £	Plant £	Total £
Y2030						
180	**Extra over 350mm high concrete ramp for steps**					
	To reduce levels 0.25 m max depth (D2030/240 - 0.27 m3), Excavated material off site (D2052/005 - 0.27 m3), Filling to excavations ne 250 mm av thick (D2070/035 - 0.16 m3), Staircases (E1040/585 - 0.351 m3), Sides of ground beams and edges of beds 250 mm to 500 mm high (E2025/092 - - 1.8 m), Stairflights 2 m wide, waist 150 mm, risers vertical 150 mm high (E2025/444 - 0.6 m), Square mesh ref A142 2.22 kg/m2 bent (E3030/215 - 1.2 m2), Trowelling (E4110/070 - 1.08 m2)	Item	61.50	109.66	37.32	208.48
182	**Extra over 500mm high concrete ramp for steps**					
	To reduce levels 0.25 m max depth (D2030/240 - 0.54 m3), Excavated material off site (D2052/005 - 0.54 m3), Filling to excavations ne 250 mm av thick (D2070/035 - 0.32 m3), Staircases (E1040/585 - 1.3 m3), Sides of ground beams and edges of beds 250 mm to 500 mm high (E2025/092 - - 1.8 m), Stairflights 2 m wide, waist 150 mm, risers vertical 150 mm high (E2025/444 - 1.2 m), Square mesh ref A142 2.22 kg/m2 bent (E3030/215 - 3.24 m2), Trowelling (E4110/070 - 2.16 m2)	Item	202.02	341.37	85.45	628.84
184	**Extra over 750mm high concrete ramp for steps**					
	To reduce levels 0.25 m max depth (D2030/240 - 0.68 m3), Excavated material off site (D2052/005 - 0.68 m3), Filling to excavations ne 250 mm av thick (D2070/035 - 0.41 m3), Staircases (E1040/585 - 1.98 m3), Sides of ground beams and edges of beds 250 mm to 500 mm high (E2025/092 - - 1.8 m), Stairflights 2 m wide, waist 150 mm, risers vertical 150 mm high (E2025/444 - 1.5 m), Square mesh ref A142 2.22 kg/m2 bent (E3030/215 - 5.4 m2), Trowelling (E4110/070 - 2.7 m2)	Item	297.22	484.72	113.65	895.59
186	**Extra over 1000mm high concrete ramp for steps**					
	To reduce levels 0.25 m max depth (D2030/240 - 0.82 m3), Excavated material off site (D2052/005 - 0.82 m3), Filling to excavations ne 250 mm av thick (D2070/035 - 0.57 m3), Staircases (E1040/585 - 3.29 m3), Sides of ground beams and edges of beds 250 mm to 500 mm high (E2025/092 - - 1.8 m), Stairflights 2 m wide, waist 150 mm, risers vertical 150 mm high (E2025/444 - 2.1 m), Square mesh ref A142 2.22 kg/m2 bent (E3030/215 - 11.34 m2), Trowelling (E4110/070 - 3.78 m2)	Item	483.59	768.89	160.56	1413.04
Y2040	**PREFABRICATED EXTERNAL RAMPS**					
200	**Supply stepless portable ramp 2000 x 700mm wide**					
	Stepless portable ramp 2000 x 770mm N&C ref: P64330.17 (N1055/010 - 1 nr)	nr	428.40			428.40
202	**Supply stepless portable ramp 2500 x 770mm wide**					
	Stepless portable folding ramp 2500 x 770mm N&C ref: P64330.21 (N1055/015 - 1 nr)	nr	656.00			656.00

© NSR 01 Aug 2008 - 31 Jul 2009   Y31

	Y20 : EXTERNAL RAMPS AND STEPS		Mat. £	Lab. £	Plant £	Total £
Y2040						
204	**Supply standard wheelchair ramp 1220mm long**					
	Standard wheelchair ramp 1220mm long (N1055/025 - 1 nr)	nr	262.44			262.44
206	**Supply standard wheelchair ramp with single folding handrail 1830mm long**					
	Standard wheelchair ramp with single folding handrail 1830mm long (N1055/030 - 1 nr)	nr	620.00			620.00
208	**Supply standard wheelchair ramp with double folding handrail 1830mm long**					
	Standard wheelchair ramp with double folding handrail 1830mm long (N1055/035 - 1 nr)	nr	780.00			780.00

© NSR 01 Aug 2008 - 31 Jul 2009   Y32

			Mat. £	Lab. £	Plant £	Total £
	Y21 : FINISHES AND SURFACE TREATMENTS (EXTERNALLY)					
Y2120	**ANTI SLIP SURFACING**					
100	**Apply anti slip tape to concrete stairs, ramps or steps 51mm wide, black**					
	Anti slip tape 51mm wide, black (M5090/100 - 1 m), extra over for application of edge sealing compound (M5090/132 - 2 m), extra over for application of floor primer (M5090/134 - 0.051 m2)	m	3.74	3.61		7.35
102	**Apply anti slip tape to concrete stairs, ramps or steps 51mm wide, coloured**					
	Anti slip tape 51mm wide, coloured (M5090/102 - 1 m), extra over for application of edge sealing compound (M5090/132 - 2 m), extra over for application of floor primer (M5090/134 - 0.051 m2)	m	4.71	3.61		8.32
104	**Apply anti slip tape to concrete stairs, ramps or steps 102mm wide, black**					
	Anti slip tape 102mm wide, black (M5090/104 - 1 m), extra over for application of edge sealing compound (M5090/132 - 2 m), extra over for application of floor primer (M5090/134 - 0.102 m2)	m	6.58	4.34		10.92
106	**Apply anti slip tape to concrete stairs, ramps or steps 102mm wide, coloured**					
	Anti slip tape 102mm wide, coloured (M5090/106 - 1 m), extra over for application of edge sealing compound (M5090/132 - 2 m), extra over for application of floor primer (M5090/134 - 0.102 m2)	m	8.50	4.34		12.84
108	**Apply anti slip tape to marble or terrazzo surfaces 25mm wide, clear**					
	Anti slip tape 25mm wide, clear (M5090/108 - 1 m), extra over for application of edge sealing compound (M5090/132 - 2 m), extra over for application of floor primer (M5090/134 - 0.025 m2)	m	3.46	3.53		6.99
110	**Apply anti slip tape to marble or terrazzo surfaces 51mm wide, clear**					
	Anti slip tape 51mm wide, clear (M5090/110 - 1 m), extra over for application of edge sealing compound (M5090/132 - 2 m), extra over for application of floor primer (M5090/134 - 0.051 m2)	m	6.03	3.61		9.64
112	**Apply anti slip tape to marble or terrazzo surfaces 102mm wide, clear**					
	Anti slip tape 102mm wide, clear (M5090/112 - 1 m), extra over for application of edge sealing compound (M5090/132 - 2 m), extra over for application of floor primer (M5090/134 - 0.102 m2)	m	11.16	4.34		15.50
114	**Apply anti slip cleats to marble or terrazzo surfaces 152 x 610mm wide, clear**					
	Anti slip cleats 152mm x 610, clear (M5090/116 - 1 nr), extra over for application of edge sealing compound (M5090/132 - 1.52 m), extra over for application of floor primer (M5090/134 - 0.093 m2)	nr	11.73	3.90		15.63

© NSR 01 Aug 2008 - 31 Jul 2009

Y33

Y21 : FINISHES AND SURFACE TREATMENTS (EXTERNALLY)			Mat. £	Lab. £	Plant £	Total £
Y2120						
120	**Supply and fix aluminium anti slip plates to metal or timber steps or stairways, 114 x 635mm**					
	Aluminium Anti slip plates 114mm x 635, screwed to metal or timber; (M5090/122 - 1 nr), extra over for application of edge sealing compound (M5090/132 - 1.5 m)	nr	43.89	5.90		49.79
122	**Supply and fix self adhesive anti slip cleats, 19 x 610mm, black**					
	Anti slip cleats 19mm x 610, black (M5090/124 - 1 nr), extra over for application of edge sealing compound (M5090/132 - 1.26 m), extra over for application of floor primer (M5090/134 - 0.012 m2)	nr	1.89	2.85		4.74
124	**Supply and fix self adhesive anti slip cleats, 19 x 610mm, coloured**					
	Anti slip cleats 19mm x 610, coloured (M5090/125 - 1 nr), extra over for application of edge sealing compound (M5090/132 - 1.26 m), extra over for application of floor primer (M5090/134 - 0.012 m2)	nr	1.86	2.85		4.71
126	**Supply and fix self adhesive anti slip cleats, 152 x 610mm, black**					
	Anti slip cleats 152mm x 610, black (M5090/126 - 1 nr), extra over for application of edge sealing compound (M5090/132 - 1.52 m), extra over for application of floor primer (M5090/134 - 0.093 m2)	nr	6.38	5.06		11.44
128	**Supply and fix self adhesive anti slip cleats, 152 x 610mm, yellow**					
	Anti slip cleats 152mm x 610, yellow (M5090/128 - 1 nr), extra over for application of edge sealing compound (M5090/132 - 1.52 m), extra over for application of floor primer (M5090/134 - 0.093 m2)	nr	7.11	5.06		12.17
130	**Supply and fix self adhesive anti slip tread, 140 x 140mm, yellow**					
	Anti slip tread 140mm x 140, yellow (M5090/130 - 1 nr), extra over for application of edge sealing compound (M5090/132 - 0.56 m), extra over for application of floor primer (M5090/134 - 0.02 m2)	nr	1.98	2.27		4.25
140	**Apply acrylic resin anti slip coating to floors, any colour**					
	Acrylic resin anti slip coating to floors, any colour (M5090/136 - 1 m2)	m2	8.93	5.76		14.69
150	**Supply and lay Takpave by Rediweld, tactile surface tiles, Dot, dash or Corduroy tiles; individual tiles**					
	Pavings level or to falls in areas ne 2 m2 (Q2610/015 - 0.18 m2)	nr	14.25	5.98		20.23
152	**Supply and lay Takpave by Rediweld, tactile surface tiles, Dot, dash or Corduroy tiles in small areas n.e 2m2**					
	Pavings level or to falls in areas ne 2 m2 (Q2610/015 - 1 m2)	m2	79.16	33.23		112.39
154	**Supply and lay Takpave by Rediweld, tactile surface tiles, Dot, dash or Corduroy tiles in areas exc 2m2**					
	Pavings level or to falls (Q2610/010 - 1 m2)	m2	79.16	27.69		106.85

© NSR 01 Aug 2008 - 31 Jul 2009   Y34

	Y21 : FINISHES AND SURFACE TREATMENTS (EXTERNALLY)		Mat. £	Lab. £	Plant £	Total £
Y2130	WALL FINISHES					
200	**Cement sand render in two coats 13mm thick, over 300mm wide**					
	Walls over 300 mm wide to brickwork, blockwork or stonework (M2010/050 - 1 m2)	m2	2.15	9.50	0.35	12.00
202	**Cement sand render in two coats 13mm thick n.e. 300mm wide**					
	Walls ne 300 mm wide to brickwork, blockwork or stonework (M2010/100 - 1 m)	m	0.64	7.91	0.29	8.84
204	**Cement sand render in two coats 13mm thick with pebbledash finish over 300mm wide**					
	Walls over 300 mm wide to brickwork, blockwork or stonework (M2014/050 - 1 m2)	m2	3.05	9.50	0.35	12.90
206	**Cement sand render in two coats 13mm thick with pebbledash finish n.e. 300mm wide**					
	Walls ne 300 mm wide to brickwork, blockwork or stonework (M2014/100 - 1 m)	m	0.91	7.91	0.29	9.11
210	**Cement sand render in two coats 13mm thick with Tyrolean finish over 300mm wide**					
	Walls over 300 mm wide to brickwork, blockwork or stonework (M2015/050 - 1 m2)	m2	13.32	14.25	0.73	28.30
212	**Cement sand render in two coats 13mm thick with Tyrolean finish n.e. 300mm wide**					
	Walls ne 300 mm wide to brickwork, blockwork or stonework (M2015/100 - 1 m)	m	4.00	13.82	0.65	18.47
220	**Extra over for angle bead**					
	External angle render bead (M2080/910 - 1 m)	m	0.92	2.37		3.29
222	**Extra over for stop bead**					
	External render stop (M2080/915 - 1 m)	m	1.24	2.37		3.61

© NSR 01 Aug 2008 - 31 Jul 2009

Y21 : FINISHES AND SURFACE TREATMENTS (EXTERNALLY)	Mat. £	Lab. £	Plant £	Total £

© NSR 01 Aug 2008 - 31 Jul 2009 — Y36

			Mat. £	Lab. £	Plant £	Total £
Y2210	**Y22 : WINDOWS, DOORS & DOORWAYS**					
	CANOPIES					
002	**Supply and fit GRP flat roofed canopy 1500 x 800mm projection**					
	Flashings 150 mm girth horizontal (H7110/100 - 1.5 m), Aprons 150 mm girth horizontal (H7110/250 - 1.5 m), GRP flat roofed canopy, 1500 x 800mm projection (PC £250) (L1410/050 - 1 nr)	nr	291.78	59.20		350.98
004	**Supply and fit GRP flat roofed canopy 2500 x 800mm projection**					
	Flashings 150 mm girth horizontal (H7110/100 - 2.5 m), Aprons 150 mm girth horizontal (H7110/250 - 2.5 m), GRP flat roofed canopy, 2500 x 800mm projection (PC £400) (L1410/055 - 1 nr)	nr	469.20	90.27		559.47
006	**Supply and fit GRP flat roofed canopy 3500 x 800mm projection**					
	Flashings 150 mm girth horizontal (H7110/100 - 3.5 m), Aprons 150 mm girth horizontal (H7110/250 - 3.5 m), GRP flat roofed canopy, 3500 x 800mm projection (PC £500) (L1410/060 - 1 nr)	nr	595.32	121.35		716.67
008	**Supply and fit GRP mono pitched canopy 1500 x 800mm projection**					
	Flashings 150 mm girth horizontal (H7110/100 - 1.5 m), Aprons 150 mm girth horizontal (H7110/250 - 1.5 m), GRP mono pitched canopy, 1500 x 800mm projection (PC £400) (L1410/065 - 1 nr)	nr	441.78	59.20		500.98
010	**Supply and fit GRP mono pitched canopy 2500 x 800mm projection**					
	Flashings 150 mm girth horizontal (H7110/100 - 2.5 m), Aprons 150 mm girth horizontal (H7110/250 - 2.5 m), GRP mono pitched canopy, 2500 x 800mm projection (PC £400) (L1410/070 - 1 nr)	nr	619.20	90.27		709.47
012	**Supply and fit GRP mono pitched canopy 3500 x 800mm projection**					
	Flashings 150 mm girth horizontal (H7110/100 - 3.5 m), Aprons 150 mm girth horizontal (H7110/250 - 3.5 m), GRP mono pitched canopy, 3500 x 800mm projection (PC £650) (L1410/075 - 1 nr)	nr	745.32	121.35		866.67
014	**Supply and fit GRP duo pitched canopy 1500 x 800mm projection**					
	Flashings 150 mm girth stepped (H7110/110 - 2.12 m), GRP duo pitched canopy, 1500 x 800mm projection (PC £400) (L1410/080 - 1 nr)	nr	434.36	66.04		500.40
016	**Supply and fit GRP duo pitched canopy 2500 x 800mm projection**					
	Flashings 150 mm girth stepped (H7110/110 - 3.53 m), GRP duo pitched canopy, 2500 x 800mm projection (PC £400) (L1410/085 - 1 nr)	nr	605.93	99.91		705.84
018	**Supply and fit GRP duo pitched canopy 3500 x 800mm projection**					
	Flashings 150 mm girth stepped (H7110/110 - 4.95 m), GRP duo pitched canopy, 3500 x 800mm projection (PC £650) (L1410/090 - 1 nr)	nr	726.33	134.01		860.34

© NSR 01 Aug 2008 - 31 Jul 2009

Y37

	Y22 : WINDOWS, DOORS & DOORWAYS		Mat. £	Lab. £	Plant £	Total £
Y2210						
020	**Supply and fit GRP cantilevered canopy 3275 x 2440mm projection**					
	Flashings 150 mm girth stepped (H7110/110 - 3.3 m), GRP cantilevered canopy, 3275 x 2440mm projection; (PC £700); flashings measured seperately (L1410/100 - 1 nr)	nr	753.92	148.08		902.00
022	**Supply and fit GRP cantilevered canopy 4875 x 2440mm projection**					
	Flashings 150 mm girth stepped (H7110/110 - 4.9 m), GRP cantilevered canopy, 4875 x 2440mm projection; (PC £950); flashings measured seperately (L1410/105 - 1 nr)	nr	1028.21	195.91		1224.12
024	**Supply and fit GRP cantilevered canopy 6475 x 2440mm projection**					
	Flashings 150 mm girth stepped (H7110/110 - 6.5 m), GRP cantilevered canopy, 6475 x 2440mm projection; (PC £950); flashings measured seperately (L1410/110 - 1 nr)	nr	1306.40	238.72		1545.12
Y2220	**FORMING NEW DOORWAYS**					
015	**Forming new single door opening through masonry wall up to 250mm thick**					
	Cutting to form openings through 215 mm brick wall 900 x 2000 mm, insert precast prestressed concrete lintel 65 mm deep, wedging and pinning to underside of existing construction, forming square jambs and sills (C2010/628 - 1 nr)	nr	30.30	347.41	23.23	400.94
020	**Forming new single door opening through block wall**					
	Cutting to form openings through 100 mm block wall 900 x 2000 mm, insert precast prestressed concrete lintel 65 mm deep, wedging and pinning to underside of existing construction, forming square jambs and sills (C2010/638 - 1 nr)	nr	17.52	188.81	7.06	213.39
025	**Forming new single door opening through metal or timber stud partition**					
	Cutting to form openings through 125 mm stud partition 900 x 2000 mm, inserting studs at head and jambs, making good structure, making good finishings (C2010/681 - 1 nr)	nr	24.24	69.28		93.52
035	**Forming new double door opening through masonry wall up to 250mm thick**					
	Cutting to form openings through 215 mm brick wall 1200 x 2000 mm, insert precast prestressed concrete lintel 65 mm deep, wedging and pinning to underside of existing construction, forming square jambs and sills (C2010/630 - 1 nr)	nr	34.87	407.83	23.23	465.93
040	**Forming new double door opening through block wall**					
	Cutting to form openings through 100 mm block wall 1200 x 2000 mm, insert precast prestressed concrete lintel 65 mm deep, wedging and pinning to underside of existing construction, forming square jambs and sills (C2010/640 - 1 nr)	nr	24.21	211.47	11.95	247.63

© NSR 01 Aug 2008 - 31 Jul 2009   Y38

	Y22 : WINDOWS, DOORS & DOORWAYS		Mat. £	Lab. £	Plant £	Total £
Y2220						
045	**Forming new double door opening through metal or timber stud partition**					
	Cutting to form openings through 100 mm stud partition 800 x 2000 mm, inserting studs at head and jambs, making good structure, making good finishings (C2010/675 - 1.5 nr)	nr	29.07	94.97		124.04
Y2230	**WIDENING EXISTING DOORWAYS**					
100	**Widen existing single doorway in masonry wall up to 250mm thick**					
	Cutting to form openings through 215 mm brick wall ne 0.50 m2 (C2010/595 - 1 nr), Cutting out defective precast reinforced concrete internal lintel and renewing with 150 x 100 x 1000 mm long precast concrete lintel (C4005/050 - 1 nr), Stripping off/removing/taking down doors in preparation for replacement (C5170/800 - 1 nr), Stripping off/removing/taking down door frames and linings, making good finishings in preparation for replacement (C5170/818 - 1 nr), Walls ne 300 mm wide to brickwork, blockwork or stonework (M2025/100 - 6 m), Bonding/jointing new to existing (M2025/150 - 6 m)	nr	20.65	254.59	14.95	290.19
105	**Widen existing single doorway in blockwork wall**					
	Cutting to form openings through 100 mm block wall ne 0.50 m2 (C2010/605 - 1 nr), Cutting out defective precast reinforced concrete internal lintel and renewing with 150 x 100 x 1000 mm long precast concrete lintel (C4005/050 - 1 nr), Stripping off/removing/taking down doors in preparation for replacement (C5170/800 - 1 nr), Stripping off/removing/taking down door frames and linings, making good finishings in preparation for replacement (C5170/818 - 1 nr), Walls ne 300 mm wide to brickwork, blockwork or stonework (M2025/100 - 6 m), Bonding/jointing new to existing (M2025/150 - 6 m)	nr	20.65	237.98	10.04	268.67
110	**Widen existing single doorway in metal or timber stud partition wall**					
	Cutting to form openings through stud partition ne 0.50 m2 (C2010/670 - 1 nr), Stripping off/removing/taking down doors in preparation for replacement (C5170/800 - 1 nr), Stripping off/removing/taking down door frames and linings, making good finishings in preparation for replacement (C5170/818 - 1 nr), Wall or partition members 50 x 75 mm (G2010/122 - 3 m)	nr	16.04	45.47		61.51
115	**Widen existing double doorway in masonry wall up to 250mm thick**					
	Cutting to form openings through 215 mm brick wall ne 0.50 m2 (C2010/595 - 1 nr), Cutting out defective precast reinforced concrete internal lintel and renewing with 150 x 100 x 2000 mm long precast concrete lintel (C4005/055 - 1 nr), Stripping off/removing/taking down doors in preparation for replacement (C5170/800 - 2 nr), Stripping off/removing/taking down door frames and linings, making good finishings in preparation for replacement (C5170/818 - 1 nr), Walls ne 300 mm wide to brickwork, blockwork or stonework (M2025/100 - 8 m), Bonding/jointing new to existing (M2025/150 - 8 m)	nr	30.97	305.54	16.24	352.75

© NSR 01 Aug 2008 - 31 Jul 2009

Y39

	Y22 : WINDOWS, DOORS & DOORWAYS		Mat. £	Lab. £	Plant £	Total £
Y2230						
120	**Widen existing double doorway in blockwork wall**					
	Cutting to form openings through 100 mm block wall ne 0.50 m2 (C2010/605 - 1 nr), Cutting out defective precast reinforced concrete internal lintel and renewing with 150 x 100 x 2000 mm long precast concrete lintel (C4005/055 - 1 nr), Stripping off/removing/taking down doors in preparation for replacement (C5170/800 - 1 nr), Stripping off/removing/taking down door frames and linings, making good finishings in preparation for replacement (C5170/818 - 1 nr), Walls ne 300 mm wide to brickwork, blockwork or stonework (M2025/100 - 8 m), Bonding/jointing new to existing (M2025/150 - 8 m)	nr	30.97	283.17	11.33	325.47
125	**Widen existing double doorway in metal or timber stud partition wall**					
	Cutting to form openings through stud partition ne 0.50 m2 (C2010/670 - 1 nr), Stripping off/removing/taking down doors in preparation for replacement (C5170/800 - 2 nr), Stripping off/removing/taking down door frames and linings, making good finishings in preparation for replacement (C5170/818 - 1 nr), Wall or partition members 50 x 75 mm (G2010/122 - 4 m)	nr	19.87	54.40		74.27
Y2240	**BLOCK UP EXISTING DOORWAYS**					
150	**Block up existing single door opening in internal masonry wall 215mm thick; plastered both sides; decorated.**					
	Filling in openings 900 x 2000 x 215 mm thick, bonding and wedging and pinning to existing (C2065/854 - 1 nr), Timber skirtings, 25 x 100 mm, cutting out mid-section ne 1 m long replacing with new jointed to existing (C5105/135 - 2 nr), Stripping off/removing/taking down doors in preparation for replacement (C5170/800 - 1 nr), Stripping off/removing/taking down door frames and linings, making good finishings in preparation for replacement (C5170/818 - 1 nr), Walls over 300 mm wide to brickwork, blockwork or stonework 1 to 5 m2 (M2025/051 - 4 m2), Bonding/jointing new to existing (M2025/150 - 6 m), Plaster general surfaces over 300 mm girth (M6015/140 - 4 m2)	nr	101.14	361.53	1.96	464.63
155	**Block up existing single door opening in external masonry wall 215mm thick; plastered internally, facing brick externally; decorated.**					
	Filling in openings 900 x 2000 x 215 mm thick, bonding and wedging and pinning to existing (C2070/854 - 1 nr), Timber skirtings, 25 x 100 mm, cutting out mid-section ne 1 m long replacing with new jointed to existing (C5105/135 - 1 nr), Stripping off/removing/taking down doors in preparation for replacement (C5170/800 - 1 nr), Stripping off/removing/taking down door frames and linings, making good finishings in preparation for replacement (C5170/818 - 1 nr), Walls over 300 mm wide to brickwork, blockwork or stonework 1 to 5 m2 (M2025/051 - 2 m2), Bonding/jointing new to existing (M2025/150 - 3 m), Plaster general surfaces over 300 mm girth (M6015/140 - 2 m2)	nr	111.98	307.51	0.98	420.47

© NSR 01 Aug 2008 - 31 Jul 2009   Y40

	Y22 : WINDOWS, DOORS & DOORWAYS		Mat. £	Lab. £	Plant £	Total £
Y2240						
160	**Block up existing single door opening in internal blockwork wall; plastered both sides; decorated.**					
	Filling in openings 900 x 2000 x 100 mm thick, bonding and wedging and pinning to existing (C2075/850 - 1 nr), Timber skirtings, 25 x 100 mm, cutting out mid-section ne 1 m long replacing with new jointed to existing (C5105/135 - 2 nr), Stripping off/removing/taking down doors in preparation for replacement (C5170/800 - 1 nr), Stripping off/removing/taking down door frames and linings, making good finishings in preparation for replacement (C5170/818 - 1 nr), Walls over 300 mm wide to brickwork, blockwork or stonework 1 to 5 m2 (M2025/051 - 4 m2), Bonding/jointing new to existing (M2025/150 - 6 m), Plaster general surfaces over 300 mm girth (M6015/140 - 4 m2)	nr	60.88	186.76	1.96	249.60
165	**Block up existing single door opening in metal or timber stud partition wall; plastered both sides; decorated.**					
	Timber skirtings, 25 x 100 mm, cutting out mid-section ne 1 m long replacing with new jointed to existing (C5105/135 - 2 nr), Stripping off/removing/taking down doors in preparation for replacement (C5170/800 - 1 nr), Stripping off/removing/taking down door frames and linings, making good finishings in preparation for replacement (C5170/818 - 1 nr), Wall or partition members 50 x 75 mm (G2010/122 - 9 m), Walls over 300 mm wide to timber 1 to 5 m2 (M2040/151 - 4 m2), Plaster general surfaces over 300 mm girth (M6015/140 - 4 m2)	nr	75.32	159.93	2.96	238.21
170	**Block up existing double door opening in internal masonry wall 215mm thick; plastered both sides; decorated.**					
	Filling in openings 1200 x 2000 x 215 mm thick, bonding and wedging and pinning to existing (C2065/878 - 1 nr), Timber skirtings, 25 x 100 mm, cutting out mid-section ne 1 m long replacing with new jointed to existing (C5105/135 - 3.6 nr), Stripping off/removing/taking down doors in preparation for replacement (C5170/800 - 2 nr), Stripping off/removing/taking down door frames and linings, making good finishings in preparation for replacement (C5170/818 - 1 nr), Walls over 300 mm wide to brickwork, blockwork or stonework 1 to 5 m2 (M2025/051 - 7.2 m2), Bonding/jointing new to existing (M2025/150 - 12 m), Plaster general surfaces over 300 mm girth (M6015/140 - 7.2 m2)	nr	143.20	535.34	3.53	682.07
175	**Block up existing double door opening in external masonry wall 215mm thick; plastered internally, facing brick externally; decorated.**					
	Filling in openings 1200 x 2000 x 215 mm thick, bonding and wedging and pinning to existing (C2070/878 - 1 nr), Timber skirtings, 25 x 100 mm, cutting out mid-section ne 1 m long replacing with new jointed to existing (C5105/135 - 3.6 nr), Stripping off/removing/taking down doors in preparation for replacement (C5170/800 - 2 nr), Stripping off/removing/taking down door frames and linings, making good finishings in preparation for replacement (C5170/818 - 1 nr), Walls over 300 mm wide to brickwork, blockwork or stonework 1 to 5 m2 (M2025/051 - 3.6 m2), Bonding/jointing new to existing (M2025/150 - 5.6 m), Plaster general surfaces over 300 mm girth (M6015/140 - 3.6 m2)	nr	157.23	451.15	1.76	610.14

© NSR 01 Aug 2008 - 31 Jul 2009   Y41

	Y22 : WINDOWS, DOORS & DOORWAYS		Mat. £	Lab. £	Plant £	Total £
Y2240						
180	**Block up existing double door opening in internal blockwork wall; plastered both sides; decorated.**					
	Filling in openings 1200 x 2000 x 100 mm thick, bonding and wedging and pinning to existing (C2075/874 - 1 nr), Timber skirtings, 25 x 100 mm, cutting out mid-section ne 1 m long replacing with new jointed to existing (C5105/135 - 3.6 nr), Stripping off/removing/taking down doors in preparation for replacement (C5170/800 - 2 nr), Stripping off/removing/taking down door frames and linings, making good finishings in preparation for replacement (C5170/818 - 1 nr), Walls over 300 mm wide to brickwork, blockwork or stonework 1 to 5 m2 (M2025/051 - 7.2 m2), Bonding/jointing new to existing (M2025/150 - 11.2 m), Plaster general surfaces over 300 mm girth (M6015/140 - 7.2 m2)	nr	89.38	307.83	3.53	400.74
185	**Block up existing double door opening in metal or timber stud partition wall; plastered both sides; decorated.**					
	Stripping off/removing/taking down doors in preparation for replacement (C5170/800 - 2 nr), Stripping off/removing/taking down door frames and linings, making good finishings in preparation for replacement (C5170/818 - 1 nr), Wall or partition members 50 x 75 mm (G2010/122 - 14 m), Walls over 300 mm wide to timber 1 to 5 m2 (M2040/151 - 7.2 m2), Plaster general surfaces over 300 mm girth (M6015/140 - 7.2 m2), Skirtings, picture rails, architraves and the like 19 x 75 mm (P2010/068 - 3.6 m)	nr	122.68	257.55	5.33	385.56
Y2250	**DOORS AND FRAMES INTERNAL**					
212	**Supply and fix 35mm thick internal plywood faced flush door, 900mm wide new ironmongery, decorated**					
	Easing doors (C5170/825 - 1 nr), Doors 35 mm thick, non standard, formed from 2134 x 914 mm blank (L2030/250 - 1 nr), Wood general surfaces over 300 mm girth (M6040/570 - 3.5 m2), Wood general surfaces over 300 mm girth (M6050/570 - 3.5 m2), Pairs butt hinges 75 mm ref H.04400.31 (P2110/056 - 1 nr), Mortice latches 75 mm ref H.00410.03 (P2110/243 - 1 nr), Sets lever furniture 95 x 38 mm ref H.02880.01, SAA finish (P2140/260 - 1 nr)	nr	82.69	127.25		209.94
214	**Supply and fix 40mm thick internal plywood faced flush door, 900mm wide, new ironmongery, decorated**					
	Easing doors (C5170/825 - 1 nr), Doors 40 mm thick, non standard, formed from 2134 x 914 mm blank (L2030/255 - 1 nr), Wood general surfaces over 300 mm girth (M6040/570 - 3.5 m2), Wood general surfaces over 300 mm girth (M6050/570 - 3.5 m2), Pairs butt hinges 75 mm ref H.04400.31 (P2110/056 - 1 nr), Mortice latches 75 mm ref H.00410.03 (P2110/243 - 1 nr), Sets lever furniture 95 x 38 mm ref H.02880.01, SAA finish (P2140/260 - 1 nr)	nr	102.21	130.13		232.34

© NSR 01 Aug 2008 - 31 Jul 2009   Y42

	Y22 : WINDOWS, DOORS & DOORWAYS		Mat. £	Lab. £	Plant £	Total £
Y2250						
216	**Supply and fit 35mm thick internal plywood faced flush door, 900mm wide with 2 vision panels approx 700 x 150mm each, new ironmongery , decorated**					
	Easing doors (C5170/825 - 1 nr), Doors 35 mm thick, non standard, formed from 2134 x 914 mm blank (L2030/250 - 1 nr), Extra for forming panel (any size for glazing) (L2030/350 - 2 nr), To wood with screwed beads and insulating strips in panes (beads and strips included elsewhere) 0.15 to 4 m2 (L4039/140 - 0.21 m2), Strips to edge of glass and the like 6 mm thick (L4090/556 - 3.4 m), Wood glazed windows and screens over 300 mm girth in panes 0.10 to 0.50 m2 (M6040/602 - 3.5 m2), Wood glazed windows and screens over 300 mm girth in panes 0.10 to 0.50 m2 (M6050/602 - 3.5 m2), Cover fillets, stops, trims, beads, nosings and the like 12 x 12 mm (P2020/100 - 3.4 m), Pairs butt hinges 75 mm ref H.04400.31 (P2110/056 - 1 nr), Mortice latches 75 mm ref H.00410.03 (P2110/243 - 1 nr), Sets lever furniture 95 x 38 mm ref H.02880.01, SAA finish (P2140/260 - 1 nr)	nr	107.82	218.66		326.48
218	**Supply and fix 40mm thick internal plywood faced flush door, 900mm wide with 2 vision panels approx 700 x 150mm each, new ironmongery , decorated**					
	Easing doors (C5170/825 - 1 nr), Doors 40 mm thick, non standard, formed from 2134 x 914 mm blank (L2030/255 - 1 nr), Extra for forming panel (any size for glazing) (L2030/350 - 2 nr), To wood with screwed beads and insulating strips in panes (beads and strips included elsewhere) ne 0.15 m2 (L4039/130 - 0.21 m2), Strips to edge of glass and the like 6 mm thick (L4090/556 - 3.4 m), Wood glazed windows and screens over 300 mm girth in panes 0.10 to 0.50 m2 (M6040/602 - 3.5 m2), Wood glazed windows and screens over 300 mm girth in panes 0.10 to 0.50 m2 (M6050/602 - 3.5 m2), Cover fillets, stops, trims, beads, nosings and the like 12 x 12 mm (P2020/100 - 3.4 m), Pairs butt hinges 75 mm ref H.04400.31 (P2110/056 - 1 nr), Mortice latches 75 mm ref H.00410.03 (P2110/243 - 1 nr), Sets lever furniture 95 x 38 mm ref H.02880.01, SAA finish (P2140/260 - 1 nr)	nr	127.34	229.15		356.49
220	**Supply and fix 44mm thick 30 min fire resisting internal plywood faced flush door, 900mm wide, new ironmongery, decorated**					
	Doors 44 mm thick, non standard, formed from 2134 x 914 mm blank (L2040/260 - 1 nr), Wood general surfaces over 300 mm girth (M6040/570 - 3.5 m2), Wood general surfaces over 300 mm girth (M6050/570 - 3.5 m2), Intumescent strips with smoke seal, white aluminium holder (P1080/160 - 5 m), Pairs butt hinges 75 mm ref H.04400.31 (P2110/056 - 1 nr), Mortice latches 75 mm ref H.00410.03 (P2110/243 - 1 nr), Sets lever furniture 95 x 38 mm ref H.02880.01, SAA finish (P2140/260 - 1 nr)	nr	142.00	162.53		304.53

© NSR 01 Aug 2008 - 31 Jul 2009   Y43

	Y22 : WINDOWS, DOORS & DOORWAYS		Mat. £	Lab. £	Plant £	Total £
Y2250						
225	**Supply and fix 44mm thick 30 min fire resisting internal plywood faced flush door, 900mm wide, with 2 vision panels approx 700 x 150mm each, new ironmongery, decorated**					
	Doors 44 mm thick, non standard, formed from 2134 x 914 mm blank (L2040/260 - 1 nr), Extra for forming panel (any size for glazing) (L2040/350 - 2 nr), To wood with screwed beads and insulating strips in panes (beads and strips included elsewhere) ne 0.15 m2 (L4039/130 - 0.21 m2), Strips to edge of glass and the like 6 mm thick (L4090/556 - 3.4 m), Wood general surfaces over 300 mm girth (M6040/570 - 3.5 m2), Wood general surfaces over 300 mm girth (M6050/570 - 3.5 m2), Intumescent strips with smoke seal, white aluminium holder (P1080/160 - 5 m), Cover fillets, stops, trims, beads, nosings and the like 12 x 12 mm (P2020/100 - 3.4 m), Pairs butt hinges 75 mm ref H.04400.31 (P2110/056 - 1 nr), Mortice latches 75 mm ref H.00410.03 (P2110/243 - 1 nr), Sets lever furniture 95 x 38 mm ref H.02880.01, SAA finish (P2140/260 - 1 nr)	nr	169.54	225.25		394.79
230	**Supply and fix 54mm thick 60 min fire resisting internal plywood faced flush door, 900mm wide, new ironmongery, decorated**					
	Doors 54 mm thick, non standard, formed from 2134 x 914 mm blank (L2045/265 - 1 nr), Wood general surfaces over 300 mm girth (M6040/570 - 3.5 m2), Wood general surfaces over 300 mm girth (M6050/570 - 3.5 m2), Intumescent strips with smoke seal, white aluminium holder (P1080/160 - 5 m), Pairs butt hinges 75 mm ref H.04400.31 (P2110/056 - 1 nr), Mortice latches 75 mm ref H.00410.03 (P2110/243 - 1 nr), Sets lever furniture 95 x 38 mm ref H.02880.01, SAA finish (P2140/260 - 1 nr)	nr	257.03	169.15		426.18
235	**Supply and fix 54mm thick 60 min fire resisting internal plywood faced flush door, 900mm wide, with 2 vision panels approx 700 x 150mm each, new ironmongery, decorated**					
	Doors 54 mm thick, non standard, formed from 2134 x 914 mm blank (L2045/265 - 1 nr), Extra for forming panel (any size for glazing) (L2045/350 - 2 nr), To wood with screwed beads and insulating strips in panes (beads and strips included elsewhere) ne 0.15 m2 (L4039/130 - 0.21 m2), Strips to edge of glass and the like 6 mm thick (L4090/556 - 3.4 m), Wood general surfaces over 300 mm girth (M6040/570 - 3.5 m2), Wood general surfaces over 300 mm girth (M6050/570 - 3.5 m2), Intumescent strips with smoke seal, white aluminium holder (P1080/160 - 5 m), Cover fillets, stops, trims, beads, nosings and the like 12 x 12 mm (P2020/100 - 3.4 m), Pairs butt hinges 75 mm ref H.04400.31 (P2110/056 - 1 nr), Mortice latches 75 mm ref H.00410.03 (P2110/243 - 1 nr), Sets lever furniture 95 x 38 mm ref H.02880.01, SAA finish (P2140/260 - 1 nr)	nr	284.57	234.75		519.32
250	**Supply and fix 44mm thick internal hardwood door type 2XGG, 900mm wide, new ironmongery, decorated**					
	Doors 838 x 1981 x 44 mm, standard type 2XG (L2015/154 - 1 nr), To wood with screwed beads and insulating strips in panes (beads and strips included elsewhere) 0.15 to 4 m2 (L4039/140 - 1.23 m2), Strips to edge of glass and the like 6 mm thick (L4090/556 - 6.5 m), Wood glazed doors over 300 mm girth in panes 0.50 to 1 m2 (M6070/642 - 3.5 m2), Cover fillets, stops, trims, beads, nosings and the like 12 x 12 mm (P2020/100 - 6.5 m), Mortice latches 75 mm ref H.00410.03 (P2110/243 - 1 nr), Pairs butt hinges 100 mm ref H.04455.21 (P2120/064 - 1.5 nr), Sets lever furniture 95 x 38 mm ref H.02880.01, SAA finish (P2140/260 - 1 nr)	nr	344.78	167.82		512.60

© NSR 01 Aug 2008 - 31 Jul 2009   Y44

	Y22 : WINDOWS, DOORS & DOORWAYS		Mat. £	Lab. £	Plant £	Total £
Y2250						
252	**Supply and fix 44mm thick internal oak door type 2XGG, 900mm wide, new ironmongery , decorated**					
	Doors 838 x 1981 x 44 mm, standard type 2XG (L2020/154 - 1 nr), To wood with screwed beads and insulating strips in panes (beads and strips included elsewhere) 0.15 to 4 m2 (L4039/140 - 1.23 m2), Strips to edge of glass and the like 6 mm thick (L4090/556 - 6.5 m), Wood glazed doors over 300 mm girth in panes 0.50 to 1 m2 (M6070/642 - 3.5 m2), Cover fillets, stops, trims, beads, nosings and the like 12 x 12 mm (P2035/100 - 6.5 m), Mortice latches 75 mm ref H.00410.03 (P2110/243 - 1 nr), Pairs butt hinges 100 mm ref H.04455.21 (P2120/064 - 1.5 nr), Sets lever furniture 95 x 38 mm ref H.02880.01, SAA finish (P2140/260 - 1 nr)	nr	417.34	173.29		590.63
260	**Extra over new internal door for sliding door gear and pelmet**					
	Sliding door pelmet Hendersons type 40P 1850 long on bracket with 2nr end caps (N1022/100 - 1 nr), Pairs butt hinges 75 mm ref H.04400.31 (P2110/056 - -1 nr), Sliding door gear J2, n.e. 1.00m ref: H0883501 (P2110/420 - 1 nr)	nr	64.94	13.54		78.48
264	**Supply and fit 35mm thick internal plywood faced flush double doors of unequal width, primary leaf 926mm wide with 2 vision panels per leaf approx 700 x 150mm each, new ironmongery , decorated**					
	Easing doors (C5170/825 - 1 nr), Doors 35 mm thick, non standard, formed from 2134 x 914 mm blank (L2030/250 - 1 nr), Doors 762 x 1981 x 35 mm (L2030/270 - 1 nr), Extra for forming panel (any size for glazing) (L2030/350 - 4 nr), To wood with screwed beads and insulating strips in panes (beads and strips included elsewhere) 0.15 to 4 m2 (L4039/140 - 0.42 m2), Strips to edge of glass and the like 6 mm thick (L4090/556 - 6.8 m), Wood glazed windows and screens over 300 mm girth in panes 0.10 to 0.50 m2 (M6040/602 - 7 m2), Wood glazed windows and screens over 300 mm girth in panes 0.10 to 0.50 m2 (M6050/602 - 7 m2), Cover fillets, stops, trims, beads, nosings and the like 12 x 12 mm (P2020/100 - 6.8 m), Pairs butt hinges 75 mm ref H.04400.31 (P2110/056 - 3 nr), Mortice latches 75 mm ref H.00410.03 (P2110/243 - 1 nr), Flush bolts 150 mm ref H.04870.11, sunk slide, BMA finish (P2130/142 - 2 nr), Sets lever furniture 95 x 38 mm ref H.02880.01, SAA finish (P2140/260 - 1 nr)	nr	221.51	423.34		644.85

© NSR 01 Aug 2008 - 31 Jul 2009

Y45

	Y22 : WINDOWS, DOORS & DOORWAYS		Mat. £	Lab. £	Plant £	Total £
Y2250						
266	**Supply and fit 40mm thick internal plywood faced flush double doors of unequal width, primary leaf 926mm wide with 2 vision panels per leaf approx 700 x 150mm each, new ironmongery, decorated**					
	Easing doors (C5170/825 - 1 nr), Doors 40 mm thick, non standard, formed from 2134 x 914 mm blank (L2030/255 - 1 nr), Doors 762 x 2040 x 40 mm (L2030/280 - 1 nr), Extra for forming panel (any size for glazing) (L2030/350 - 4 nr), To wood with screwed beads and insulating strips in panes (beads and strips included elsewhere) 0.15 to 4 m2 (L4039/140 - 0.42 m2), Strips to edge of glass and the like 6 mm thick (L4090/556 - 6.8 m), Wood glazed windows and screens over 300 mm girth in panes 0.10 to 0.50 m2 (M6040/602 - 7 m2), Wood glazed windows and screens over 300 mm girth in panes 0.10 to 0.50 m2 (M6050/602 - 7 m2), Cover fillets, stops, trims, beads, nosings and the like 12 x 12 mm (P2020/100 - 6.8 m), Pairs butt hinges 75 mm ref H.04400.31 (P2110/056 - 3 nr), Mortice latches 75 mm ref H.00410.03 (P2110/243 - 1 nr), Flush bolts 150 mm ref H.04870.11, sunk slide, BMA finish (P2130/142 - 2 nr), Sets lever furniture 95 x 38 mm ref H.02880.01, SAA finish (P2140/260 - 1 nr)	nr	243.13	429.54		672.67
268	**Supply and fit 44mm thick 30 min fire resisting internal plywood faced flush double doors of unequal width, primary leaf 926mm wide with 2 vision panels per leaf approx 700 x 150mm each, new ironmongery, decorated**					
	Easing doors (C5170/825 - 1 nr), Extra for forming panel (any size for glazing) (L2030/350 - 4 nr), Doors 44 mm thick, non standard, formed from 2134 x 914 mm blank (L2040/260 - 1 nr), Doors 762 x 1981 x 44 mm (L2040/290 - 1 nr), To wood with screwed beads and insulating strips in panes (beads and strips included elsewhere) 0.15 to 4 m2 (L4039/140 - 0.42 m2), Strips to edge of glass and the like 6 mm thick (L4090/556 - 6.8 m), Wood glazed windows and screens over 300 mm girth in panes 0.10 to 0.50 m2 (M6040/602 - 7 m2), Wood glazed windows and screens over 300 mm girth in panes 0.10 to 0.50 m2 (M6050/602 - 7 m2), Cover fillets, stops, trims, beads, nosings and the like 12 x 12 mm (P2020/100 - 6.8 m), Pairs butt hinges 75 mm ref H.04400.31 (P2110/056 - 3 nr), Mortice latches 75 mm ref H.00410.03 (P2110/243 - 1 nr), Flush bolts 150 mm ref H.04870.11, sunk slide, BMA finish (P2130/142 - 2 nr), Sets lever furniture 95 x 38 mm ref H.02880.01, SAA finish (P2140/260 - 1 nr)	nr	321.59	437.17		758.76

© NSR 01 Aug 2008 - 31 Jul 2009   Y46

	Y22 : WINDOWS, DOORS & DOORWAYS		Mat. £	Lab. £	Plant £	Total £
Y2250						
270	**Supply and fit 54mm thick 60 min fire resisting internal plywood faced flush double doors of unequal width, primary leaf 926mm wide with 2 vision panels per leaf approx 700 x 150mm each, new ironmongery , decorated**					
	Easing doors (C5170/825 - 1 nr), Extra for forming panel (any size for glazing) (L2030/350 - 4 nr), Doors 54 mm thick, non standard, formed from 2134 x 914 mm blank (L2045/265 - 1 nr), Doors 762 x 1981 x 54 mm (L2045/330 - 1 nr), To wood with screwed beads and insulating strips in panes (beads and strips included elsewhere) 0.15 to 4 m2 (L4039/140 - 0.42 m2), Strips to edge of glass and the like 6 mm thick (L4090/556 - 6.8 m), Wood glazed windows and screens over 300 mm girth in panes 0.10 to 0.50 m2 (M6040/602 - 7 m2), Wood glazed windows and screens over 300 mm girth in panes 0.10 to 0.50 m2 (M6050/602 - 7 m2), Cover fillets, stops, trims, beads, nosings and the like 12 x 12 mm (P2020/100 - 6.8 m), Pairs butt hinges 75 mm ref H.04400.31 (P2110/056 - 3 nr), Mortice latches 75 mm ref H.00410.03 (P2110/243 - 1 nr), Flush bolts 150 mm ref H.04870.11, sunk slide, BMA finish (P2130/142 - 2 nr), Sets lever furniture 95 x 38 mm ref H.02880.01, SAA finish (P2140/260 - 1 nr)	nr	536.37	445.37		981.74
272	**Supply and fix 44mm thick internal hardwood double doors type 2XGG, of unequal width, primary leaf 926mm wide, new ironmongery , decorated**					
	Doors 44 mm thick, non-standard, one panel 12 mm plywood square or moulded (L2015/050 - 2 m2), Doors 838 x 1981 x 44 mm, standard type 2XG (L2015/154 - 1 nr), To wood with screwed beads and insulating strips in panes (beads and strips included elsewhere) 0.15 to 4 m2 (L4039/140 - 2.4 m2), Strips to edge of glass and the like 6 mm thick (L4090/556 - 12 m), Wood glazed doors over 300 mm girth in panes 0.50 to 1 m2 (M6070/642 - 7 m2), Cover fillets, stops, trims, beads, nosings and the like 12 x 12 mm (P2020/100 - 7 m), Mortice latches 75 mm ref H.00410.03 (P2110/243 - 1 nr), Pairs butt hinges 100 mm ref H.04455.21 (P2120/064 - 3 nr), Flush bolts 150 mm ref H.04870.11, sunk slide, BMA finish (P2130/142 - 2 nr), Sets lever furniture 95 x 38 mm ref H.02880.01, SAA finish (P2140/260 - 1 nr)	nr	1133.52	322.58		1456.10
274	**Supply and fix 44mm thick internal oak double doors type 2XGG, of unequal width, primary leaf 926mm wide, new ironmongery , decorated**					
	Doors 44 mm thick, non-standard, one panel 12 mm plywood square or moulded (L2020/050 - 2 m2), Doors 838 x 1981 x 44 mm, standard type 2XG (L2020/154 - 1 nr), To wood with screwed beads and insulating strips in panes (beads and strips included elsewhere) 0.15 to 4 m2 (L4039/140 - 2.4 m2), Strips to edge of glass and the like 6 mm thick (L4090/556 - 12 m), Wood glazed doors over 300 mm girth in panes 0.50 to 1 m2 (M6070/642 - 7 m2), Cover fillets, stops, trims, beads, nosings and the like 12 x 12 mm (P2035/100 - 12 m), Mortice latches 75 mm ref H.00410.03 (P2110/243 - 1 nr), Pairs butt hinges 100 mm ref H.04455.21 (P2120/064 - 1.5 nr), Flush bolts 150 mm ref H.04870.11, sunk slide, BMA finish (P2130/142 - 2 nr), Sets lever furniture 95 x 38 mm ref H.02880.01, SAA finish (P2140/260 - 1 nr)	nr	1411.57	344.77		1756.34

© NSR 01 Aug 2008 - 31 Jul 2009

Y47

	Y22 : WINDOWS, DOORS & DOORWAYS		Mat. £	Lab. £	Plant £	Total £
Y2250						
300	**Supply and fit 32 x 100mm softwood single door lining, new architraves, decorated**					
	Jambs and heads 32 x 100 mm (L2065/452 - 5 m), Plugging at 450 mm centres (L2068/667 - 5 m), Screwing at 450 mm centres (L2068/672 - 5 m), Wood general surfaces in isolated surfaces ne 300 mm girth (M6040/580 - 5 m), Wood general surfaces in isolated surfaces ne 300 mm girth (M6050/580 - 5 m), Skirtings, picture rails, architraves and the like 19 x 50 mm (P2010/064 - 10 m)	nr	41.60	108.35		149.95
305	**Supply and fit 32 x 150mm softwood single door lining, new architraves, decorated**					
	Jambs and heads 32 x 150 mm (L2065/456 - 5 m), Plugging at 450 mm centres (L2068/667 - 5 m), Screwing at 450 mm centres (L2068/672 - 5 m), Wood general surfaces in isolated surfaces ne 300 mm girth (M6040/580 - 5 m), Wood general surfaces in isolated surfaces ne 300 mm girth (M6050/580 - 5 m), Skirtings, picture rails, architraves and the like 19 x 50 mm (P2010/064 - 10 m)	nr	51.30	108.35		159.65
310	**Supply and fit 62 x 75mm hardwood single door frame, new architraves, decorated**					
	Jambs and heads 62 x 75 mm (L2062/482 - 5 m), Plugging at 450 mm centres (L2068/667 - 5 m), Screwing at 450 mm centres (L2068/672 - 5 m), Wood general surfaces in isolated surfaces ne 300 mm girth (M6070/580 - 10 m), Skirtings, picture rails, architraves and the like 19 x 50 mm (P2020/064 - 10 m)	nr	78.65	119.55		198.20
312	**Supply and fit 62 x 75mm oak single door frame, new architraves, decorated**					
	Jambs and heads 62 x 75 mm (L2063/482 - 5 m), Plugging at 450 mm centres (L2068/667 - 5 m), Screwing at 450 mm centres (L2068/672 - 5 m), Pelleting at 450 mm centres (L2068/691 - 5 m), Wood general surfaces in isolated surfaces ne 300 mm girth (M6070/580 - 10 m), Skirtings, picture rails, architraves and the like 19 x 50 mm (P2035/064 - 10 m)	nr	128.35	165.25		293.60
320	**Supply and fit 32 x 100mm softwood double door lining, new architraves, decorated**					
	Jambs and heads 32 x 100 mm (L2065/452 - 6 m), Plugging at 450 mm centres (L2068/667 - 6 m), Screwing at 450 mm centres (L2068/672 - 6 m), Wood general surfaces in isolated surfaces ne 300 mm girth (M6040/580 - 6 m), Wood general surfaces in isolated surfaces ne 300 mm girth (M6050/580 - 6 m), Skirtings, picture rails, architraves and the like 19 x 50 mm (P2010/064 - 12 m)	nr	49.92	130.02		179.94
322	**Supply and fit 32 x 150mm softwood double door lining, new architraves, decorated**					
	Jambs and heads 32 x 150 mm (L2065/456 - 6 m), Plugging at 450 mm centres (L2068/667 - 6 m), Screwing at 450 mm centres (L2068/672 - 6 m), Wood general surfaces in isolated surfaces ne 300 mm girth (M6040/580 - 6 m), Wood general surfaces in isolated surfaces ne 300 mm girth (M6050/580 - 6 m), Skirtings, picture rails, architraves and the like 19 x 50 mm (P2010/064 - 12 m)	nr	61.56	130.02		191.58

© NSR 01 Aug 2008 - 31 Jul 2009   Y48

	Y22 : WINDOWS, DOORS & DOORWAYS		Mat. £	Lab. £	Plant £	Total £
Y2250						
324	**Supply and fit 62 x 75mm hardwood double door frame, new architraves, decorated**					
	Jambs and heads 62 x 75 mm (L2062/482 - 6 m), Plugging at 450 mm centres (L2068/667 - 6 m), Screwing at 450 mm centres (L2068/672 - 6 m), Wood general surfaces in isolated surfaces ne 300 mm girth (M6070/580 - 12 m), Skirtings, picture rails, architraves and the like 19 x 50 mm (P2020/064 - 12 m)	nr	94.38	143.46		237.84
326	**Supply and fit 62 x 75mm oak double door frame, new architraves, decorated**					
	Jambs and heads 62 x 75 mm (L2063/482 - 6 m), Plugging at 450 mm centres (L2068/667 - 6 m), Screwing at 450 mm centres (L2068/672 - 6 m), Pelleting at 450 mm centres (L2068/691 - 6 m), Wood general surfaces in isolated surfaces ne 300 mm girth (M6070/580 - 12 m), Skirtings, picture rails, architraves and the like 19 x 50 mm (P2035/064 - 12 m)	nr	154.02	198.30		352.32
Y2260	**DOORS AND FRAMES EXTERNAL**					
340	**Supply and fit 40mm thick external plywood faced flush door, 900mm wide, new ironmongery, decorated**					
	Extra for weatherboards 50 x 75 mm, screwing (L2010/220 - 1 m), Doors 40 mm thick, non standard, formed from 2134 x 914 mm blank (L2035/255 - 1 nr), Wood general surfaces in isolated surfaces ne 300 mm girth (M6035/580 - 5 m), Wood general surfaces over 300 mm girth (M6040/570 - 2 m2), Wood general surfaces over 300 mm girth (M6135/570 - 2 m2), Pairs butt hinges 75 mm ref H.04400.31 (P2110/056 - 1.5 nr), Mortice sash locks 75 mm ref H.00600.02 (P2110/194 - 1 nr), Sets lever furniture 152 x 38 mm ref H.02875.01, SAA finish (P2140/262 - 1 nr)	nr	127.45	146.27		273.72
341	**Supply and fit 44mm thick external plywood faced flush door, 900mm wide, new ironmongery, decorated**					
	Extra for weatherboards 50 x 75 mm, screwing (L2010/220 - 1 m), Doors 44 mm thick, non standard, formed from 2134 x 914 mm blank (L2035/260 - 1 nr), Wood general surfaces over 300 mm girth (M6040/570 - 2 m2), Wood general surfaces over 300 mm girth (M6135/570 - 2 m2), Wood general surfaces in isolated surfaces ne 300 mm girth (M6220/580 - 5 m), Pairs butt hinges 75 mm ref H.04400.31 (P2110/056 - 1.5 nr), Mortice sash locks 75 mm ref H.00600.02 (P2110/194 - 1 nr), Sets lever furniture 152 x 38 mm ref H.02875.01, SAA finish (P2140/262 - 1 nr)	nr	194.54	140.50		335.04

© NSR 01 Aug 2008 - 31 Jul 2009

Y49

	Y22 : WINDOWS, DOORS & DOORWAYS		Mat. £	Lab. £	Plant £	Total £
Y2260						
342	**Supply and fit 40mm thick external plywood faced flush door, 900mm wide with 2 vision panels approx 700 x 150mm each, new ironmongery, decorated** Easing doors (C5170/825 - 1 nr), Doors 40 mm thick, non standard, formed from 2134 x 914 mm blank (L2035/255 - 1 nr), Extra for forming panel (any size for glazing) (L2035/350 - 1 nr), To wood with screwed beads and insulating strips in panes (beads and strips included elsewhere) 0.15 to 4 m2 (L4039/140 - 0.21 m2), Strips to edge of glass and the like 6 mm thick (L4090/556 - 3.4 m), Wood glazed windows and screens over 300 mm girth in panes 0.10 to 0.50 m2 (M6040/602 - 3.5 m2), Wood glazed windows and screens over 300 mm girth in panes 0.10 to 0.50 m2 (M6050/602 - 3.5 m2), Pairs butt hinges 100 mm ref H.04415.11, heavy duty (P2110/062 - 1.5 nr), Mortice sash locks 75 mm ref H.00600.02 (P2110/194 - 1 nr), Sets lever furniture 152 x 38 mm ref H.02875.01, SAA finish (P2140/262 - 1 nr)	nr	150.41	199.41		349.82
343	**Supply and fit 44mm thick external plywood faced flush door, 900mm wide with 2 vision panels approx 700 x 150mm each, new ironmongery, decorated** Easing doors (C5170/825 - 1 nr), Doors 44 mm thick, non standard, formed from 2134 x 914 mm blank (L2035/260 - 1 nr), Extra for forming panel (any size for glazing) (L2035/350 - 1 nr), To wood with screwed beads and insulating strips in panes (beads and strips included elsewhere) 0.15 to 4 m2 (L4039/140 - 0.21 m2), Strips to edge of glass and the like 6 mm thick (L4090/556 - 3.4 m), Wood glazed windows and screens over 300 mm girth in panes 0.10 to 0.50 m2 (M6040/602 - 3.5 m2), Wood glazed windows and screens over 300 mm girth in panes 0.10 to 0.50 m2 (M6050/602 - 3.5 m2), Pairs butt hinges 100 mm ref H.04415.11, heavy duty (P2110/062 - 1.5 nr), Mortice sash locks 75 mm ref H.00600.02 (P2110/194 - 1 nr), Sets lever furniture 152 x 38 mm ref H.02875.01, SAA finish (P2140/262 - 1 nr)	nr	218.41	202.29		420.70
344	**Supply and fix 44mm thick 30 min fire resisting external plywood faced flush door, 900mm wide, new ironmongery inc panic bolt, decorated** Doors 44 mm thick, non standard, formed from 2134 x 914 mm blank (L2050/260 - 1 nr), Wood general surfaces over 300 mm girth (M6040/570 - 1.8 m2), Wood general surfaces over 300 mm girth (M6050/570 - 1.8 m2), Wood general surfaces over 300 mm girth (M6135/570 - 1.8 m2), Wood general surfaces over 300 mm girth (M6145/570 - 1.8 m2), Intumescent strips with smoke seal, white aluminium holder (P1080/160 - 5 m), Pairs butt hinges 100 mm ref H.04415.11, heavy duty (P2110/062 - 1.5 nr), Panic bolts for single doors ref H.04030.01 (P2110/170 - 1 nr), Mortice locks 75 mm ref H.00605.11 (P2110/192 - 1 nr), Sets lever furniture 152 x 38 mm ref H.02875.01, SAA finish (P2140/262 - 1 nr)	nr	254.54	203.72		458.26

© NSR 01 Aug 2008 - 31 Jul 2009   Y50

	Y22 : WINDOWS, DOORS & DOORWAYS		Mat. £	Lab. £	Plant £	Total £
Y2260						
345	**Supply and fix 54mm thick 60 min fire resisting external plywood faced flush door, 900mm wide, new ironmongery including panic bolt, decorated**					
	Doors 54 mm thick, non standard, formed from 2134 x 914 mm blank (L2055/265 - 1 nr), Wood general surfaces over 300 mm girth (M6040/570 - 1.8 m2), Wood general surfaces over 300 mm girth (M6050/570 - 1.8 m2), Wood general surfaces over 300 mm girth (M6135/570 - 1.8 m2), Wood general surfaces over 300 mm girth (M6145/570 - 1.8 m2), Intumescent strips with smoke seal, white aluminium holder (P1080/160 - 5 m), Pairs butt hinges 100 mm ref H.04415.11, heavy duty (P2110/062 - 1.5 nr), Panic bolts for single doors ref H.04030.01 (P2110/170 - 1 nr), Mortice sash locks 75 mm ref H.00600.02 (P2110/194 - 1 nr), Sets lever furniture 152 x 38 mm ref H.02875.01, SAA finish (P2140/262 - 1 nr)	nr	366.96	210.34		577.30
346	**Supply and fix 44mm thick external hardwood door type 2XGG, 900mm wide, new ironmongery , decorated**					
	Doors 838 x 1981 x 44 mm, standard type 2XG (L2015/154 - 1 nr), To wood with screwed beads and insulating strips in panes (beads and strips included elsewhere) 0.15 to 4 m2 (L4039/140 - 1.23 m2), Strips to edge of glass and the like 6 mm thick (L4090/556 - 6.5 m), Wood glazed doors over 300 mm girth in panes 0.50 to 1 m2 (M6070/642 - 3.5 m2), Cover fillets, stops, trims, beads, nosings and the like 12 x 12 mm (P2020/100 - 6.5 m), Mortice sash locks 75 mm ref H.00600.02 (P2110/194 - 1 nr), Pairs butt hinges 100 mm ref H.04455.21 (P2120/064 - 1.5 nr), Sets lever furniture 152 x 38 mm ref H.02875.01, SAA finish (P2140/262 - 1 nr)	nr	350.42	168.97		519.39
347	**Supply and fix 44 mm thick external oak door type 2XGG, 900mm wide, new ironmongery , decorated**					
	Doors 838 x 1981 x 44 mm, standard type 2XG (L2020/154 - 1 nr), To wood with screwed beads and insulating strips in panes (beads and strips included elsewhere) 0.15 to 4 m2 (L4039/140 - 1.23 m2), Strips to edge of glass and the like 6 mm thick (L4090/556 - 6.5 m), Wood glazed doors over 300 mm girth in panes 0.50 to 1 m2 (M6070/642 - 3.5 m2), Cover fillets, stops, trims, beads, nosings and the like 12 x 12 mm (P2035/100 - 6.5 m), Mortice sash locks 75 mm ref H.00600.02 (P2110/194 - 1 nr), Pairs butt hinges 100 mm ref H.04455.21 (P2120/064 - 1.5 nr), Sets lever furniture 152 x 38 mm ref H.02875.01, SAA finish (P2140/262 - 1 nr)	nr	422.99	174.44		597.43

© NSR 01 Aug 2008 - 31 Jul 2009   Y51

	Y22 : WINDOWS, DOORS & DOORWAYS		Mat. £	Lab. £	Plant £	Total £
Y2260						
348	**Supply and fit 40 mm thick external plywood faced flush double doors of unequal width, primary leaf 926mm wide with 2 vision panels per leaf approx 700 x 150mm each, new ironmongery, decorated**					
	Easing doors (C5170/825 - 1 nr), Extra for forming panel (any size for glazing) (L2030/350 - 4 nr), Doors 40 mm thick, non standard, formed from 2134 x 914 mm blank (L2035/255 - 1 nr), Doors 762 x 2040 x 40 mm (L2035/280 - 1 nr), To wood with screwed beads and insulating strips in panes (beads and strips included elsewhere) 0.15 to 4 m2 (L4039/140 - 0.42 m2), Strips to edge of glass and the like 6 mm thick (L4090/556 - 6.8 m), Wood glazed windows and screens over 300 mm girth in panes 0.10 to 0.50 m2 (M6040/602 - 7 m2), Wood glazed windows and screens over 300 mm girth in panes 0.10 to 0.50 m2 (M6050/602 - 7 m2), Cover fillets, stops, trims, beads, nosings and the like 12 x 12 mm (P2020/100 - 6.8 m), Pairs butt hinges 75 mm ref H.04400.31 (P2110/056 - 3 nr), Mortice latches 75 mm ref H.00410.03 (P2110/243 - 1 nr), Flush bolts 150 mm ref H.04870.11, sunk slide, BMA finish (P2130/142 - 2 nr), Sets lever furniture 95 x 38 mm ref H.02880.01, SAA finish (P2140/260 - 1 nr)	nr	276.67	432.42		709.09
349	**Supply and fit 62 x 100mm softwood external single door frame with low level threshold, decorated**					
	Jambs and heads 62 x 100 mm (L2060/484 - 5 m), Plugging at 450 mm centres (L2068/667 - 5 m), Screwing at 450 mm centres (L2068/672 - 5 m), Pointing frames (L2075/605 - 5 m), Wood general surfaces in isolated surfaces ne 300 mm girth (M6040/580 - 5 m), Wood general surfaces in isolated surfaces ne 300 mm girth (M6135/580 - 5 m), Sills Raintite ne 1 m long (P2140/472 - 1 nr)	nr	70.69	110.89		181.58
350	**Supply and fit 62 x 100mm softwood external double door frame with low level threshold, decorated**					
	Jambs and heads 62 x 100 mm (L2060/484 - 6 m), Plugging at 450 mm centres (L2068/667 - 6 m), Screwing at 450 mm centres (L2068/672 - 6 m), Pointing frames (L2075/605 - 6 m), Wood general surfaces in isolated surfaces ne 300 mm girth (M6040/580 - 6 m), Wood general surfaces in isolated surfaces ne 300 mm girth (M6135/580 - 6 m), Sills Raintite ne 1 m long (P2140/472 - 2 nr)	nr	100.50	139.98		240.48
355	**Supply and fit 62 x 100mm hardwood external single door frame with low level threshold, decorated**					
	Jambs and heads 62 x 100 mm (L2062/484 - 5 m), Plugging at 450 mm centres (L2068/667 - 5 m), Screwing at 450 mm centres (L2068/672 - 5 m), Pelleting at 450 mm centres (L2068/691 - 5 m), Pointing frames (L2075/605 - 5 m), Wood general surfaces in isolated surfaces ne 300 mm girth (M6070/580 - 5 m), Wood general surfaces in isolated surfaces ne 300 mm girth (M6155/580 - 5 m), Sills Raintite ne 1 m long (P2140/472 - 1 nr)	nr	113.39	149.39		262.78

© NSR 01 Aug 2008 - 31 Jul 2009   Y52

	Y22 : WINDOWS, DOORS & DOORWAYS		Mat. £	Lab. £	Plant £	Total £
Y2260						
356	**Supply and fit 62 x 100mm hardwood external double door frame with low level threshold, decorated**					
	Jambs and heads 62 x 100 mm (L2062/484 - 6 m), Plugging at 450 mm centres (L2068/667 - 6 m), Screwing at 450 mm centres (L2068/672 - 6 m), Pelleting at 450 mm centres (L2068/691 - 6 m), Pointing frames (L2075/605 - 6 m), Wood general surfaces in isolated surfaces ne 300 mm girth (M6070/580 - 6 m), Wood general surfaces in isolated surfaces ne 300 mm girth (M6155/580 - 6 m), Sills Raintite ne 1 m long (P2140/472 - 2 nr)	nr	151.74	186.18		337.92
360	**Supply and fit 62 x 100mm oak external single door frame with low level threshold, decorated**					
	Jambs and heads 62 x 100 mm (L2063/484 - 5 m), Plugging at 450 mm centres (L2068/667 - 5 m), Screwing at 450 mm centres (L2068/672 - 5 m), Pelleting at 450 mm centres (L2068/691 - 5 m), Pointing frames (L2075/605 - 5 m), Wood general surfaces in isolated surfaces ne 300 mm girth (M6070/580 - 5 m), Wood general surfaces in isolated surfaces ne 300 mm girth (M6155/580 - 5 m), Sills Raintite ne 1 m long (P2140/472 - 1 nr)	nr	138.94	156.24		295.18
365	**Supply and fit 62 x 100mm oak external double door frame with low level threshold, decorated**					
	Jambs and heads 62 x 100 mm (L2063/484 - 6 m), Plugging at 450 mm centres (L2068/667 - 6 m), Screwing at 450 mm centres (L2068/672 - 6 m), Pelleting at 450 mm centres (L2068/691 - 6 m), Pointing frames (L2075/605 - 6 m), Wood general surfaces in isolated surfaces ne 300 mm girth (M6070/580 - 6 m), Wood general surfaces in isolated surfaces ne 300 mm girth (M6155/580 - 6 m), Sills Raintite ne 1 m long (P2140/472 - 2 nr)	nr	182.40	194.40		376.80
Y2270	**WINDOWS**					
370	**Remove plywood or similar panel from window and reglaze in safety glass n.e 0.50m2**					
	Removing plywood, insulation board or other sheet lining (C2010/370 - 1 m2), To wood with screwed beads and insulating strips in panes (beads and strips included elsewhere) 0.15 to 4 m2 (L4044/140 - 1 m2), Strips to edge of glass and the like 6 mm thick (L4090/556 - 4 m), Cover fillets, stops, trims, beads, nosings and the like 13 x 13 mm (P2020/102 - 4 m)	m2	91.24	69.88		161.12
375	**Remove transome from window and reglaze, window size n.e. 1m2**					
	Stripping off/removing/taking down casements in preparation for replacement (C5165/726 - 1 nr), To wood with screwed beads (beads included elsewhere) in panes 0.15 to 4 m2 (L4016/100 - 0.75 m2), Strips to edge of glass and the like 6 mm thick (L4090/556 - 4 m), Removing and preparing for renewal wood rebates glazed with screwed beads (L4091/597 - 4 m), Cover fillets, stops, trims, beads, nosings and the like 13 x 13 mm (P2020/102 - 4 m)	nr	32.62	77.28		109.90

© NSR 01 Aug 2008 - 31 Jul 2009   Y53

	Y22 : WINDOWS, DOORS & DOORWAYS		Mat. £	Lab. £	Plant £	Total £
Y2270						
376	**Remove transome from window and reglaze, window size 1 - 2m2**					
	Stripping off/removing/taking down casements in preparation for replacement (C5165/726 - 1 nr), To wood with screwed beads (beads included elsewhere) in panes 0.15 to 4 m2 (L4016/100 - 1.75 m2), Strips to edge of glass and the like 6 mm thick (L4090/556 - 6 m), Removing and preparing for renewal wood rebates glazed with screwed beads (L4091/597 - 6 m), Cover fillets, stops, trims, beads, nosings and the like 13 x 13 mm (P2020/102 - 6 m)	nr	69.65	142.89		212.54
377	**Remove transome from window and reglaze, window size exc 2m2**					
	Stripping off/removing/taking down casements in preparation for replacement (C5165/726 - 1 nr), To wood with screwed beads (beads included elsewhere) in panes 0.15 to 4 m2 (L4016/100 - 1 m2), Strips to edge of glass and the like 6 mm thick (L4090/556 - 3 m), Removing and preparing for renewal wood rebates glazed with screwed beads (L4091/597 - 3 m), Cover fillets, stops, trims, beads, nosings and the like 13 x 13 mm (P2020/102 - 3 m)	m2	38.97	80.72		119.69
378	**Extra over for fitting new softwood transome in new position**					
	Sills, mullions and transomes 50 x 50 mm (L2065/500 - 1 m), Cover fillets, stops, trims, beads, nosings and the like 12 x 12 mm (P2020/100 - 2 m)	m	2.75	11.38		14.13
379	**Extra over for fitting new hardwood transome in new position**					
	Sills, mullions and transomes 50 x 50 mm (L2066/500 - 1 m), Cover fillets, stops, trims, beads, nosings and the like 12 x 12 mm (P2020/100 - 2 m)	m	6.00	12.97		18.97
380	**Extra over for fitting new oak transome in new position**					
	Sills, mullions and transomes 50 x 50 mm (L2067/500 - 1 m), Cover fillets, stops, trims, beads, nosings and the like 12 x 12 mm (P2035/100 - 2 m)	m	8.99	13.83		22.82
Y2280	**IRONMONGERY**					
400	**Remove existing threshold strip**					
	Draught strips screwing (C2015/945 - 1 m)	Item		2.59		2.59
401	**Supply and fit Phlexicare RX100 threshold strip, 915mm wide**					
	Threshold strip, easy access, 915mm RX100 SAA ref: H.08540.21 (P2140/474 - 1 nr)	Item	15.23	8.64		23.87
402	**Supply and fit Phlexicare RX100 threshold strip, 1000mm wide**					
	Threshold strip, easy access, 1000mm RX100 SAA ref: H.08540.31 (P2140/475 - 1 nr)	Item	17.58	8.64		26.22

© NSR 01 Aug 2008 - 31 Jul 2009   Y54

	Y22 : WINDOWS, DOORS & DOORWAYS		Mat. £	Lab. £	Plant £	Total £
Y2280						
403	**Supply and fit Phlexicare RX100 threshold strip, 1219mm wide**					
	Threshold strip, easy access, 1219mm RX100 SAA ref: H.08540.51 (P2140/476 - 1 nr)	Item	31.58	8.64		40.22
405	**Remove existing hinges and replace with Ball bearing butt hinges 102mm polished brass**					
	Pairs butt hinges 100 mm steel (C2015/610 - 1 nr), Pairs ball bearing butt hinges 102 mm ref H.04427.02 (P2120/066 - 1 nr)	pr	6.15	7.49		13.64
406	**Remove existing hinges and replace with Ball bearing butt hinges 102mm stainless steel**					
	Pairs butt hinges 100 mm steel (C2015/610 - 1 nr), Pairs ball bearing butt hinges 102 mm ref H.04427.03 (P2110/064 - 1 nr)	pr	6.69	7.49		14.18
408	**Supply and fit R.A.D.A.R lockset to door**					
	R.A.D.A.R. Lockset ref H.01485.02 (P2110/222 - 1 nr)	Item	111.81	16.70		128.51
409	Extra over for red cover plate to R.A.D.A.R lockset					
	R.A.D.A.R. Coverplate ref H.01485.04, Coloured (P2110/224 - 1 nr)	Item	16.00	4.32		20.32
410	**Supply and fit R.A.D.A.R rimlock to door**					
	R.A.D.A.R. Rimlock ref H.01485.51 (P2110/226 - 1 nr)	Item	92.50	13.82		106.32
411	**Remove existing lever handles and replace with coloured steel cored nylon lever handles and roses, Normbau or similar**					
	Sets lever furniture (C2015/785 - 1 nr), Normbau steel cored lever set ref:0516.01 (P2160/350 - 1 nr)	Item	40.38	5.19		45.57
412	**Remove existing lever handles and replace with coloured steel cored nylon lever handles with backplate, Normbau or similar**					
	Sets lever furniture (C2015/785 - 1 nr), Normbau steel cored lever set on backplate ref:0623 (P2160/352 - 1 nr)	Item	46.68	7.78		54.46
413	**Remove existing lever handles and replace with coloured solid nylon lever handles and roses, Normbau or similar**					
	Sets lever furniture (C2015/785 - 1 nr), Normbau solid nylon lever set ref:0246.08 (P2160/354 - 1 nr), Normbau solid nylon lever set rose ref: 0247.10 (P2160/358 - 1 nr)	Item	20.70	6.92		27.62
414	**Remove existing lever handles and replace with coloured solid nylon lever handles with backplate, Normbau or similar**					
	Sets lever furniture (C2015/785 - 1 nr), Normbau solid nylon lever set ref:0246.08 (P2160/354 - 1 nr), Normbau solid nylon lever set backplate ref: 0347 (P2160/360 - 1 nr)	Item	28.50	9.22		37.72
415	**Remove existing lever handles and replace with combi lever handles and roses, Normbau or similar**					
	Sets lever furniture (C2015/785 - 1 nr), Normbau Combi lever set ref:2840.01 (P2160/340 - 1 nr), Normbau Combi rose ref: 0390.03 (P2160/344 - 1 nr)	Item	94.56	8.07		102.63

© NSR 01 Aug 2008 - 31 Jul 2009   Y55

	Y22 : WINDOWS, DOORS & DOORWAYS		Mat. £	Lab. £	Plant £	Total £
Y2280						
416	Remove existing lever handles and replace with combi lever handles with backplates, Normbau or similar					
	Sets lever furniture (C2015/785 - 1 nr), Normbau Combi lever set ref:2840.01 (P2160/340 - 1 nr), Normbau Combi backplate set ref: 0405.05 (P2160/346 - 1 nr)	Item	101.21	9.22		110.43
417	extra over for fixing lock/latch and lever furniture in new position, incl making good door and frame					
	Mortice locks (C2020/735 - 1 nr), Sets lever furniture (C2020/785 - 1 nr), Filling keyholes with softwood plug and making good both sides (C5170/873 - 1 nr)	Item	0.20	17.57		17.77
418	Remove existing escutcheon and replace with combi escutcheon set, Normbau or similar					
	Escutcheons (C2015/780 - 2 nr), Normbau Combi Escutcheons ref: 0245.05 (P2160/342 - 1 nr)	Item	7.27	2.89		10.16
419	Remove existing escutcheon and replace with combi escutcheon and thumb turn, Normbau or similar					
	Escutcheons (C2015/780 - 2 nr), Normbau Combi escutcheon and thumb turn set ref: 0245.03 (P2160/348 - 1 nr)	Item	16.27	4.04		20.31
420	Provide coloured steel cored nylon lever handles and roses to new door in leu of Standard SAA lever furniture, Normbau or similar					
	Sets lever furniture 95 x 38 mm ref H.02880.01, SAA finish (P2140/260 - -1 nr), Normbau steel cored lever set ref:0516.01 (P2160/350 - 1 nr)	Item	30.04	-2.59		27.45
422	Provide coloured steel cored nylon lever handles with backplate to new door in leu of Standard SAA lever furniture, Normbau or similar					
	Sets lever furniture 95 x 38 mm ref H.02880.01, SAA finish (P2140/260 - -1 nr), Normbau steel cored lever set on backplate ref:0623 (P2160/352 - 1 nr)	Item	36.34			36.34
426	Provide coloured solid nylon lever handles and roses to new door in leu of Standard SAA lever furniture, Normbau or similar					
	Sets lever furniture 95 x 38 mm ref H.02880.01, SAA finish (P2140/260 - -1 nr), Normbau solid nylon lever set ref:0246.08 (P2160/354 - 1 nr), Normbau solid nylon lever set rose ref: 0247.10 (P2160/358 - 1 nr)	Item	10.36	-0.86		9.50
428	Provide coloured solid nylon lever handles with backplate to new door in leu of Standard SAA lever furniture, Normbau or similar					
	Sets lever furniture 95 x 38 mm ref H.02880.01, SAA finish (P2140/260 - -1 nr), Normbau solid nylon lever set ref:0246.08 (P2160/354 - 1 nr), Normbau solid nylon lever set backplate ref: 0347 (P2160/360 - 1 nr)	Item	18.16	1.44		19.60

© NSR 01 Aug 2008 - 31 Jul 2009

Y56

	Y22 : WINDOWS, DOORS & DOORWAYS		Mat. £	Lab. £	Plant £	Total £
Y2280						
440	**Reposition existing D handles**					
	Pull handles (C2015/860 - 1 nr), Filling holes caused by decay with proprietary wood filler and sanding smooth (C5170/867 - 1 nr), Pull handles (P2175/860 - 1 nr)	Item	4.90	9.22		14.12
442	**Remove existing D handles**					
	Pull handles (C2015/860 - 1 nr)	Item		0.86		0.86
444	**Supply and fit coloured solid nylon D handle 23 x 160mm Normbau or similar**					
	D pull handles 23 x 160 mm ref Normbau, solid nylon (P2160/370 - 1 nr)	Item	33.08	3.74		36.82
446	**Supply and fit coloured solid nylon D handle 31 x 175mm Normbau or similar**					
	D pull handles 31 x 175 mm ref Normbau, solid nylon (P2160/372 - 1 nr)	Item	56.27	3.74		60.01
448	**Supply and fit coloured solid nylon D handle 31 x 300mm Normbau or similar**					
	D pull handles 31 x 300 mm ref Normbau, solid nylon (P2160/374 - 1 nr)	Item	89.04	3.74		92.78
450	**Supply and fit coloured solid nylon D handle 34 x 200mm Normbau or similar**					
	D pull handles 34 x 200 mm ref Normbau, solid nylon (P2160/376 - 1 nr)	Item	68.21	3.74		71.95
452	**Supply and fit coloured steel cored nylon D handle 34 x 200mm Normbau or similar**					
	D pull handles 34 x 200 mm ref Normbau, steel cored nylon (P2160/378 - 1 nr)	Item	78.54	3.74		82.28
454	**Supply and fit coloured steel cored nylon D handle 34 x 300mm Normbau or similar**					
	D pull handles 34 x 300 mm ref Normbau, steel cored nylon (P2160/380 - 1 nr)	Item	97.51	3.74		101.25
456	**Supply and fit coloured steel cored nylon D handle 34 x 400mm Normbau or similar**					
	D pull handles 34 x 400 mm ref Normbau, steel cored nylon (P2160/382 - 1 nr)	Item	111.60	3.74		115.34
458	**Supply and fit coloured steel cored nylon back to back D handle 34 x 200mm Normbau or similar**					
	D pull handles, back to back 34 x 220 mm ref Normbau, steel cored nylon (P2160/384 - 1 pr)	Item	110.38	4.32		114.70

© NSR 01 Aug 2008 - 31 Jul 2009   Y57

	Y22 : WINDOWS, DOORS & DOORWAYS		Mat. £	Lab. £	Plant £	Total £
Y2290	DOOR AND WINDOW CONTROLS					
540	Remove existing door closer and prepare to recieve new.					
	Overhead door closers (C2020/660 - 1 nr)	nr		10.08		10.08
544	Supply and fix door closer (Briton series 2100)					
	Briton 2100 series overhead door closers (P2110/102 - 1 nr)	nr	209.10	21.60		230.70
550	Supply and fix electromagnetic door closer (Briton series 996) including flush mounted wall magnet and connection to power supply					
	Briton 996 series electromagnetic door closer (P2110/126 - 1 nr), Briton 996 series flush mounted wall magnet (P2110/127 - 1 nr), Heating and power socket, switch socket and the like, single switch, 2.5 mm2 single core cables (V9020/208 - 1 nr)	nr	772.26	58.40		830.66
552	Supply and fix low energy door closer (Briton 2500 series) including connection to power supply					
	Briton 2500 series low energy door closer (P2110/130 - 1 nr), Heating and power socket, switch socket and the like, single switch, 2.5 mm2 single core cables (V9020/208 - 1 nr)	nr	2574.10	51.20		2625.30
554	Extra over low energy door closer for narrow style push pad, including connection to power supply					
	Briton 2500 series narrow style push pad (not inc wiring) (P2110/131 - 1 nr), Heating and power socket, switch socket and the like, single switch, 2.5 mm2 single core cables (V9020/208 - 0.5 nr)	nr	258.24	28.88		287.12
556	Extra over low energy door closer for large plate style push pad, including connection to power supply					
	Briton 2500 series large plate style push pad (not inc wiring) (P2110/132 - 1 nr), Heating and power socket, switch socket and the like, single switch, 2.5 mm2 single core cables (V9020/208 - 0.5 nr)	nr	401.56	28.88		430.44
558	Extra over low energy door closer for 3 position auto key switch, including connection to power supply					
	Briton 2500 series 3 position auto key switch (P2110/133 - 1 nr), Heating and power socket, switch socket and the like, single switch, 2.5 mm2 single core cables (V9020/208 - 0.5 nr)	nr	218.11	22.54		240.65
560	Extra over low energy door closer for 4 position auto key switch, including connection to power supply					
	Briton 2500 series 4 position auto key switch (P2110/134 - 1 nr), Heating and power socket, switch socket and the like, single switch, 2.5 mm2 single core cables (V9020/208 - 0.5 nr)	nr	258.24	22.54		280.78
562	Extra over low energy door closer for morning entry key switch, including connection to power supply					
	Briton 2500 series morning entry key switch (not inc wiring) (P2110/135 - 1 nr), Heating and power socket, switch socket and the like, single switch, 2.5 mm2 single core cables (V9020/208 - 0.5 nr)	nr	258.24	22.54		280.78

© NSR 01 Aug 2008 - 31 Jul 2009

Y58

	Y22 : WINDOWS, DOORS & DOORWAYS		Mat. £	Lab. £	Plant £	Total £
Y2290						
564	**Extra over low energy door closer for motion sensor, including connection to power supply**					
	Briton 2500 series motion sensor (not inc wiring) (P2110/136 - 1 nr), Heating and power socket, switch socket and the like, single switch, 2.5 mm2 single core cables (V9020/208 - 0.5 nr)	nr	453.16	22.54		475.70
566	**Extra over low energy door closer for 340mm safety sensor kit , including connection to power supply**					
	Briton 2500 series safety sensor kit 340mm (P2110/137 - 1 nr)	nr	716.63	8.46		725.09
568	**Extra over low energy door closer for 700mm safety sensor kit , including connection to power supply**					
	Briton 2500 series safety sensor kit 700mm (P2110/138 - 1 nr)	nr	814.09	8.46		822.55
600	**Supply and fix remote control electrical window opener to sash window (not incl electrical supply)**					
	Electrical window opener; Compact chain activator with infra red receiver and remote control; ref: H3405008 (P2110/144 - 1 nr)	nr	353.80	17.28		371.08
605	**Supply and fix switch operated 24volt DC electrical window opener to sash window (not incl electrical supply)**					
	CN1 electrical window opener, 24vdc, single chain with 250mm travel and CP02 transformer; ref: H3405006 (P2110/140 - 1 nr)	nr	187.50	17.28		204.78
610	**Supply and fix switch operated 230volt AC electrical window opener to sash window (not incl electrical supply)**					
	CN1 electrical window opener, 230v, single chain with 250mm travel; ref: H3405007 (P2110/142 - 1 nr)	nr	155.90	17.28		173.18

© NSR 01 Aug 2008 - 31 Jul 2009

Y59

Y22 : WINDOWS, DOORS & DOORWAYS	Mat. £	Lab. £	Plant £	Total £

			Mat. £	Lab. £	Plant £	Total £
	Y24 : INTERNAL RAMPS AND STEPS					
Y2440	**ANCILLARY WORKS**					
100	**Demolish existing internal 1.00 metre wide reinforced concrete steps and ramp; approx 450mm high; approx 3nr steps; ramp n.e. 7 metres, protect existing structure and make good** Protection to floors, 12 mm (C1073/943 - 6 m2), To reduce levels 0.25 m max depth (D2030/240 - 2.5 m3), Extra over excavation irrespective of depth for breaking out reinforced concrete (D2030/760 - 2.5 m3), Excavated material off site (D2052/005 - 2.5 m3), Walls over 300 mm wide to concrete 1 to 5 m2 (M2030/061 - 7 m2)	Item	68.25	371.29	263.46	703.00
102	**Demolish existing internal 1.00 metre wide reinforced concrete steps and ramp; approx 750mm high; approx 5nr steps; Ramp n.e. 12 metres, protect existing structure and make good** Protection to floors, 12 mm (C1073/943 - 6 m2), To reduce levels 0.25 m max depth (D2030/240 - 6.15 m3), Extra over excavation irrespective of depth for breaking out reinforced concrete (D2030/760 - 6.15 m3), Excavated material off site (D2052/005 - 6.15 m3), Walls over 300 mm wide to concrete 1 to 5 m2 (M2030/061 - 12 m2)	Item	80.70	781.94	645.87	1508.51
104	**Demolish existing internal 1.00 metre wide reinforced concrete steps and ramp; over 5nr steps; Ramp exc. 12 metres, protect existing structure and make good** Protection to floors, 12 mm (C1073/943 - 1 m2), To reduce levels 0.25 m max depth (D2030/240 - 0.7 m3), Extra over excavation irrespective of depth for breaking out reinforced concrete (D2030/760 - 0.7 m3), Excavated material off site (D2052/005 - 0.7 m3), Walls over 300 mm wide to concrete (M2030/060 - 1 m2)	m2	10.96	86.82	73.33	171.11
106	**Demolish existing internal 2.00 metre wide reinforced concrete steps and ramp; approx 450mm high; approx 3nr steps; ramp n.e. 7 metres, protect existing structure and make good** Protection to floors, 12 mm (C1073/943 - 12 m2), To reduce levels 0.25 m max depth (D2030/240 - 5 m3), Extra over excavation irrespective of depth for breaking out reinforced concrete (D2030/760 - 5 m3), Excavated material off site (D2052/005 - 5 m3), Walls over 300 mm wide to brickwork, blockwork or stonework (M2025/050 - 7 m2), Bonding/jointing new to existing (M2025/150 - 14 m)	Item	120.82	753.63	523.91	1398.36
108	**Demolish existing internal 2.00 metre wide reinforced concrete steps and ramp; approx 750mm high; approx 5nr steps; Ramp n.e. 12 metres, protect existing structure and make good** Protection to floors, 12 mm (C1073/943 - 12 m2), To reduce levels 0.25 m max depth (D2030/240 - 12.3 m3), Extra over excavation irrespective of depth for breaking out reinforced concrete (D2030/760 - 12.3 m3), Excavated material off site (D2052/005 - 12.3 m3), Walls over 300 mm wide to concrete (M2030/060 - 6 m2)	Item	116.58	1359.70	1283.81	2760.09

	Y24 : INTERNAL RAMPS AND STEPS		Mat. £	Lab. £	Plant £	Total £
Y2440						
110	**Demolish existing internal 2.00 metre wide reinforced concrete steps and ramp; over 5nr steps; Ramp exc. 12 metres, protect existing structure and make good**					
	Protection to floors, 12 mm (C1073/943 - 1 m2), To reduce levels 0.25 m max depth (D2030/240 - 0.7 m3), Extra over excavation irrespective of depth for breaking out reinforced concrete (D2030/760 - 0.7 m3), Excavated material off site (D2052/005 - 0.7 m3), Walls over 300 mm wide to concrete (M2030/060 - 1 m2)	m2	10.96	86.82	73.33	171.11
Y2450	**RAMPS**					
200	**Form reinforced concrete ramp between existing walls, n.e. 200mm high; 1800mm wide at gradient of ave 1:15; ensuring 1800mm long landings at top and bottom; including screeded finish (floor covering and handrails measured elsewhere)**					
	Filling to excavations ne 250 mm av thick (D2070/035 - 1.29 m3), Compacting filling, blinding with sand (D2082/080 - 8.64 m2), Slabs ne 150 mm thick (E1040/355 - 1.19 m3), Square mesh ref A142 2.22 kg/m2 (E3030/210 - 8.64 m2), Trowelling to falls (E4110/072 - 8.64 m2), Floors 25 mm thick level or to falls only ne 15 degrees from horizontal to concrete (M1010/250 - 8.64 m2), Extra for steel trowel finish (M1010/480 - 8.64 m2)	nr	262.42	284.53	33.00	579.95
202	**Form reinforced concrete ramp between existing walls, n.e. 350mm high; 1800mm wide at gradient of 1:15; include for 1800mm long landings at top and bottom; including screeded finish (floor covering and handrails measured elsewhere)**					
	Filling to excavations ne 250 mm av thick (D2070/035 - 1.26 m3), Compacting filling, blinding with sand (D2082/080 - 12.69 m2), Beds 150 mm to 450 mm thick (E1040/335 - 2.79 m3), Square mesh ref A142 2.22 kg/m2 (E3030/210 - 12.69 m2), Trowelling to falls (E4110/072 - 12.69 m2), Floors 25 mm thick level or to falls only ne 15 degrees from horizontal to concrete (M1010/250 - 12.69 m2), Extra for steel trowel finish (M1010/480 - 12.69 m2)	nr	477.30	437.07	43.56	957.93
204	**Form reinforced concrete ramp between existing walls, n.e. 500mm high; 1800mm wide at gradient of 1:15; include for 1800mm long landings at top and bottom; including screeded finish (floor covering and handrails measured elsewhere)**					
	Filling to excavations ne 250 mm av thick (D2070/035 - 1.67 m3), Compacting filling, blinding with sand (D2082/080 - 16.74 m2), Beds 150 mm to 450 mm thick (E1040/335 - 4.99 m3), Square mesh ref A142 2.22 kg/m2 (E3030/210 - 16.74 m2), Trowelling to falls (E4110/072 - 16.74 m2), Floors 25 mm thick level or to falls only ne 15 degrees from horizontal to concrete (M1010/250 - 16.74 m2), Extra for steel trowel finish (M1010/480 - 16.74 m2)	nr	776.01	630.22	65.64	1471.87

	Y24 : INTERNAL RAMPS AND STEPS		Mat. £	Lab. £	Plant £	Total £
Y2450						
206	Form reinforced concrete ramp abutting existing wall, n.e. 200mm high; 1800mm wide at gradient of ave 1:15; ensuring 1800mm long landings at top and bottom; Outer face veneered plywood lined with hardwood cill; including screeded finish (floor covering and handrails measured elsewhere)					
	Filling to excavations ne 250 mm av thick (D2070/035 - 1.25 m3), Compacting filling, blinding with sand (D2082/080 - 8.64 m2), Slabs ne 150 mm thick (E1040/355 - 1.71 m3), Sides of ground beams and edges of beds ne 250 mm high (E2030/090 - 5 m), Square mesh ref A142 2.22 kg/m2 (E3030/210 - 8.64 m2), Trowelling to falls (E4110/072 - 8.64 m2), Walls 19 mm thick ne 300 mm wide (K1110/214 - 3 m), Floors 25 mm thick level or to falls only ne 15 degrees from horizontal to concrete (M1010/250 - 8.64 m2), Extra for steel trowel finish (M1010/480 - 8.64 m2), Skirtings, picture rails, architraves and the like 25 x 100 mm (P2020/076 - 4 m)	nr	389.84	495.43	36.62	921.89
208	Form reinforced concrete ramp abutting existing wall, n.e. 350mm high; 1800mm wide at gradient of ave 1:15; ensuring 1800mm long landings at top and bottom; Outer face veneered plywood lined with hardwood cill; including screeded finish (floor covering and handrails measured elsewhere)					
	Filling to excavations ne 250 mm av thick (D2070/035 - 1.27 m3), Compacting filling, blinding with sand (D2082/080 - 12.69 m2), Slabs ne 150 mm thick (E1040/355 - 2.79 m3), Sides of ground beams and edges of beds ne 250 mm high (E2030/090 - 6 m), Square mesh ref A142 2.22 kg/m2 (E3030/210 - 12.69 m2), Trowelling to falls (E4110/072 - 12.69 m2), Walls 19 mm thick ne 300 mm wide (K1110/214 - 6 m), Floors 25 mm thick level or to falls only ne 15 degrees from horizontal to concrete (M1010/250 - 12.69 m2), Extra for steel trowel finish (M1010/480 - 12.69 m2), Skirtings, picture rails, architraves and the like 25 x 100 mm (P2020/076 - 6 m)	nr	592.70	706.76	48.68	1348.14
210	Form reinforced concrete ramp abutting existing wall, n.e. 500mm high; 1800mm wide at gradient of ave 1:15; ensuring 1800mm long landings at top and bottom; Outer face veneered plywood lined with hardwood cill; including screeded finish (floor covering and handrails measured elsewhere)					
	Filling to excavations ne 250 mm av thick (D2070/035 - 1.67 m3), Compacting filling, blinding with sand (D2082/080 - 16.74 m2), Slabs ne 150 mm thick (E1040/355 - 4.99 m3), Sides of ground beams and edges of beds ne 250 mm high (E2030/090 - 8 m), Square mesh ref A142 2.22 kg/m2 (E3030/210 - 16.74 m2), Trowelling to falls (E4110/072 - 16.74 m2), Walls 19 mm thick ne 300 mm wide (K1110/214 - 8 m), Floors 25 mm thick level or to falls only ne 15 degrees from horizontal to concrete (M1010/250 - 16.74 m2), Extra for steel trowel finish (M1010/480 - 16.74 m2), Skirtings, picture rails, architraves and the like 25 x 100 mm (P2020/076 - 8 m)	nr	929.37	1004.79	74.57	2008.73
212	Extra over concrete ramps for facing in 215mm facing brick PC £375/1000					
	Walls 215 mm thick, facework one side (F1034/104 - 1 m2), Walls 19 mm thick ne 300 mm wide (K1110/214 - -3 m)	m2	22.77	22.16	1.33	46.26
214	Extra over for brick on edge coping to edge of ramp					
	Brick on edge copings 225 x 112 mm horizontal, set weathering (F1032/400 - 1 m)	m	6.96	10.29	0.38	17.63

© NSR 01 Aug 2008 - 31 Jul 2009   Y63

Y24 : INTERNAL RAMPS AND STEPS			Mat. £	Lab. £	Plant £	Total £
Y2450						
216	**Extra over 350mm high concrete ramp for steps**					
	Filling to excavations ne 250 mm av thick (D2070/035 - 0.16 m3), Staircases (E1040/585 - 0.351 m3), Sides of ground beams and edges of beds 250 mm to 500 mm high (E2025/092 - -1.8 m), Stairflights 2 m wide, waist 150 mm, risers vertical 150 mm high (E2025/444 - 0.6 m), Square mesh ref A142 2.22 kg/m2 bent (E3030/215 - 1.2 m2), Trowelling (E4110/070 - 1.08 m2)	Item	61.50	108.79	21.24	191.53
218	**Extra over 500mm high concrete ramp for steps**					
	Filling to excavations ne 250 mm av thick (D2070/035 - 0.32 m3), Staircases (E1040/585 - 1.3 m3), Sides of ground beams and edges of beds 250 mm to 500 mm high (E2025/092 - -1.8 m), Stairflights 2 m wide, waist 150 mm, risers vertical 150 mm high (E2025/444 - 1.2 m), Square mesh ref A142 2.22 kg/m2 bent (E3030/215 - 3.24 m2), Trowelling (E4110/070 - 2.16 m2)	Item	202.02	339.61	53.29	594.92
230	**Extra over for 75mm softwood skirting board to edge of ramp, decorated**					
	Wood general surfaces in isolated surfaces ne 300 mm girth (M6040/580 - 1 m), Wood general surfaces in isolated surfaces ne 300 mm girth (M6045/580 - 1 m), Skirtings, picture rails, architraves and the like 19 x 75 mm (P2010/068 - 1 m)	m	2.89	8.35		11.24
232	**Extra over for 150mm softwood skirting board to edge of ramp, decorated**					
	Wood general surfaces in isolated surfaces ne 300 mm girth (M6040/580 - 1 m), Wood general surfaces in isolated surfaces ne 300 mm girth (M6045/580 - 1 m), Skirtings, picture rails, architraves and the like 19 x 150 mm (P2010/070 - 1 m)	m	4.34	8.92		13.26
234	**Extra over for 75mm hardwood skirting board to edge of ramp, decorated**					
	Wood general surfaces in isolated surfaces ne 300 mm girth (M6070/580 - 1 m), Skirtings, picture rails, architraves and the like 19 x 75 mm (P2020/068 - 1 m)	m	2.93	6.33		9.26
236	**Extra over for 150mm hardwood skirting board to edge of ramp, decorated**					
	Wood general surfaces in isolated surfaces ne 300 mm girth (M6070/580 - 1 m), Skirtings, picture rails, architraves and the like 19 x 150 mm (P2020/070 - 1 m)	m	4.97	6.91		11.88
238	**Extra over for 75mm oak skirting board to edge of ramp, decorated**					
	Wood general surfaces in isolated surfaces ne 300 mm girth (M6070/580 - 1 m), Skirtings, picture rails, architraves and the like 19 x 75 mm (P2035/068 - 1 m)	m	7.32	6.33		13.65

Y24 : INTERNAL RAMPS AND STEPS			Mat. £	Lab. £	Plant £	Total £
Y2450 240	**Extra over for 150mm oak skirting board to edge of ramp, decorated** Wood general surfaces in isolated surfaces ne 300 mm girth (M6070/580 - 1 m), Skirtings, picture rails, architraves and the like 19 x 150 mm (P2035/070 - 1 m)	m	13.74	6.91		20.65

Y24 : INTERNAL RAMPS AND STEPS		Mat. £	Lab. £	Plant £	Total £

© NSR 01 Aug 2008 - 31 Jul 2009 — Y66

			Mat. £	Lab. £	Plant £	Total £
	Y25 : FIXTURES AND FITTINGS					
Y2510	**INTERNAL HANDRAILS**					
002	**Remove existing handrail**					
	Removing handrails and brackets (C2010/340 - 1 m)	m		1.94		1.94
004	**Remove existing handrail and refix**					
	Timber handrails and brackets, resecuring (C5105/175 - 1 m), Plugging (isolated) (L2068/669 - 2 nr), Screwing (isolated) (L2068/674 - 2 nr)	m	0.33	6.32		6.65
010	**Provide and fix 50mm dia softwood handrail to wall on nylon coated elbow brackets, decorated to contrast with background**					
	Wood general surfaces in isolated surfaces ne 300 mm girth (M6040/580 - 1 m), Isolated handrails and grab rails 50 x 50 mm (P2010/190 - 1 m), Handrail Brackets (P2110/086 - 2 nr)	m	6.84	18.72		25.56
015	**Provide and fix 50mm dia hardwood handrail to wall on nylon coated elbow brackets, 2 coats lacquer**					
	Wood general surfaces in isolated surfaces ne 300 mm girth (M6070/580 - 1 m), Isolated handrails and grab rails 50 x 50 mm (P2020/190 - 1 m), Handrail Brackets (P2110/086 - 2 nr)	m	6.92	18.43		25.35
020	**Provide and fix 50mm dia iroko handrail to wall on nylon coated elbow brackets, 2 coats lacquer**					
	Wood general surfaces in isolated surfaces ne 300 mm girth (M6070/580 - 1 m), Isolated handrails and grab rails 50 x 50 mm (P2025/190 - 1 m), Handrail Brackets (P2110/086 - 2 nr)	m	7.47	18.43		25.90
025	**Provide and fix 50mm dia oak handrail to wall on nylon coated elbow brackets, 2 coats lacquer**					
	Wood general surfaces in isolated surfaces ne 300 mm girth (M6070/580 - 1 m), Isolated handrails and grab rails 50 x 50 mm (P2035/190 - 1 m), Handrail Brackets (P2110/086 - 2 nr)	m	14.62	18.43		33.05
Y2520	**SHELVING AND WALL CUPBOARDS**					
100	**Remove existing shelving, make good**					
	Removing isolated shelves and bearers ne 300 mm wide, making good finishings (C2010/322 - 1 m)	m	0.43	2.52		2.95
102	**Supply and fix MDF shelving 200mm wide, decorated**					
	Wood general surfaces over 300 mm girth (M6040/570 - 0.419 m2), Isolated shelves and worktops 18 x 200 mm (P2053/154 - 1 m)	m	4.15	9.89		14.04
104	**Supply and fix Plywood shelving 200mm wide, decorated**					
	Hard building board general surfaces over 300 mm girth (M6040/540 - 0.418 m2), Isolated shelves and worktops 18 x 200 mm, 12 x 18 mm wrot softwood lipping (P2055/266 - 1 m)	m	9.25	11.85		21.10

© NSR 01 Aug 2008 - 31 Jul 2009   Y67

	Y25 : FIXTURES AND FITTINGS		Mat. £	Lab. £	Plant £	Total £
Y2520						
106	**Supply and fix softwood shelving 200mm wide, decorated**					
	Wood general surfaces over 300 mm girth (M6040/570 - 0.419 m2), Isolated shelves and worktops 19 x 200 mm (P2010/154 - 1 m)	m	7.24	9.89		17.13
108	**Supply and fix hardwood shelving 200mm wide, decorated**					
	Wood general surfaces over 300 mm girth (M6070/570 - 0.419 m2), Isolated shelves and worktops 19 x 200 mm (P2020/154 - 1 m)	m	6.70	8.90		15.60
110	**Supply and fix oak shelving 200mm wide, decorated**					
	Wood general surfaces over 300 mm girth (M6070/570 - 0.419 m2), Isolated shelves and worktops 19 x 200 mm (P2035/154 - 1 m)	m	18.41	8.90		27.31
114	**Supply and fix N&C drop down counter unit**					
	Supply and fit drop down counter unit N&C ref:P6820420 (N1022/150 - 1 nr)	nr	552.34	8.64		560.98
116	**Supply and fix N&C drop down counter unit with induction loop**					
	Supply and fit drop down counter unit with induction loop N&C ref:P6820421 (N1022/155 - 1 nr)	nr	720.34	8.64		728.98
120	**Reposition existing shelving to correct height n.e. 1000mm long**					
	Removing isolated shelves and bearers ne 300 mm wide, making good finishings (C2010/322 - 1 m), Plugging (isolated) (P2075/669 - 4 nr), Screwing (isolated) (P2075/674 - 4 nr)	nr	0.75	9.40		10.15
122	**Reposition existing shelving to correct height 1000 - 2000mm long**					
	Removing isolated shelves and bearers ne 300 mm wide, making good finishings (C2010/322 - 2 m), Plugging (isolated) (P2075/669 - 6 nr), Screwing (isolated) (P2075/674 - 6 nr)	nr	1.34	15.36		16.70
130	**Remove wall fixed storage cupboard**					
	Stripping off/removing/taking down fittings, setting aside for re-use (C6210/051 - 1 nr)	nr		8.62		8.62
132	**Reposition storage cupboard fixed to wall**					
	Taking down wall units, resecuring components and refixing (C6210/200 - 1 nr)	nr	1.95	27.69		29.64
140	**Remove cupboard hinges and replace with new to open 180 deg**					
	Taking off kitchen unit doors, renewing hinges and refitting (C6210/290 - 1 pr)	pr	3.68	2.88		6.56
142	**Remove cupboard handle and replace with new contrasting coloured handle**					
	Drawer/cupboard knobs or handles (C2015/855 - 1 nr), Cupboard handles (P2150/300 - 1 nr)	nr	4.50	2.88		7.38

© NSR 01 Aug 2008 - 31 Jul 2009   Y68

			Mat. £	Lab. £	Plant £	Total £
	Y26 : FINISHES & SURFACE TREATMENTS (INTERNALLY)					
Y2610	**FLOOR FINISHES**					
002	**Repair cracks to screed n.e 50mm wide**					
	Repairing cracks up to 50 mm wide, in floor screed, raking out, apply bonding agent and filling (C6022/360 - 1 m)	m	0.83	17.62		18.45
004	**Patch repair 25mm screed to floor in small areas n.e. 1m2**					
	Stripping off/removing/taking down floors 25 mm thick to concrete in preparation for replacement ne 1 m2 (C6010/064 - 1 m2), Floors 25 mm thick level or to falls only ne 15 degrees from horizontal to concrete ne 1 m2 (M1010/252 - 1 m2), Extra for steel trowel finish (M1010/480 - 1 m2)	nr	4.66	23.45	1.92	30.03
006	**Patch repair 25mm screed to floor in small areas 1 - 5 m2**					
	Stripping off/removing/taking down floors 25 mm thick to concrete in preparation for replacement 1 to 5 m2 (C6010/062 - 1 m2), Floors 25 mm thick level or to falls only ne 15 degrees from horizontal to concrete 1 to 5 m2 (M1010/251 - 1 m2), Extra for steel trowel finish (M1010/480 - 1 m2)	m2	4.66	21.78	1.77	28.21
008	**Replace 25mm screed to floor in areas exc 5 m2**					
	Stripping off/removing/taking down floors 25 mm thick to concrete in preparation for replacement (C6010/060 - 1 m2), Floors 25 mm thick level or to falls only ne 15 degrees from horizontal to concrete (M1010/250 - 1 m2), Extra for steel trowel finish (M1010/480 - 1 m2)	m2	4.66	18.69	1.48	24.83
010	**Patch repair 50mm screed to floor in small areas n.e. 1m2**					
	Stripping off/removing/taking down floors 50 mm thick to concrete in preparation for replacement ne 1 m2 (C6010/104 - 1 m2), Floors 50 mm thick level or to falls only ne 15 degrees from horizontal to concrete ne 1 m2 (M1010/282 - 1 m2), Extra for steel trowel finish (M1010/480 - 1 m2)	nr	9.32	34.16	2.79	46.27
012	**Patch repair 50mm screed to floor in small areas 1 - 5m2**					
	Stripping off/removing/taking down floors 50 mm thick to concrete in preparation for replacement 1 to 5 m2 (C6010/102 - 1 m2), Floors 50 mm thick level or to falls only ne 15 degrees from horizontal to concrete 1 to 5 m2 (M1010/281 - 1 m2), Extra for steel trowel finish (M1010/480 - 1 m2)	m2	9.32	32.65	2.56	44.53
014	**Replace 50mm screed to floor in areas exc 5 m2**					
	Stripping off/removing/taking down floors 50 mm thick to concrete in preparation for replacement (C6010/100 - 1 m2), Floors 50 mm thick level or to falls only ne 15 degrees from horizontal to concrete (M1010/280 - 1 m2), Extra for steel trowel finish (M1010/480 - 1 m2)	m2	9.32	29.13	2.21	40.66
020	**Raise level of existing concrete floor to eliminate step by 25mm**					
	Floors level or to falls only not exceeding 15 degrees from horizontal to existing concrete (C6020/300 - 1 m2), Floors 25 mm thick level or to falls only ne 15 degrees from horizontal to concrete (M1010/250 - 1 m2), Extra for steel trowel finish (M1010/480 - 1 m2)	m2	5.08	18.26	0.65	23.99

© NSR 01 Aug 2008 - 31 Jul 2009 — Y69

	Y26 : FINISHES & SURFACE TREATMENTS (INTERNALLY)		Mat. £	Lab. £	Plant £	Total £
Y2610						
022	**Raise level of existing concrete floor to eliminate step by 50mm**					
	Floors level or to falls only not exceeding 15 degrees from horizontal to existing concrete (C6020/300 - 1 m2), Floors 50 mm thick level or to falls only ne 15 degrees from horizontal to concrete (M1010/280 - 1 m2), Extra for steel trowel finish (M1010/480 - 1 m2)	m2	9.74	26.97	1.15	37.86
024	**Raise level of existing concrete floor to eliminate step by 75mm**					
	Floors level or to falls only not exceeding 15 degrees from horizontal to existing concrete (C6020/300 - 1 m2), Floors 75 mm thick level or to falls only ne 15 degrees from horizontal to concrete (M1010/290 - 1 m2), Extra for steel trowel finish (M1010/480 - 1 m2)	m2	14.40	30.13	1.34	45.87
Y2620	**VINYL FLOOR COVERINGS**					
010	**Remove existing sheet vinyl floor covering in small areas n.e. 5m2, prepare to receive new floor covering**					
	Stripping off/removing/taking down floors to cement-sand in preparation for replacement 1 to 5 m2 (C6095/802 - 1 m2)	m2		5.17		5.17
012	**Remove existing sheet vinyl floor covering in areas exceeding 5m2, prepare to receive new floor covering**					
	Stripping off/removing/taking down floors to cement-sand in preparation for replacement (C6095/800 - 1 m2)	m2		3.88		3.88
014	**Remove existing vinyl tile floor covering in small areas n.e. 5m2, prepare to receive new floor covering**					
	Stripping off/removing/taking down floors to cement-sand in preparation for replacement 1 to 5 m2 (C6090/752 - 1 m2)	m2		10.34		10.34
016	**Remove existing vinyl tile floor covering in areas exceeding 5m2, prepare to receive new floor covering**					
	Stripping off/removing/taking down floors to cement-sand in preparation for replacement (C6090/750 - 1 m2)	m2		8.40		8.40
018	**Supply and lay new safety vinyl flooring (Altro 2.5mm classic D25) in small areas n.e 5m2**					
	Floors over 300 mm wide level or to falls to screed 1 to 5 m2 (M5028/051 - 1 m2)	m2	29.13	8.93		38.06
020	**Supply and lay new safety vinyl flooring (Altro 2.5mm classic D25) in areas exceeding 5m2**					
	Floors over 300 mm wide level or to falls to screed (M5028/050 - 1 m2)	m2	29.13	8.06		37.19
022	**Supply and lay new safety vinyl flooring (Altro 2.5mm classic D25) n.e 300mm wide**					
	Floors ne 300 mm wide level or to falls to screed (M5028/100 - 1 m)	m	8.74	4.61		13.35

© NSR 01 Aug 2008 - 31 Jul 2009

Y70

Y26 : FINISHES & SURFACE TREATMENTS (INTERNALLY)			Mat. £	Lab. £	Plant £	Total £
Y2620						
024	**Supply and lay new safety vinyl flooring (Altro 2.5mm classic D25) n.e 300mm wide to stairs**					
	Floors ne 300 mm wide level or vertical to stair treads or risers (M5028/112 - 1 m)	m	8.74	5.76		14.50
026	**Extra over for 100mm high self coved skirting**					
	Self coved skirtings 100 mm high (M5028/350 - 1 m)	m	5.97	2.59		8.56
028	**Extra over for 150mm high self coved skirting**					
	Self coved skirtings 150 mm high (M5028/355 - 1 m)	m	7.33	3.17		10.50
030	**Supply and lay new safety vinyl flooring (Altro 3.5mm Atlas 40) in small areas n.e 5m2**					
	Floors over 300 mm wide level or to falls to screed 1 to 5 m2 (M5029/051 - 1 m2)	m2	35.33	10.37		45.70
032	**Supply and lay new safety vinyl flooring (Altro 3.5mm Atlas 40) in areas exceeding 5m2**					
	Floors over 300 mm wide level or to falls to screed (M5029/050 - 1 m2)	m2	35.33	9.50		44.83
034	**Supply and lay new safety vinyl flooring (Altro 3.5mm Atlas 40) n.e 300mm wide**					
	Floors ne 300 mm wide level or to falls to screed (M5029/100 - 1 m)	m	10.60	5.18		15.78
036	**Supply and lay new safety vinyl flooring (Altro 3.5mm Atlas 40) n.e 300mm wide to stairs**					
	Floors ne 300 mm wide level or vertical to stair treads or risers (M5029/112 - 1 m)	m	10.60	6.34		16.94
038	**Extra over for 100mm high self coved skirting**					
	Self coved skirtings 100 mm high (M5029/350 - 1 m)	m	6.59	3.17		9.76
040	**Extra over for 150mm high self coved skirting**					
	Self coved skirtings 150 mm high (M5029/355 - 1 m)	m	8.26	4.61		12.87
Y2630	**CARPET FLOOR COVERINGS**					
100	**Remove existing edge fixed carpet in small areas n.e. 5m2, prepare to receive new floor covering**					
	Stripping off/removing/taking down floors to cement-sand in preparation for replacement 1 to 5 m2 (C6092/802 - 1 m2)	m2		3.88		3.88

© NSR 01 Aug 2008 - 31 Jul 2009

Y71

	Y26 : FINISHES & SURFACE TREATMENTS (INTERNALLY)		Mat. £	Lab. £	Plant £	Total £
Y2630						
102	**Remove existing edge fixed carpet in areas exceeding 5m2, prepare to receive new floor covering**					
	Stripping off/removing/taking down floors to cement-sand in preparation for replacement (C6092/800 - 1 m2)	m2		2.80		2.80
104	**Remove existing carpet tiling in small areas n.e. 5m2, prepare to receive new floor covering**					
	Stripping off/removing/taking down floors to cement-sand in preparation for replacement 1 to 5 m2 (C6085/752 - 1 m2)	m2		10.34		10.34
106	**Remove existing carpet tiling in areas exceeding 5m2, prepare to receive new floor covering**					
	Stripping off/removing/taking down floors to cement-sand in preparation for replacement (C6085/750 - 1 m2)	m2		8.40		8.40
110	**Supply and lay new carpet to screed (Phlexicare Sandringham or Buckingham Carpet) in small areas n.e. 5m2**					
	Floors over 300 mm wide level or to falls to screed 1 to 5 m2 (M5036/051 - 1 m2)	m2	14.70	7.49		22.19
112	**Supply and lay new carpet to screed (Phlexicare Sandringham or Buckingham Carpet) in areas exceeding 5m2**					
	Floors over 300 mm wide level or to falls to screed (M5036/050 - 1 m2)	m2	14.70	5.76		20.46
114	**Supply and lay new carpet to screed (Phlexicare Sandringham or Buckingham Carpet) n.e. 300mm wide**					
	Floors ne 300 mm wide level or to falls to screed (M5036/100 - 1 m)	m	4.41	3.46		7.87
116	**Supply and lay new carpet to timber (Phlexicare Sandringham or Buckingham Carpet) in small areas n.e. 5m2**					
	Floors over 300 mm wide level or to falls to timber 1 to 5 m2 (M5036/061 - 1 m2)	m2	14.70	7.49		22.19
118	**Supply and lay new carpet to timber (Phlexicare Sandringham or Buckingham Carpet) in areas exceeding 5m2**					
	Floors over 300 mm wide level or to falls to timber (M5036/060 - 1 m2)	m2	14.70	5.76		20.46
120	**Supply and lay new carpet to timber (Phlexicare Sandringham or Buckingham Carpet) n.e. 300mm wide**					
	Floors ne 300 mm wide level or to falls to timber (M5036/110 - 1 m)	m	4.41	3.46		7.87
122	**Supply and lay new carpet to stairs (Phlexicare Sandringham or Buckingham Carpet)**					
	Floors ne 300 mm wide level or vertical to stair treads or risers (M5036/112 - 1 m)	m	4.41	4.61		9.02

© NSR 01 Aug 2008 - 31 Jul 2009   Y72

	Y26 : FINISHES & SURFACE TREATMENTS (INTERNALLY)		Mat. £	Lab. £	Plant £	Total £
Y2630						
124	**Extra over for underlay**					
	Extra for underlay (M5036/175 - 1 m2)	m2	5.20	1.44		6.64
Y2640	**ENTRANCE MATTING**					
200	**Supply and lay intermmediate carpet mat (Seton or similar) 600 x 900mm**					
	Seton - Intermediate carpet mats 600 x 900 (M5040/010 - 1 nr)	nr	49.90	2.88		52.78
202	**Supply and lay intermmediate carpet mat (Seton or similar) 900 x 1500mm**					
	Seton - Intermediate carpet mats 900 x 1500 (M5040/015 - 1 nr)	nr	115.20	3.46		118.66
204	**Supply and lay indoor carpet mat (Seton or similar) 600 x 900mm**					
	Seton - Indoor carpet mats 600 x 900 (M5040/030 - 1 nr)	nr	48.30	2.88		51.18
206	**Supply and lay indoor carpet mat (Seton or similar) 900 x 1500mm**					
	Seton - Indoor carpet mats 900 x 1500 (M5040/035 - 1 nr)	nr	107.40	3.46		110.86
208	**Supply and lay outdoor carpet mat (Seton or similar) 600 x 900mm**					
	Seton - Outdoor carpet mats 600 x 900 (M5040/020 - 1 nr)	nr	48.30	2.88		51.18
210	**Supply and lay outdoor carpet mat (Seton or similar) 900 x 1500mm**					
	Seton - Outdoor carpet mats 900 x 1500 (M5040/025 - 1 nr)	nr	107.40	3.46		110.86
212	**Supply and lay Primary barrier matting, exterior (Gradus Topguard plain or similar)**					
	Gradus Primary barrier matting - Exterior use, Topguard plain (M5040/040 - 1 m2)	m2	258.16	4.32		262.48
214	**Supply and lay Primary barrier matting, interior (Gradus Boardwalk or similar)**					
	Gradus Primary barrier matting - Interior use, Boardwalk (M5040/042 - 1 m2)	m2	180.25	4.32		184.57
216	**Supply and lay Secondary barrier matting, interior (Gradus Boulevard 5000 or similar)**					
	Gradus secondary barrier matting - Boulevard 5000 (M5040/044 - 1 m2)	m2	36.48	4.32		40.80
218	**Supply and lay Secondary barrier matting, interior (Gradus Boulevard 6000 or similar)**					
	Gradus secondary barrier matting - Boulevard 6000 (M5040/046 - 1 m2)	m2	42.68	4.32		47.00

© NSR 01 Aug 2008 - 31 Jul 2009

Y73

	Y26 : FINISHES & SURFACE TREATMENTS (INTERNALLY)		Mat. £	Lab. £	Plant £	Total £
Y2640						
220	**Remove existing matwell frame and mat from concrete floor, fill recess in to existing floor level**					
	Remove existing matwell frame and mat. (C2010/140 - 1 nr), Floors 25 mm thick level or to falls only ne 15 degrees from horizontal to concrete ne 1 m2 (M1010/252 - 1 m2), Bonding/jointing new to existing (M1010/490 - 4 m)	m2	4.66	38.22	0.78	43.66
222	**Remove existing matwell frame and mat from timber floor, fill recess in to existing floor level**					
	Remove existing matwell frame and mat. (C2010/140 - 1 nr), Floors 25 mm thick area ne 1 m2 to timber (C5155/715 - 1 nr)	m2	28.37	16.33		44.70
230	**Form new matwell in concrete floor including aluminium matwell frame 1m2 (barrier matting measured elsewhere)**					
	100 mm thick (C1072/740 - 4 m), Stripping off/removing/taking down floors 25 mm thick to concrete in preparation for replacement ne 1 m2 (C6010/064 - 1 m2), Gradus Aluminium matwell edging, plugging and screwing (M5080/858 - 4 m)	nr	51.88	44.80	5.38	102.06
232	**Form new matwell in concrete floor including Simbrass matwell frame 1m2 (barrier matting measured elsewhere)**					
	100 mm thick (C1072/740 - 4 m), Stripping off/removing/taking down floors 25 mm thick to concrete in preparation for replacement ne 1 m2 (C6010/064 - 1 m2), Gradus brass matwell edging, plugging and screwing (M5080/928 - 4 m)	nr	90.88	44.80	5.38	141.06
240	**Form new matwell in timber floor including aluminium matwell frame 1m2 (barrier matting measured elsewhere)**					
	Stripping off/removing/taking down butt jointed floor boards, any thickness, denail joists in preparation for replacement areas ne 1 m2 (C5140/504 - 1 m2), Floors 25 mm thick area ne 1 m2 to timber (C5155/715 - 1 nr), Floor members 50 x 125 mm (G2010/078 - 4 m), Gradus Aluminium matwell edging, plugging and screwing (M5080/858 - 4 m)	nr	101.17	51.83		153.00
242	**Form new matwell in timber floor including simbrass matwell frame 1m2 (barrier matting measured elsewhere)**					
	Stripping off/removing/taking down butt jointed floor boards, any thickness, denail joists in preparation for replacement areas ne 1 m2 (C5140/504 - 1 m2), Floors 25 mm thick area ne 1 m2 to timber (C5155/715 - 1 nr), Floor members 50 x 125 mm (G2010/078 - 4 m), Gradus brass matwell edging, plugging and screwing (M5080/928 - 4 m)	nr	140.17	51.83		192.00
244	**Form new matwell in concrete floor including aluminium matwell frame 1-3m2 (barrier matting measured elsewhere)**					
	100 mm thick (C1072/740 - 7 m), Stripping off/removing/taking down floors 25 mm thick to concrete in preparation for replacement ne 1 m2 (C6010/064 - 3 m2), Gradus Aluminium matwell edging, plugging and screwing (M5080/858 - 7 m)	nr	90.79	89.45	10.84	191.08

© NSR 01 Aug 2008 - 31 Jul 2009   Y74

Y26 : FINISHES & SURFACE TREATMENTS (INTERNALLY)			Mat. £	Lab. £	Plant £	Total £
Y2640						
246	**Form new matwell in concrete floor including Simbrass matwell frame 1-3m2 (barrier matting measured elsewhere)**					
	100 mm thick (C1072/740 - 7 m), Stripping off/removing/taking down floors 25 mm thick to concrete in preparation for replacement ne 1 m2 (C6010/064 - 3 m2), Gradus brass matwell edging, plugging and screwing (M5080/928 - 7 m)	nr	159.04	89.45	10.84	259.33
248	**Form new matwell in timber floor including aluminium matwell frame 1-3m2 (barrier matting measured elsewhere)**					
	Stripping off/removing/taking down butt jointed floor boards, any thickness, denail joists in preparation for replacement areas ne 1 m2 (C5140/504 - 3 m2), Floors 25 mm thick over 300 mm wide to timber areas 1 to 5 m2 (C5155/707 - 3 m2), Floor members 50 x 125 mm (G2010/078 - 7 m), Gradus Aluminium matwell edging, plugging and screwing (M5080/858 - 7 m)	nr	212.51	107.84		320.35
250	**Form new matwell in timber floor including simbrass matwell frame 1-3m2 (barrier matting measured elsewhere)**					
	Stripping off/removing/taking down butt jointed floor boards, any thickness, denail joists in preparation for replacement areas ne 1 m2 (C5140/504 - 3 m2), Floors 19 mm thick over 300 mm wide to timber areas 1 to 5 m2 (C5155/682 - 3 m2), Floor members 50 x 125 mm (G2010/078 - 7 m), Gradus brass matwell edging, plugging and screwing (M5080/928 - 7 m)	nr	259.19	106.31		365.50
Y2650	**EDGINGS TO STEPS AND STAIRS (INTERNAL)**					
300	**Supply and fit pre-formed nosing with coloured slip resistant inserts - Gradus type AS11**					
	Stair edgings, A range, type AS11, plugging and screwing (M5085/100 - 1 m)	m	12.18	5.76		17.94
302	**Supply and fit pre-formed nosing with coloured slip resistant inserts - Gradus type AS12**					
	Stair edgings, A range, type AS12, plugging and screwing (M5085/101 - 1 m)	m	17.90	5.76		23.66
304	**Supply and fit pre-formed nosing with coloured slip resistant inserts - Gradus type AR10**					
	Stair edgings, A range, type AR10, plugging and screwing (M5085/103 - 1 m)	m	23.52	5.76		29.28
306	**Supply and fit pre-formed nosing with coloured slip resistant inserts - Gradus type AR32**					
	Stair edgings, A range, type AR32, plugging and screwing (M5085/105 - 1 m)	m	19.64	5.76		25.40
308	**Supply and fit pre-formed nosing with coloured slip resistant inserts - Gradus type ELA5140**					
	Stair edgings, A range, type Elite ELA5140 , plugging and screwing (M5085/106 - 1 m)	m	10.69	5.76		16.45

© NSR 01 Aug 2008 - 31 Jul 2009   Y75

	Y26 : FINISHES & SURFACE TREATMENTS (INTERNALLY)		Mat. £	Lab. £	Plant £	Total £
Y2650						
310	**Supply and fit pre-formed nosing with coloured slip resistant inserts - Gradus type ELA11750**					
	Stair edgings, A range, type Elite ELA11750 , plugging and screwing (M5085/107 - 1 m)	m	22.52	5.76		28.28
314	**Supply and fit pre-formed nosing with coloured slip resistant inserts - Gradus type G11**					
	Stair edgings, G range, type G11, plugging and screwing (M5085/108 - 1 m)	m	14.45	5.76		20.21
320	**Supply and fit pre-formed nosing with coloured slip resistant inserts - Gradus type G12**					
	Stair edgings, G range, type G12, plugging and screwing (M5085/110 - 1 m)	m	21.20	5.76		26.96
322	**Supply and fit pre-formed nosing with coloured slip resistant inserts - Gradus type G54**					
	Stair edgings, G range, type G54, plugging and screwing (M5085/112 - 1 m)	m	57.32	6.91		64.23
324	**Supply and fit pre-formed nosing with coloured slip resistant inserts - Gradus type CT71**					
	Stair edgings, CT range, type CT71, plugging and screwing (M5085/116 - 1 m)	m	25.99	5.76		31.75
326	**Supply and fit pre-formed nosing with coloured slip resistant inserts - Gradus type CT712**					
	Stair edgings, CT range, type CT712, plugging and screwing (M5085/114 - 1 m)	m	30.33	5.76		36.09
328	**Supply and fit pre-formed nosing with coloured slip resistant inserts - Gradus Hardnose type RN511**					
	Stair edgings, Hardnose range, type RN511, plugging and screwing (M5085/118 - 1 m)	m	12.44	5.76		18.20
330	**Supply and fit pre-formed nosing with coloured slip resistant inserts - Gradus Hardnose type RN727**					
	Stair edgings, Hardnose range, type RN727, plugging and screwing (M5085/120 - 1 m)	m	23.51	5.76		29.27
332	**Supply and fit pre-formed nosing with coloured slip resistant inserts - Gradus Hardnose type Elite ELR1150**					
	Stair edgings, Hardnose range, type Elite ELR1150 , plugging and screwing (M5085/128 - 1 m)	m	9.50	5.76		15.26
334	**Supply and fit pre-formed nosing with coloured slip resistant inserts - Gradus Hardnose type DDA1/ DDA2**					
	Stair edgings, Hardnose range, type DDA1/DDA2, plugging and screwing (M5085/122 - 1 m)	m	13.39	5.76		19.15

© NSR 01 Aug 2008 - 31 Jul 2009   Y76

	Y26 : FINISHES & SURFACE TREATMENTS (INTERNALLY)		Mat. £	Lab. £	Plant £	Total £
Y2650						
336	**Supply and fit pre-formed nosing with coloured slip resistant inserts - Gradus Hardnose type DDA3/ DDA4**					
	Stair edgings, Hardnose range, type DDA3/DDA4, plugging and screwing (M5085/123 - 1 m)	m	17.09	5.76		22.85
340	**Supply and fit pre-formed nosing with coloured slip resistant inserts - Gradus Bronze type BG511**					
	Stair edgings, Bronze BG range, type BG511, plugging and screwing (M5085/132 - 1 m)	m	39.35	5.76		45.11
342	**Supply and fit pre-formed nosing with coloured slip resistant inserts - Gradus Bronze type BG532**					
	Stair edgings, Bronze BG range, type BG532, plugging and screwing (M5085/134 - 1 m)	m	55.24	6.91		62.15
344	**Supply and fit pre-formed nosing with coloured slip resistant inserts - Gradus Bronze Elite type ELB1140**					
	Stair edgings, Bronze Elite range, type ELB1140, plugging and screwing (M5085/136 - 1 m)	m	36.38	5.76		42.14
346	**Supply and fit pre-formed nosing with coloured slip resistant inserts - Gradus Bronze Elite type ELB72740**					
	Stair edgings, Bronze Elite range, type ELB72740, plugging and screwing (M5085/137 - 1 m)	m	60.09	6.91		67.00
350	**Supply and fit pre-formed nosing with coloured slip resistant inserts - Gradus trims types TA5, TA5B, TA5C**					
	Stair edgings, Trim, types TA5, TA5B, TA5C, plugging and screwing (M5085/140 - 1 m)	m	9.41	5.76		15.17
352	**Supply and fit pre-formed nosing with coloured slip resistant inserts - Gradus trims types TA5, TA5B, TA5C, corners**					
	Stair edgings, Trim, type TA5 corners (M5085/141 - 1 nr)	nr	6.32	1.44		7.76
354	**Supply and fit pre-formed nosing with coloured slip resistant inserts - Gradus trim type BT42**					
	Stair edgings, Trim, type BT42, plugging and screwing (M5085/142 - 1 m)	m	21.23	5.76		26.99
356	**Supply and fit pre-formed nosing with coloured slip resistant inserts - Gradus trim type BT42 corners**					
	Stair edgings, Trim, type BT42 corners (M5085/143 - 1 nr)	nr	25.00	1.44		26.44
358	**Supply and fit pre-formed nosing with coloured slip resistant inserts - Gradus trim type HT38**					
	Stair edgings, Trim, type HT38, plugging and screwing (M5085/144 - 1 m)	m	5.06	5.76		10.82
360	**Supply and fit pre-formed nosing with coloured slip resistant inserts - Gradus trim type HT38 corners**					
	Stair edgings, Trim, type HT38 corners (M5085/145 - 1 nr)	nr	6.32	1.44		7.76

© NSR 01 Aug 2008 - 31 Jul 2009   Y77

Y26 : FINISHES & SURFACE TREATMENTS (INTERNALLY)			Mat. £	Lab. £	Plant £	Total £
Y2660	ANTI SLIP SURFACING					
400	Apply anti slip tape to concrete stairs, ramps or steps 51mm wide, black					
	Anti slip tape 51mm wide, black (M5090/100 - 1 m)	m	2.56	1.73		4.29
402	Apply anti slip tape to concrete stairs, ramps or steps 51mm wide, coloured					
	Anti slip tape 51mm wide, coloured (M5090/102 - 1 m)	m	3.53	1.73		5.26
404	Apply anti slip tape to concrete stairs, ramps or steps 102mm wide, black					
	Anti slip tape 102mm wide, black (M5090/104 - 1 m)	m	5.12	2.30		7.42
406	Apply anti slip tape to concrete stairs, ramps or steps 102mm wide, coloured					
	Anti slip tape 102mm wide, coloured (M5090/106 - 1 m)	m	7.04	2.30		9.34
408	Apply anti slip tape to marble or terrazzo surfaces 25mm wide, clear					
	Anti slip tape 25mm wide, clear (M5090/108 - 1 m)	m	2.42	1.73		4.15
410	Apply anti slip tape to marble or terrazzo surfaces 51mm wide, clear					
	Anti slip tape 51mm wide, clear (M5090/110 - 1 m)	m	4.85	1.73		6.58
412	Apply anti slip tape to marble or terrazzo surfaces 102mm wide, clear					
	Anti slip tape 102mm wide, clear (M5090/112 - 1 m)	m	9.70	2.30		12.00
414	Apply anti slip cleats to marble or terrazzo surfaces 152 x 610mm wide, clear					
	Anti slip cleats 152mm x 610, clear (M5090/116 - 1 nr)	nr	10.54	2.30		12.84
416	Apply water resistant anti slip tape to shower or changing areas or the like 25mm wide.					
	Anti slip tape 25mm wide, water resistant (M5090/118 - 1 m)	m	2.54	1.73		4.27
418	Apply water resistant anti slip tape to shower or changing areas or the like 50mm wide.					
	Anti slip tape 50mm wide, water resistant (M5090/120 - 1 m)	m	5.06	1.73		6.79
420	Supply and fix aluminium anti slip plates to metal or timber steps or stairways, 114 x 635mm					
	Aluminium Anti slip plates 114mm x 635, screwed to metal or timber; (M5090/122 - 1 nr)	nr	43.22	4.61		47.83
422	Supply and fix self adhesive anti slip cleats, 19 x 610mm, black					
	Anti slip cleats 19mm x 610, black (M5090/124 - 1 nr)	nr	1.26	1.73		2.99
424	Supply and fix self adhesive anti slip cleats, 19 x 610mm, coloured					
	Anti slip cleats 19mm x 610, coloured (M5090/125 - 1 nr)	nr	1.23	1.73		2.96

© NSR 01 Aug 2008 - 31 Jul 2009   Y78

Y26 : FINISHES & SURFACE TREATMENTS (INTERNALLY)			Mat. £	Lab. £	Plant £	Total £
Y2660						
426	**Supply and fix self adhesive anti slip cleats, 152 x 610mm, black**					
	Anti slip cleats 152mm x 610, black (M5090/126 - 1 nr)	nr	5.19	3.46		8.65
428	**Supply and fix self adhesive anti slip cleats, 152 x 610mm, yellow**					
	Anti slip cleats 152mm x 610, yellow (M5090/128 - 1 nr)	nr	5.92	3.46		9.38
430	**Supply and fix self adhesive anti slip tread, 140 x 140mm, yellow**					
	Anti slip tread 140mm x 140, yellow (M5090/130 - 1 nr)	nr	1.62	1.73		3.35
440	**Apply acrylic resin anti slip coating to floors, any colour**					
	Acrylic resin anti slip coating to floors, any colour (M5090/136 - 1 m2)	m2	8.93	5.76		14.69
442	**Apply anti slip coating to marble, granite, terrazzo or tiled areas**					
	Antil slip treatment to tiled areas (M5090/138 - 1 m2)	m2	1.60	2.30		3.90
Y2670	**WALL FINISHES**					
500	**Supply and fix ceramic tiling over 300mm wide PC £25.00/m2**					
	Walls over 300 mm wide to plaster (M4070/300 - 1 m2)	m2	27.48	40.28		67.76
502	**Supply and fix ceramic tiling over 300mm wide n.e. 5m2 PC £25.00/m2**					
	Walls over 300 mm wide to plaster 1 to 5 m2 (M4070/301 - 1 m2)	m2	27.48	45.31		72.79
504	**Supply and fix ceramic tiling n.e 300mm wide PC £25.00/m2**					
	Walls ne 300 mm wide to plaster (M4070/310 - 1 m)	m	8.24	20.14		28.38
506	**Extra over for fair edge tiles**					
	Extra over for fair edge tiles (M4070/320 - 1 m)	m	7.09	1.44		8.53
510	**Plaster skim to walls over 300mm wide**					
	Walls over 300 mm wide to cement/sand or plaster (M2023/070 - 1 m2)	m2	0.87	4.75	0.18	5.80
512	**Plaster skim to walls n.e. 300mm wide**					
	Walls ne 300 mm wide to cement/sand or plaster (M2023/115 - 1 m)	m	0.26	4.75	0.18	5.19
514	**Plaster browning and skim to walls 13mm thick over 300mm wide**					
	Walls over 300 mm wide to brickwork, blockwork or stonework (M2025/050 - 1 m2)	m2	2.74	11.08	0.43	14.25

© NSR 01 Aug 2008 - 31 Jul 2009   Y79

Y26 : FINISHES & SURFACE TREATMENTS (INTERNALLY)			Mat. £	Lab. £	Plant £	Total £
Y2670						
516	**Plaster browning and skim to walls 13mm thick n.e. 300mm wide**					
	Walls ne 300 mm wide to brickwork, blockwork or stonework (M2025/100 - 1 m)	m	0.82	8.71	0.34	9.87
520	**Plaster bonding and skim to walls 13mm thick over 300mm wide**					
	Walls over 300 mm wide to concrete (M2035/060 - 1 m2)	m2	3.26	11.87	0.46	15.59
522	**Plaster bonding and skim to walls 13mm thick n.e. 300mm wide**					
	Walls ne 300 mm wide to concrete (M2035/110 - 1 m)	m	0.98	11.08	0.43	12.49
530	**Extra over for angle bead**					
	Angle beads (M2080/900 - 1 m)	m	0.92	2.37		3.29

© NSR 01 Aug 2008 - 31 Jul 2009   Y80

			Mat. £	Lab. £	Plant £	Total £
	Y27 : DECORATING (INTERNALLY)					
Y2710	**WALLPAPER**					
010	**Replace lining paper to walls**					
	Walls and columns area over 0.5 m2 to plaster (M5230/200 - 1 m2), Stripping off/removing/taking down paper from plastered walls in preparation for replacement (M6205/050 - 1 m2)	m2	0.70	8.64	0.20	9.54
012	**Replace lining paper to ceilings**					
	Ceilings and beams area over 0.5 m2 to plaster (M5230/400 - 1 m2), Stripping off/removing/taking down paper from plastered ceilings in preparation for replacement (M6205/100 - 1 m2)	m2	0.70	10.37	0.22	11.29
014	**Replace wallpaper PC £9.00/roll to walls**					
	Walls and columns area over 0.5 m2 to plaster (M5250/200 - 1 m2), Stripping off/removing/taking down paper from plastered walls in preparation for replacement (M6205/050 - 1 m2)	m2	2.24	9.50	0.20	11.94
016	**Replace wallpaper PC £9.00/roll to ceilings**					
	Ceilings and beams area over 0.5 m2 to plaster (M5250/400 - 1 m2), Stripping off/removing/taking down paper from plastered ceilings in preparation for replacement (M6205/100 - 1 m2)	m2	2.24	11.24	0.22	13.70
018	**Replace wallpaper PC £13.00/roll to walls**					
	Walls and columns area over 0.5 m2 to plaster (M5260/200 - 1 m2), Stripping off/removing/taking down paper from plastered walls in preparation for replacement (M6205/050 - 1 m2)	m2	3.12	9.50	0.20	12.82
020	**Replace wallpaper PC £13.00/roll to ceilings**					
	Ceilings and beams area over 0.5 m2 to plaster (M5260/400 - 1 m2), Stripping off/removing/taking down paper from plastered ceilings in preparation for replacement (M6205/100 - 1 m2)	m2	3.12	11.24	0.22	14.58
022	**Replace wallpaper PC £18.00/roll to walls**					
	Walls and columns area over 0.5 m2 to plaster (M5270/200 - 1 m2), Stripping off/removing/taking down paper from plastered walls in preparation for replacement (M6205/050 - 1 m2)	m2	4.19	9.50	0.20	13.89
024	**Replace wallpaper PC £18.00/roll to ceilings**					
	Ceilings and beams area over 0.5 m2 to plaster (M5270/400 - 1 m2), Stripping off/removing/taking down paper from plastered ceilings in preparation for replacement (M6205/100 - 1 m2)	m2	4.19	11.24	0.22	15.65
Y2720	**EMULSION PAINT: NEW SURFACES**					
100	**3 coats emulsion over 300mm girth to plaster**					
	Plaster general surfaces over 300 mm girth (M6015/140 - 1 m2)	m2	1.23	5.18		6.41

© NSR 01 Aug 2008 - 31 Jul 2009   Y81

	Y27 : DECORATING (INTERNALLY)		Mat. £	Lab. £	Plant £	Total £
Y2720						
102	**3 coats emulsion not exceeding 300mm girth to plaster**					
	Plaster general surfaces in isolated surfaces ne 300 mm girth (M6015/150 - 1 m)	m	0.37	2.59		2.96
106	**3 coats emulsion in areas not exceeding 0.5 m2 to plaster**					
	All isolated areas ne 0.5 m2 irrespective of girth (M6015/780 - 1 nr)	nr	0.62	5.47		6.09
108	**3 coats emulsion over 300mm girth to textured surfaces**					
	Textured general surfaces over 300 mm girth (M6015/710 - 1 m2)	m2	1.69	6.05		7.74
110	**3 coats emulsion not exceeding 300mm girth to textured surfaces**					
	Textured general surfaces in isolated surfaces ne 300 mm girth (M6015/720 - 1 m)	m	0.56	2.88		3.44
112	**3 coats emulsion over 300mm girth to paper covered surfaces**					
	Paper covered general surfaces over 300 mm girth (M6015/740 - 1 m2)	m2	1.23	4.61		5.84
114	**3 coats emulsion not exceeding 300mm girth to paper covered surfaces**					
	Paper covered general surfaces in isolated surfaces ne 300 mm girth (M6015/750 - 1 m)	m	0.37	2.59		2.96
Y2730	**EMULSION PAINT : REPAINTING**					
120	**Repaint two coats over 300mm girth to plaster**					
	Plaster general surfaces over 300 mm girth (M6010/140 - 1 m2), Plaster general surfaces over 300 mm girth (M6220/140 - 1 m2)	m2	1.12	4.32		5.44
122	**Repaint two coats not exceeding 300mm girth to plaster**					
	Plaster general surfaces in isolated surfaces ne 300 mm girth (M6010/150 - 1 m), Plaster general surfaces in isolated surfaces ne 300 mm girth (M6220/150 - 1 m)	m	0.34	2.88		3.22
124	**Repaint two coats in areas not exceeding 0.5 m2**					
	All isolated areas ne 0.5 m2 irrespective of girth (M6010/780 - 1 nr), All isolated areas ne 0.50 m2 irrespective of girth (M6220/780 - 1 nr)	nr	0.66	5.18		5.84
126	**Repaint two coats over 300mm girth to textured surfaces**					
	Textured general surfaces over 300 mm girth (M6010/710 - 1 m2), Plaster general surfaces over 300 mm girth (M6220/140 - 1 m2)	m2	1.45	5.76		7.21

	Y27 : DECORATING (INTERNALLY)		Mat. £	Lab. £	Plant £	Total £
Y2730						
128	**Repaint two coats not exceeding 300mm girth to textured surfaces**					
	Textured general surfaces in isolated surfaces ne 300 mm girth (M6010/720 - 1 m), Plaster general surfaces in isolated surfaces ne 300 mm girth (M6220/150 - 1 m)	m	0.44	2.60		3.04
130	**Repaint two coats over 300mm girth to paper covered surfaces**					
	Paper covered general surfaces over 300 mm girth (M6010/740 - 1 m2), Paper covered general surfaces over 300 mm girth (M6210/740 - 1 m2)	m2	1.15	5.47		6.62
132	**Repaint two coats not exceeding 300mm girth to paper covered surfaces**					
	Paper covered general surfaces in isolated surfaces ne 300 mm girth (M6010/750 - 1 m), Paper covered general surfaces in isolated surfaces ne 300 mm girth (M6210/750 - 1 m)	m	0.34	3.45		3.79
Y2740	**GLOSS PAINT: NEW SURFACES**					
200	**Primer, undercoat, gloss over 300mm girth**					
	Wood general surfaces over 300 mm girth (M6040/570 - 1 m2)	m2	1.67	11.23		12.90
202	**Primer, undercoat, gloss not exceeding 300mm girth**					
	Wood general surfaces in isolated surfaces ne 300 mm girth (M6040/580 - 1 m)	m	0.50	3.74		4.24
204	**Primer, undercoat, gloss in areas not exceeding 0.5 m2**					
	All isolated areas ne 0.5 m2 irrespective of girth (M6040/780 - 1 nr)	nr	1.25	10.37		11.62
210	**Primer, undercoat, gloss to wood glazed doors over 300mm girth**					
	Wood glazed doors over 300 mm girth in panes 0.50 to 1 m2 (M6040/642 - 1 m2)	m2	0.97	13.82		14.79
212	**Primer, undercoat, gloss to metal glazed doors over 300mm girth**					
	Metal glazed doors over 300 mm girth in panes 0.50 to 1 m2 (M6040/242 - 1 m2)	m2	1.50	13.82		15.32
214	**Primer, undercoat, gloss to metal panel type radiators over 300mm girth**					
	Metal panel type radiators over 300 mm girth (M6040/320 - 1 m2)	m2	2.74	9.22		11.96
216	**Primer, undercoat, gloss to metal column type radiators over 300mm girth**					
	Metal column type radiators over 300 mm girth (M6040/330 - 1 m2)	m2	2.99	16.70		19.69

© NSR 01 Aug 2008 - 31 Jul 2009   Y83

	Y27 : DECORATING (INTERNALLY)		Mat. £	Lab. £	Plant £	Total £
Y2750	**GLOSS PAINT : REPAINTING**					
300	**Repaint, undercoat, gloss over 300mm girth**					
	Wood general surfaces over 300 mm girth (M6035/570 - 1 m2), Wood general surfaces over 300 mm girth (M6220/570 - 1 m2)	m2	1.43	10.36		11.79
302	**Repaint, undercoat, gloss not exceeding 300mm girth**					
	Wood general surfaces in isolated surfaces ne 300 mm girth (M6035/580 - 1 m), Wood general surfaces in isolated surfaces ne 300 mm girth (M6220/580 - 1 m)	m	0.42	3.45		3.87
304	**Repaint, undercoat, gloss in areas not exceeding 0.5 m2**					
	All isolated areas ne 0.5 m2 irrespective of girth (M6035/780 - 1 nr), All isolated areas ne 0.50 m2 irrespective of girth (M6220/780 - 1 nr)	nr	0.72	9.79		10.51
310	**Repaint, undercoat, gloss to wood glazed doors over 300mm girth**					
	Wood glazed doors over 300 mm girth in panes 0.50 to 1 m2 (M6035/642 - 1 m2), Wood glazed doors over 300 mm girth in panes 0.50 to 1 m2 (M6220/642 - 1 m2)	m2	0.96	13.53		14.49
312	**Repaint, undercoat, gloss to metal glazed doors over 300mm girth**					
	Metal glazed doors over 300 mm girth in panes 0.50 to 1 m2 (M6035/242 - 1 m2), Metal glazed doors over 300 mm girth in panes 0.50 to 1 m2 (M6220/242 - 1 m2)	m2	1.10	12.38		13.48
314	**Repaint, undercoat, gloss to metal panel type radiators over 300mm girth**					
	Metal panel type radiators over 300 mm girth (M6035/320 - 1 m2), Metal panel type radiators over 300 mm girth (M6220/320 - 1 m2)	m2	1.69	9.51		11.20
316	**Repaint, undercoat, gloss to metal column type radiators over 300mm girth**					
	Metal column type radiators over 300 mm girth (M6035/330 - 1 m2), Metal column type radiators over 300 mm girth (M6220/330 - 1 m2)	m2	2.07	16.13		18.20

© NSR 01 Aug 2008 - 31 Jul 2009   Y84

			Mat. £	Lab. £	Plant £	Total £
	Y28 : DECORATING (EXTERNALLY)					
Y2840	**GLOSS PAINT**					
300	**Primer, undercoat, two gloss finishing coats over 300mm girth**					
	Wood general surfaces over 300 mm girth (M6135/570 - 1 m2)	m2	2.12	14.40		16.52
302	**Primer, undercoat, two gloss finishing coats not exceeding 300mm girth**					
	Wood general surfaces in isolated surfaces ne 300 mm girth (M6135/580 - 1 m)	m	0.64	4.90		5.54
304	**Primer, undercoat, two gloss finishing coats in areas not exceeding 0.5 m2**					
	All isolated areas ne 0.5 m2 irrespective of girth (M6135/780 - 1 nr)	nr	1.07	15.26		16.33
320	**Primer, undercoat, two gloss finishing coats to glazed windows and screens in panes not exceeding 0.10 m2**					
	Wood glazed windows and screens over 300 mm girth in panes ne 0.10 m2 (M6135/600 - 1 m2)	m2	1.70	29.66		31.36
322	**Primer, undercoat, two gloss finishing coats to glazed windows and screens in panes 0.10 to 0.50 m2**					
	Wood glazed windows and screens over 300 mm girth in panes 0.10 to 0.50 m2 (M6135/602 - 1 m2)	m2	1.49	22.46		23.95
324	**Primer, undercoat, two gloss finishing coats to glazed windows and screens in panes 0.50 to 1 m2**					
	Wood glazed windows and screens over 300 mm girth in panes 0.50 to 1 m2 (M6135/604 - 1 m2)	m2	1.27	19.01		20.28
326	**Primer, undercoat, two gloss finishing coats to glazed windows and screens in panes over 1 m2**					
	Wood glazed windows and screens over 300 mm girth in panes over 1 m2 (M6135/606 - 1 m2)	m2	1.10	16.99		18.09
Y2842	**GLOSS PAINT : REPAINTING**					
350	**Repaint, undercoat, two gloss finishing coats over 300mm girth**					
	Wood general surfaces over 300 mm girth (M6130/570 - 1 m2), Wood general surfaces over 300 mm girth (M6210/570 - 1 m2)	m2	1.77	12.68		14.45
352	**Repaint, undercoat, two gloss finishing coats not exceeding 300mm girth**					
	Wood general surfaces in isolated surfaces ne 300 mm girth (M6130/580 - 1 m), Wood general surfaces in isolated surfaces ne 300 mm girth (M6210/580 - 1 m)	m	0.53	4.32		4.85

© NSR 01 Aug 2008 - 31 Jul 2009

Y85

	Y28 : DECORATING (EXTERNALLY)		Mat. £	Lab. £	Plant £	Total £
Y2842						
354	**Repaint, undercoat, two gloss finishing coats in areas not exceeding 0.5 m2**					
	All isolated areas ne 0.5 m2 irrespective of girth (M6130/780 - 1 nr), All isolated areas ne 0.50 m2 irrespective of girth (M6220/780 - 1 nr)	nr	0.97	12.09		13.06
370	**Repaint, undercoat, two gloss finishing coats to glazed windows and screens in panes not exceeding 0.10 m2**					
	Wood glazed windows and screens over 300 mm girth in panes ne 0.10 m2 (M6130/600 - 1 m2), Wood glazed windows and screens over 300 mm girth in panes ne 0.10 m2 (M6220/600 - 1 m2)	m2	1.36	26.21		27.57
372	**Repaint, undercoat, two gloss finishing coats to glazed windows and screens in panes 0.10 to 0.50 m2**					
	Wood glazed windows and screens over 300 mm girth in panes 0.10 to 0.50 m2 (M6130/602 - 1 m2), Wood glazed windows and screens over 300 mm girth in panes 0.10 to 0.50 m2 (M6220/602 - 1 m2)	m2	1.44	19.59		21.03
374	**Repaint, undercoat, two gloss finishing coats to glazed windows and screens in panes 0.50 to 1 m2**					
	Wood glazed windows and screens over 300 mm girth in panes 0.50 to 1 m2 (M6130/604 - 1 m2), Wood glazed windows and screens over 300 mm girth in panes 0.50 to 1 m2 (M6220/604 - 1 m2)	m2	1.26	16.99		18.25
376	**Repaint, undercoat, two gloss finishing coats to glazed windows and screens in panes over 1 m2**					
	Wood glazed windows and screens over 300 mm girth in panes over 1 m2 (M6130/606 - 1 m2), Wood glazed windows and screens over 300 mm girth in panes over 1 m2 (M6220/606 - 1 m2)	m2	1.13	14.97		16.10

© NSR 01 Aug 2008 - 31 Jul 2009　　　　Y86

			Mat. £	Lab. £	Plant £	Total £
	Y29 : SIGNAGE AND WAYFINDING (INTERNALLY)					
Y2910	**SIGNAGE**					
010	**Supply and fix self adhesive vinyl sign 297 x 210mm**					
	Self adhesive vinyl sign size 297 x 210mm (N1510/100 - 1 nr)	nr	13.40	1.44		14.84
012	**Supply and fix self adhesive vinyl sign 297 x 420mm**					
	Self adhesive vinyl sign size 297 x 420mm (N1510/102 - 1 nr)	nr	23.40	1.44		24.84
014	**Supply and fix self adhesive rigid plastic sign 297 x 210mm**					
	Self adhesive rigid plastic sign size 297 x 210mm (N1510/104 - 1 nr)	nr	22.90	1.44		24.34
016	**Supply and fix self adhesive rigid plastic sign 297 x 420mm**					
	Self adhesive rigid plastic sign size 297 x 420mm (N1510/106 - 1 nr)	nr	34.10	1.44		35.54
018	**Supply and fix self adhesive tactile braille sign 75 x 100mm**					
	Self adhesive tactile Braille sign size 75 x 100mm (N1510/110 - 1 nr)	nr	16.56	1.44		18.00
020	**Supply and fix self adhesive tactile braille sign 150 x 150mm**					
	Self adhesive tactile Braille sign size 150 x 150mm (N1510/112 - 1 nr)	nr	21.60	1.44		23.04
022	**Supply and fix self adhesive tactile braille sign 150 x 225mm**					
	Self adhesive tactile Braille sign size 150 x 225mm (N1510/114 - 1 nr)	nr	24.80	1.73		26.53
024	**Supply and fix self adhesive tactile braille sign 150 x 300mm**					
	Self adhesive tactile Braille sign size 150 x 300mm (N1510/116 - 1 nr)	nr	28.90	1.73		30.63
026	**Supply and fix self adhesive tactile braille Photoluminescent sign 150 x 450mm**					
	Self adhesive tactile Braille photoluminescent sign size 125 x 450mm (N1510/118 - 1 nr)	nr	40.50	1.73		42.23
028	**Supply and fix self adhesive tactile braille Photoluminescent sign 150 x 150mm**					
	Self adhesive tactile Braille photoluminescent sign size 150 x 150mm (N1510/120 - 1 nr)	nr	28.90	1.44		30.34
030	**Supply and fix self adhesive tactile braille Photoluminescent sign 150 x 225mm**					
	Self adhesive tactile Braille photoluminescent sign size 150 x 225mm (N1510/122 - 1 nr)	nr	33.30	1.44		34.74

© NSR 01 Aug 2008 - 31 Jul 2009

Y87

	Y29 : SIGNAGE AND WAYFINDING (INTERNALLY)		Mat. £	Lab. £	Plant £	Total £
Y2910						
032	**Supply and fix self adhesive tactile braille Photoluminescent sign 150 x 300mm**					
	Self adhesive tactile Braille photoluminescent sign size 150 x 300mm (N1510/124 - 1 nr)	nr	37.75	1.73		39.48
034	**Supply and fix self adhesive tactile braille Photoluminescent sign 150 x 450mm**					
	Self adhesive tactile Braille photoluminescent sign size 150 x 450mm (N1510/126 - 1 nr)	nr	43.00	1.73		44.73
036	**Supply and fix rigid plastic sign 297 x 210mm**					
	1.5mm rigid plastic sign size 297 x 210mm ; screw fixed to wall (N1510/140 - 1 nr)	nr	19.60	7.78		27.38
038	**Supply and fix rigid plastic sign 297 x 420mm**					
	1.5mm rigid plastic sign size 297 x 420mm; screw fixed to wall (N1510/150 - 1 nr)	nr	30.30	7.78		38.08
040	**Supply and fix non reflective steel sign 152 x 305mm**					
	Non reflective steel sign size 152 x 305mm screw fixed to wall (N1510/160 - 1 nr)	nr	43.70	7.78		51.48
042	**Supply and fix non reflective steel sign 457 x 305mm**					
	Non reflective steel sign size 457 x 305mm screw fixed to wall (N1510/170 - 1 nr)	nr	57.30	7.78		65.08
Y2930	**AUXILIARY AIDS**					
400	**Portable induction loop**					
	Supply only Portable induction loop (W1210/050 - 1 nr)	nr	239.00			239.00
402	**Fixed counter induction loop**					
	Supply and install fixed counter induction loop, inc connection to mains (W1210/055 - 1 nr)	nr	169.00	12.69		181.69
404	**Fixed auditoriam induction loop (150m2)**					
	Supply and install fixed auditorium induction loop, 150m2 ref: AIL100F inc connection to mains (W1210/060 - 1 nr)	nr	900.00	236.88		1136.88
406	**Fixed auditoriam induction loop (250m2)**					
	Supply and install fixed auditorium induction loop, 250m2 ref: AIL250F inc connection to mains (W1210/065 - 1 nr)	nr	1200.00	236.88		1436.88
408	**Fixed auditoriam induction loop (450m2)**					
	Supply and install fixed auditorium induction loop, 450m2 ref: AIL450F inc connection to mains (W1210/070 - 1 nr)	nr	1500.00	236.88		1736.88

© NSR 01 Aug 2008 - 31 Jul 2009   Y88

	Y29 : SIGNAGE AND WAYFINDING (INTERNALLY)		Mat. £	Lab. £	Plant £	Total £
Y2940	**FILMS AND COATINGS FOR GLASS**					
500	**Apply safety film to glass in areas n.e. 0.15m2**					
	In panes n.e. 0.15m2 (L4088/270 - 1 m2)	m2	15.60	40.32		55.92
502	**Apply safety film to glass in areas 0.15 - 0.50m2**					
	In panes 0.15 - 0.50m (L4088/280 - 1 m2)	m2	15.60	25.92		41.52
504	**Apply safety film to glass in areas 0.50 - 1.00m2**					
	In panes 0.50 - 1.00m (L4088/290 - 1 m2)	m2	15.60	15.84		31.44
506	**Apply safety film to glass in areas exceeding 1.00m2**					
	In panes exceeding 1.00m2 (L4088/300 - 1 m2)	m2	15.60	11.52		27.12
508	**Apply glass highlighting strip 500 x 150mm wide**					
	Glass highlighting strip 500 x 150mm (L4088/305 - 1 nr)	nr	15.93	1.44		17.37
Y2950	**TEMPORARY BARRIER SYSTEMS**					
600	**Free standing barrier system with single beam 1.2m long**					
	Free standing barrier system - Supply only Tensaguide or similar solid beam system, Single silver or grey post (N1023/100 - 2 nr), Free standing barrier system - Supply only Tensaguide or similar solid beam system, silver or grey beam (N1023/104 - 1 nr)	nr	430.60			430.60
602	**Free standing barrier system with single beam 2.4m long**					
	Free standing barrier system - Supply only Tensaguide or similar solid beam system, Single silver or grey post (N1023/100 - 3 nr), Free standing barrier system - Supply only Tensaguide or similar solid beam system, silver or grey beam (N1023/104 - 2 nr)	nr	674.50			674.50
604	**Free standing barrier system with dual beam 1.2m long**					
	Free standing barrier system - Supply only Tensaguide or similar solid beam system, dual silver or grey post (N1023/102 - 2 nr), Free standing barrier system - Supply only Tensaguide or similar solid beam system, silver or grey beam (N1023/104 - 2 nr)	nr	494.00			494.00
606	**Free standing barrier system with dual beam 2.4m long**					
	Free standing barrier system - Supply only Tensaguide or similar solid beam system, dual silver or grey post (N1023/102 - 3 nr), Free standing barrier system - Supply only Tensaguide or similar solid beam system, silver or grey beam (N1023/104 - 4 nr)	nr	798.20			798.20
608	**Extra over for single writing surface**					
	Free standing barrier system - Supply only Tensaguide or similar solid beam system, single writing surface (N1023/106 - 1 nr)	nr	147.80			147.80
610	**Free standing webbing and post barrier system**					
	Free standing barrier system - Supply only webbing and post system comprising 1 post with retractable webbing and 1 linking post (up to 2.3ml) (N1023/108 - 1 nr)	nr	218.35			218.35

© NSR 01 Aug 2008 - 31 Jul 2009   Y89

Y29 : SIGNAGE AND WAYFINDING (INTERNALLY)		Mat. £	Lab. £	Plant £	Total £

© NSR 01 Aug 2008 - 31 Jul 2009   Y90

			Mat. £	Lab. £	Plant £	Total £
	Y30 : SANITARY CONVENIENCES					
Y3010	**SANITARY APPLIANCES**					
005	**Reposition existing urinal inc alter existing pipework as necessary**					
	Removing urinal bowls, setting aside for re-use, making good finishings (C2010/431 - 1 nr), Basins, waste, plugging and screwing (N1365/600 - 1 nr), Pipes 15 mm nominal size, screwing to timber (S1220/100 - 1 m)	nr	9.15	69.97		79.12
010	**Renew existing urinal**					
	Removing urinal bowls, making good finishings (C2010/430 - 1 nr), Basins, waste, (PC £180.00), plugging and screwing (N1363/600 - 1 nr), Sealing around sanitary appliances (P2240/100 - 0.6 m), Pipes 32 mm nominal size (KP104P), screwing to timber (R1120/150 - 0.5 m), Extra over for fittings with two ends (R1120/170 - 1 nr), Traps 32 mm nominal size (632.125 S type outlet), joint to waste outlet (R1160/805 - 1 nr)	nr	196.69	82.28		278.97
012	**Supply and fix new urinal**					
	Basins, waste, (PC £180.00), plugging and screwing (N1363/600 - 1 nr), Sealing around sanitary appliances (P2240/100 - 0.6 m), Pipes 32 mm nominal size (KP104P), screwing to timber (R1120/150 - 0.5 m), Extra over for fittings with two ends (R1120/170 - 1 nr), Traps 32 mm nominal size (632.125 S type outlet), joint to waste outlet (R1160/805 - 1 nr)	nr	194.92	62.20		257.12
014	**Extra over for new flush pipe**					
	Flush pipes and spreader to one basin, plugging and screwing (N1363/650 - 1 nr)	nr	45.34	11.58		56.92
020	**Reposition existing wash hand basin inc alter existing pipework as necessary**					
	Removing basins and brackets, setting aside for re-use, making good finishings (C2010/426 - 1 nr), Basins 560 x 405 mm, waste, plug and chain, pair of taps, brackets, plugging and screwing (N1340/400 - 1 nr), Pipes 15 mm nominal size, screwing to timber (S1220/100 - 2 m), Extra over for fittings with two ends (S1220/152 - 2 nr)	nr	18.89	108.62		127.51

© NSR 01 Aug 2008 - 31 Jul 2009

Y91

	Y30 : SANITARY CONVENIENCES		Mat. £	Lab. £	Plant £	Total £
Y3010						
022	**Renew washbasin and brackets, chrome plated taps, 'waste, overflow and chain, adapt supply and disposal pipework, silicone mastic sealant**					
	Removing basins and brackets, making good finishings (C2010/425 - 1 nr), Stripping out ne 25 mm plastics or copper pipework (C7550/500 - 1 m), Stripping out 25 to 50 mm plastics or copper pipework (C7550/502 - 0.5 m), Basins 560 x 405 mm, chromium plated waste, plug and chain, pair of chromium plated taps, brackets (PC £105.00), plugging and screwing (N1339/415 - 1 nr), Sealing around sanitary appliances (P2240/100 - 1.2 m), Pipes 32 mm nominal size (KP104P), screwing to timber (R1120/150 - 0.5 m), Extra over for fittings with two ends (R1120/170 - 1 nr), Bottle traps 32 mm nominal size (611.125 P type outlet), joint to waste outlet (R1160/830 - 1 nr), Pipes 15 mm nominal size, screwing to timber (S1220/100 - 1 m), Extra over for connections to appliances and equipment, straight tap connector (S1220/131 - 1 nr), Fixed to surface, single core 4 mm2, screwing to timber (V9015/090 - 1 m), Labelled "SAFETY ELECTRICAL EARTH DO NOT REMOVE" (V9205/010 - 2 nr)	nr	141.31	143.63		284.94
024	**Replace washbasin and pedestal, chrome plated taps, waste, overflow and chain, adapt supply and disposal pipework, silicone mastic sealant**					
	Removing basins and pedestals, making good finishings (C2010/427 - 1 nr), Stripping out ne 25 mm plastics or copper pipework (C7550/500 - 1 m), Stripping out 25 to 50 mm plastics or copper pipework (C7550/502 - 0.5 m), Basins 560 x 405 mm, chromium plated waste, plug and chain, pair of chromium plated taps, pedestal, (PC £115.00), plugging and screwing (N1339/417 - 1 nr), Sealing around sanitary appliances (P2240/100 - 1.2 m), Pipes 32 mm nominal size (KP104P), screwing to timber (R1120/150 - 0.5 m), Extra over for fittings with two ends (R1120/170 - 1 nr), Bottle traps 32 mm nominal size (611.125 P type outlet), joint to waste outlet (R1160/830 - 1 nr), Pipes 15 mm nominal size, screwing to timber (S1220/100 - 1 m), Extra over for connections to appliances and equipment, straight tap connector (S1220/131 - 1 nr), Fixed to surface, single core 4 mm2, screwing to timber (V9015/090 - 1 m), Labelled "SAFETY ELECTRICAL EARTH DO NOT REMOVE" (V9205/010 - 2 nr)	nr	150.23	149.42		299.65
026	**Convert basin or bath taps to lever operated pillar taps**					
	Basin taps, Simplicity conversion kits N&C ref:P63491.25 (N1310/144 - 1 pr)	pr	14.68	4.44		19.12
028	**Replace basin taps with chrome plated lever operated pillar taps**					
	Pillar taps, Novalever N&C ref:P63000.33 (N1310/122 - 1 nr)	nr	51.44	6.65		58.09
030	**Replace basin taps with chrome plated non concussive pillar taps**					
	Pillar taps, Timed-flow Pillar tap N&C ref:S23320.01 (N1310/124 - 1 nr)	nr	123.78	6.65		130.43
032	**Replace basin mounted taps with pulse 8 automatic taps including plug in power supply**					
	Pulse 8 Automatic tap, Basin mounted with plug in power supply N&C ref:P68201.34 (N1310/166 - 1 nr)	nr	422.00	8.87		430.87

© NSR 01 Aug 2008 - 31 Jul 2009   Y92

Y30 : SANITARY CONVENIENCES			Mat. £	Lab. £	Plant £	Total £
Y3010						
034	**Replace wall mounted taps with pulse 8 automatic taps including plug in power supply**					
	Pulse 8 Automatic tap, Wall mounted with plug in power supply N&C ref:P68201.34 (N1310/168 - 1 nr)	nr	423.30	11.09		434.39
044	**Removing WC pan and high level cistern, new low level WC pan and cistern, supply pipework, overflow pipe, seat and cover**					
	Removing WC and cistern, making good finishings (C2010/439 - 1 nr), Stripping out ne 25 mm plastics or copper pipework (C7550/500 - 2 m), WC pans (PC £65.00), screwing (N1359/555 - 1 nr), Seats and covers PC £20.00 (N1359/565 - 1 nr), Low level syphonic plastic cisterns and covers, flush pipe and clips (PC £50.00), plugging and screwing (N1359/585 - 1 nr), Cutting or forming holes for pipes ne 55 mm nominal size in common brickwork 102 mm thick, making good (P3110/054 - 1 nr), Cutting or forming holes for pipes ne 55 mm nominal size in facing brickwork 102 mm thick, making good (P3110/060 - 1 nr), Pipes 19 mm nominal size (500.75), screwing to timber (R1110/050 - 1.5 m), Extra over for connections to appliances and equipment, bent connector (502.75.90) (R1110/070 - 1 nr), Pipes 15 mm nominal size, plugging and screwing to masonry (S1220/105 - 1.5 m), Extra over for connections to pipe ends, bent connector screwed joint to steel (S1220/129 - 1 nr), Extra over for fittings with two ends (S1220/152 - 1 nr)	nr	175.97	166.00	8.25	350.22
050	**Replace existing WC seat with new seat and cover PC 65.00**					
	Removing WC seats (C2010/441 - 1 nr), Disabled Seats and covers PC £65.00 (N1359/566 - 1 nr)	nr	65.00	10.04		75.04
052	**Replace existing WC seat with new raised seat and cover PC 95.00**					
	Removing WC seats (C2010/441 - 1 nr), Disabled Raised Seats and covers PC £95.00 (N1359/567 - 1 nr)	nr	95.00	11.72		106.72
060	**Supply and fit Doc M Toilet Package, standard white**					
	Olympus Economy Doc M Toilet Package N&C ref:P68300.50 (N1367/060 - 1 nr)	nr	722.06	178.09		900.15
062	**Supply and fit Doc M Toilet Package, Olympus Classic with coloured rails**					
	Olympus Classic Doc M Toilet Package N&C ref:P68300.48 (N1367/050 - 1 nr)	nr	738.45	178.09		916.54
064	**Supply and fit Doc M Toilet Package, Olympus Pentland Blue Close Coupled.**					
	Olympus Pentland Blue Close Coupled Doc M Toilet Package N&C ref:P68300.51 (N1367/070 - 1 nr)	nr	481.30	171.18		652.48
066	**Supply and fit Doc M Toilet Package, Olympus Ambulant.**					
	Olympus Ambulant Doc M Toilet Package N&C ref:P68300.49 (N1367/072 - 1 nr)	nr	380.03	55.47		435.50
080	**Replace bath taps with chrome plated lever operated pillar taps**					
	Bath taps, Novalever N&C ref:P63000.31 (N1310/152 - 1 nr)	nr	56.69	9.61		66.30

© NSR 01 Aug 2008 - 31 Jul 2009 — Y93

	Y30 : SANITARY CONVENIENCES		Mat. £	Lab. £	Plant £	Total £
Y3010						
100	**Remove existing bath**					
	Removing baths, making good finishings (C2010/445 - 1 nr)	nr	3.80	38.15		41.95
102	**Supply and fix Windsor powered bath, series 1 manual traverse 1700 x 745mm, chrome plated lever taps, waste, overflow and chain, adapt supply and disposal pipework, GRP panels, silicone mastic sealant**					
	Stripping off/removing/taking down to plaster walls in preparation for replacement 1 to 5 m2 (C6065/602 - 0.7 m2), Stripping out ne 25 mm plastics or copper pipework (C7550/500 - 1 m), Stripping out 25 to 50 mm plastics or copper pipework (C7550/502 - 1 m), Walls ne 300 mm wide to plaster (M4070/310 - 2.5 m), Powered baths 1700 x 745 x 440 mm, Windsor series 1 manual traverse, complete with built in hydraulic seat lift, chromium plated waste, overflow, plug and chain, GRP front panel and pair of chromium plated taps N&C ref:P68085 (N1320/260 - 1 nr), Sealing around sanitary appliances (P2240/100 - 2.5 m), Pipes 40 mm nominal size (KP204P), screwing to timber (R1120/200 - 1 m), Extra over for fittings with two ends (R1120/220 - 1 nr), Pipes 15 mm nominal size, screwing to timber (S1220/100 - 1 m), Extra over for connections to appliances and equipment, straight tap connector (S1220/131 - 1 nr), Fixed to surface, single core 4 mm2, screwing to timber (V9015/090 - 1 m), Labelled "SAFETY ELECTRICAL EARTH DO NOT REMOVE" (V9205/010 - 2 nr)	nr	5121.32	240.56	0.34	5362.22
104	**Supply and fix Windsor powered bath, series 2 powered traverse; 1700 x 745mm, chrome plated lever taps, waste, overflow and chain, adapt supply and disposal pipework, GRP panels, silicone mastic sealant**					
	Stripping off/removing/taking down to plaster walls in preparation for replacement 1 to 5 m2 (C6065/602 - 0.7 m2), Stripping out ne 25 mm plastics or copper pipework (C7550/500 - 1 m), Stripping out 25 to 50 mm plastics or copper pipework (C7550/502 - 1 m), Walls ne 300 mm wide to plaster (M4070/310 - 2.5 m), Powered baths 1700 x 745 x 440 mm, Windsor series 2 Power traverse, complete with built in hydraulic seat lift, chromium plated waste, overflow, plug and chain, GRP front panel and pair of chromium plated lever taps N&C ref:P68085 (N1320/262 - 1 nr), Sealing around sanitary appliances (P2240/100 - 2.5 m), Pipes 40 mm nominal size (KP204P), screwing to timber (R1120/200 - 1 m), Extra over for fittings with two ends (R1120/220 - 1 nr), Pipes 15 mm nominal size, screwing to timber (S1220/100 - 1 m), Extra over for connections to appliances and equipment, straight tap connector (S1220/131 - 1 nr), Fixed to surface, single core 4 mm2, screwing to timber (V9015/090 - 1 m), Labelled "SAFETY ELECTRICAL EARTH DO NOT REMOVE" (V9205/010 - 2 nr)	nr	5121.32	240.56	0.34	5362.22

© NSR 01 Aug 2008 - 31 Jul 2009 — Y94

	Y30 : SANITARY CONVENIENCES		Mat. £	Lab. £	Plant £	Total £
Y3010						
106	**Supply and fix Windsor powered bath, series 3 powered traverse; 1700 x 745mm, chrome plated lever taps, waste, overflow and chain, adapt supply and disposal pipework, GRP panels, silicone mastic sealant**					
	Stripping off/removing/taking down to plaster walls in preparation for replacement 1 to 5 m2 (C6065/602 - 0.7 m2), Stripping out ne 25 mm plastics or copper pipework (C7550/500 - 1 m), Stripping out 25 to 50 mm plastics or copper pipework (C7550/502 - 1 m), Walls ne 300 mm wide to plaster (M4070/310 - 2.5 m), Powered baths 1700 x 745 x 440 mm, Windsor series 3 Power traverse, complete with built in hydraulic seat lift, powered leg raiser, chromium plated waste, overflow, plug and chain, GRP front panel and pair of chromium plated lever taps N&C ref:P68085 (N1320/264 - 1 nr), Sealing around sanitary appliances (P2240/100 - 2.5 m), Pipes 40 mm nominal size (KP204P), screwing to timber (R1120/200 - 1 m), Extra over for fittings with two ends (R1120/220 - 1 nr), Pipes 15 mm nominal size, screwing to timber (S1220/100 - 1 m), Extra over for connections to appliances and equipment, straight tap connector (S1220/131 - 1 nr), Fixed to surface, single core 4 mm2, screwing to timber (V9015/090 - 1 m), Labelled "SAFETY ELECTRICAL EARTH DO NOT REMOVE" (V9205/010 - 2 nr)	nr	6038.32	240.56	0.34	6279.22
108	**Supply and fix Claremont Heavy duty cut away bath; 1700 x 780mm, chrome plated lever taps, waste, overflow and chain, adapt supply and disposal pipework, GRP panels, silicone mastic sealant**					
	Stripping off/removing/taking down to plaster walls in preparation for replacement 1 to 5 m2 (C6065/602 - 0.7 m2), Stripping out ne 25 mm plastics or copper pipework (C7550/500 - 1 m), Stripping out 25 to 50 mm plastics or copper pipework (C7550/502 - 1 m), Walls ne 300 mm wide to plaster (M4070/310 - 2.5 m), Cut away bath 1700 x 780mm, Claremont, heavy duty N&C ref:P68065.01, lever action pop up waste and overflow, plug and chain and pair of chromium plated lever taps (N1320/266 - 1 nr), Sealing around sanitary appliances (P2240/100 - 2.5 m), Pipes 40 mm nominal size (KP204P), screwing to timber (R1120/200 - 1 m), Extra over for fittings with two ends (R1120/220 - 1 nr), Pipes 15 mm nominal size, screwing to timber (S1220/100 - 1 m), Extra over for connections to appliances and equipment, straight tap connector (S1220/131 - 1 nr), Fixed to surface, single core 4 mm2, screwing to timber (V9015/090 - 1 m), Labelled "SAFETY ELECTRICAL EARTH DO NOT REMOVE" (V9205/010 - 2 nr)	nr	794.27	228.99	0.34	1023.60

© NSR 01 Aug 2008 - 31 Jul 2009　　　　　　　　　　　　　　　　　　　　　　　　Y95

	Y30 : SANITARY CONVENIENCES		Mat. £	Lab. £	Plant £	Total £
Y3010						
110	**Supply and fix Shallow bath; 1700 x 700mm, chrome plated lever taps, waste, overflow and chain, adapt supply and disposal pipework, Plastic faced hardboard panels, silicone mastic sealant**					
	Stripping off/removing/taking down to plaster walls in preparation for replacement 1 to 5 m2 (C6065/602 - 0.7 m2), Stripping out ne 25 mm plastics or copper pipework (C7550/500 - 1 m), Stripping out 25 to 50 mm plastics or copper pipework (C7550/502 - 1 m), Walls ne 300 mm wide to plaster (M4070/310 - 2.5 m), Shallow bath 1700 x 780mm, with twin hand grips, deep enamelled pressed steel, N&C ref:P68002.51, lever action pop up waste and overflow, plug and chain and pair of chromium plated lever taps (N1320/268 - 1 nr), Side and end panels, plastic faced hardboard with aluminium trim (N1320/286 - 1 nr), Sealing around sanitary appliances (P2240/100 - 2.5 m), Pipes 40 mm nominal size (KP204P), screwing to timber (R1120/200 - 1 m), Extra over for fittings with two ends (R1120/220 - 1 nr), Pipes 15 mm nominal size, screwing to timber (S1220/100 - 1 m), Extra over for connections to appliances and equipment, straight tap connector (S1220/131 - 1 nr), Fixed to surface, single core 4 mm2, screwing to timber (V9015/090 - 1 m), Labelled "SAFETY ELECTRICAL EARTH DO NOT REMOVE" (V9205/010 - 2 nr)	nr	513.79	251.45	0.34	765.58
112	**Supply and fix Duet walk-in bath; 1660 x 740mm, chrome plated lever taps, waste, overflow and chain, adapt supply and disposal pipework, Plastic faced hardboard panels, silicone mastic sealant**					
	Stripping off/removing/taking down to plaster walls in preparation for replacement 1 to 5 m2 (C6065/602 - 0.7 m2), Stripping out ne 25 mm plastics or copper pipework (C7550/500 - 1 m), Stripping out 25 to 50 mm plastics or copper pipework (C7550/502 - 1 m), Walls ne 300 mm wide to plaster (M4070/310 - 2.5 m), Walk-in bath 1660 x 740mm, Duet, N&C ref:P68085.60, front panel, lever action pop up waste and overflow, plug and chain and pair of chromium plated lever taps (N1320/270 - 1 nr), Sealing around sanitary appliances (P2240/100 - 2.5 m), Pipes 40 mm nominal size (KP204P), screwing to timber (R1120/200 - 1 m), Extra over for fittings with two ends (R1120/220 - 1 nr), Pipes 15 mm nominal size, screwing to timber (S1220/100 - 1 m), Extra over for connections to appliances and equipment, straight tap connector (S1220/131 - 1 nr), Fixed to surface, single core 4 mm2, screwing to timber (V9015/090 - 1 m), Labelled "SAFETY ELECTRICAL EARTH DO NOT REMOVE" (V9205/010 - 2 nr)	nr	3013.82	228.99	0.34	3243.15

© NSR 01 Aug 2008 - 31 Jul 2009   Y96

Y30 : SANITARY CONVENIENCES			Mat. £	Lab. £	Plant £	Total £
Y3010						
114	**Supply and fix Duet walk-in bath with hydrotherapy unit; 1660 x 740mm, chrome plated lever taps, waste, overflow and chain, adapt supply and disposal pipework, Plastic faced hardboard panels, silicone mastic sealant**					
	Stripping off/removing/taking down to plaster walls in preparation for replacement 1 to 5 m2 (C6065/602 - 0.7 m2), Stripping out ne 25 mm plastics or copper pipework (C7550/500 - 1 m), Stripping out 25 to 50 mm plastics or copper pipework (C7550/502 - 1 m), Walls ne 300 mm wide to plaster (M4070/310 - 2.5 m), Walk-in bath 1660 x 740mm, with hydrotherapy unit, Duet, N&C ref:P68085.61, front panel, lever action pop up waste and overflow, plug and chain and pair of chromium plated lever taps (N1320/272 - 1 nr), Sealing around sanitary appliances (P2240/100 - 2.5 m), Pipes 40 mm nominal size (KP204P), screwing to timber (R1120/200 - 1 m), Extra over for fittings with two ends (R1120/220 - 1 nr), Pipes 15 mm nominal size, screwing to timber (S1220/100 - 1 m), Extra over for connections to appliances and equipment, straight tap connector (S1220/131 - 1 nr), Fixed to surface, single core 4 mm2, screwing to timber (V9015/090 - 1 m), Labelled "SAFETY ELECTRICAL EARTH DO NOT REMOVE" (V9205/010 - 2 nr)	nr	3848.82	228.99	0.34	4078.15
120	**Supply and install Bathing aids; patient showering device, as Phlexicare shower easy ref:P68245.11 or similar**					
	Bathing aids; Phlexicare 'Shower easy', N&C ref:P68245.11, plugging and screwing (N1320/320 - 1 nr)	nr	350.30	4.32		354.62
122	**Supply and install Bathing aids; wall mounted trapeze pull Phlexicare ref:P68040.01 or similar**					
	Bathing aids; Wall mounted trapeze pull N&C ref:P68040.01, plugging and screwing (N1320/322 - 1 nr)	nr	117.65	7.20		124.85
124	**Supply and install Bathing aids; trapeze ceiling lift Phlexicare ref:P68022.01 or similar**					
	Bathing aids; Trapeze ceiling lift N&C ref:P68022.01, plugging and screwing (N1320/324 - 1 nr)	nr	73.77	11.52		85.29
126	**Supply and install Bathing aids; bath seat Phlexicare ref:P68026.51 or similar**					
	Bathing aids; Bath seat N&C ref:P68026.51, (N1320/326 - 1 nr)	nr	56.73			56.73
150	**Remove existing shower base.**					
	Removing shower base, making good finishings (C2010/447 - 1 nr)	nr	3.80	20.08		23.88
152	**Supply and fit 900 x 900 x 30mm GRP slip resistant low edge profile shower tray (Not incl waste)**					
	Trays 900 x 900 mm, GRP, N&C Phlexicare ref:P68214.21 (Not inc waste fitting) (N1330/306 - 1 nr)	nr	214.22	53.55		267.77
154	**Supply and fit 1000 x 1000 x 30mm GRP slip resistant low edge profile shower tray (Not incl waste)**					
	Trays 1000 x 1000 mm, GRP, N&C Phlexicare ref:P68214.23 (Not inc waste fitting) (N1330/308 - 1 nr)	nr	257.68	53.55		311.23

© NSR 01 Aug 2008 - 31 Jul 2009   Y97

	Y30 : SANITARY CONVENIENCES		Mat. £	Lab. £	Plant £	Total £
Y3010						
156	**Supply and fit 1300 x 720 x 30mm GRP slip resistant low edge profile shower tray (Not incl waste)**					
	Trays 1300 x 720 mm, GRP, N&C Phlexicare ref:P68214.25 (Not inc waste fitting) (N1330/310 - 1 nr)	nr	244.41	53.55		297.96
158	**Supply and fit 1500 x 840 x 30mm GRP slip resistant low edge profile shower tray (Not incl waste)**					
	Trays 1500 x 840 mm, GRP, N&C Phlexicare ref:P68205.39 (Not inc waste fitting) (N1330/312 - 1 nr)	nr	331.56	53.55		385.11
160	**Supply and fit Phlexicare level deck system (option 1) comprising level deck tray, bi-fold doors, shower curtain and track. (not incl floor covering or waste fitting)**					
	Shower cubicle, Phlexicare Level Dec system option 1 ref:68218.95 (N1330/370 - 1 nr)	nr	658.62	46.08		704.70
162	**Supply and fit Phlexicare level deck system (option 2/3) comprising level deck tray, bi-fold doors, shower curtain and track. (not incl floor covering or waste fitting)**					
	Shower cubicle, Phlexicare Level Dec system option 2/3 ref:68218.96 - 98 (N1330/371 - 1 nr)	nr	725.86	46.08		771.94
164	**Supply and fit Phlexicare level deck system (option 4/5) comprising level deck tray, bi-fold doors, shower curtain and track. (not incl floor covering or waste fitting)**					
	Shower cubicle, Phlexicare Level Dec system option 4/5 ref:68218.99 - P68219.02 (N1330/372 - 1 nr)	nr	742.72	46.08		788.80
170	**Supply and fit wet floor gully; 130mm overall depth; inc connection to waste pipe and 2ml of 19mm waste pipe**					
	Shower gully for wet floor ref:S23705.11 (N1337/100 - 1 nr), Pipes 19 mm nominal size (500.75), screwing to timber (R1110/050 - 2 m), Extra over for connections to appliances and equipment, straight connector (511.75) (R1110/065 - 1 nr)	nr	58.47	28.17		86.64
172	**Supply and fit pumped shallow wet floor gully; 45mm overall depth; incl connection to waste pipe, 2ml of 19mm waste pipe and electrical connection**					
	Pumped waste for wet floor ref:P68215.89 (N1337/102 - 1 nr), Pipes 19 mm nominal size (500.75), screwing to timber (R1110/050 - 2 m), Extra over for connections to appliances and equipment, straight connector (511.75) (R1110/065 - 1 nr), Heating and power socket, switch socket and the like, single switch, 2.5 mm2 twin core and earth cables (V9020/210 - 1 nr), Extra for 16 x 16 mm pvc mini-trunking (V9020/255 - 10 m), Surface fused connection units (K330 WHI), 1 gang plastic box, plugging and screwing to masonry (V9090/825 - 1 nr)	nr	451.01	99.36		550.37

© NSR 01 Aug 2008 - 31 Jul 2009   Y98

	Y30 : SANITARY CONVENIENCES		Mat. £	Lab. £	Plant £	Total £
Y3010						
174	**Supply and fit Phlexicare pumped shallow waste system; 40mm overall depth; incl connection to waste pipe, 2ml of 19mm waste pipe and electrical connection**					
	Pumped waste system ref:P68214.90 (N1337/104 - 1 nr), Shallow bottom waste exit kit ref:P68214.93 (N1337/110 - 1 nr), Pipes 19 mm nominal size (500.75), screwing to timber (R1110/050 - 2 m), Extra over for connections to appliances and equipment, straight connector (511.75) (R1110/065 - 1 nr), Heating and power socket, switch socket and the like, single switch, 2.5 mm2 twin core and earth cables (V9020/210 - 1 nr), Extra for 16 x 16 mm pvc mini-trunking (V9020/255 - 10 m), Flush fused connection units (K330 WHI), 1 gang steel box, plugging and screwing to masonry (V9090/820 - 1 nr)	nr	361.76	107.73		469.49
176	**Supply and fit top waste exit kit and filter free pump; incl connection to waste pipe, 2ml of 19mm waste pipe and electrical connection**					
	Phlexicare filter free pump with Top waste exit kit N&C ref:P68216.04 (N1337/106 - 1 nr), Pipes 19 mm nominal size (500.75), screwing to timber (R1110/050 - 2 m), Extra over for connections to appliances and equipment, straight connector (511.75) (R1110/065 - 1 nr), Heating and power socket, switch socket and the like, single switch, 2.5 mm2 twin core and earth cables (V9020/210 - 1 nr), Extra for 16 x 16 mm pvc mini-trunking (V9020/255 - 10 m), Flush fused connection units (K330 WHI), 1 gang steel box, plugging and screwing to masonry (V9090/820 - 1 nr)	nr	379.76	90.99		470.75
178	**Supply and fit Phlexicare filter free pump and auto gully; incl connection to waste pipe, 2ml of 19mm waste pipe and electrical connection**					
	Phlexicare bottom exit pump wit auto gully N&C ref:P68216.05 (N1337/112 - 1 nr), Pipes 19 mm nominal size (500.75), screwing to timber (R1110/050 - 2 m), Extra over for connections to appliances and equipment, straight connector (511.75) (R1110/065 - 1 nr), Heating and power socket, switch socket and the like, single switch, 2.5 mm2 twin core and earth cables (V9020/210 - 1 nr), Extra for 16 x 16 mm pvc mini-trunking (V9020/255 - 10 m), Flush fused connection units (K330 WHI), 1 gang steel box, plugging and screwing to masonry (V9090/820 - 1 nr)	nr	369.26	99.36		468.62
180	**Supply and fit bottom waste; 90mm overall depth; incl connection to waste pipe and 2ml of 32mm waste pipe**					
	Bottom waste exit kit ref:P68214.92 (N1337/108 - 1 nr), Pipes 32 mm nominal size (KP104P), screwing to timber (R1120/150 - 2 m), Extra over for fittings with two ends (R1120/170 - 1 nr)	nr	38.30	32.02		70.32
190	**Supply and fix folding shower seat**					
	Folding shower seat, Classic N&C ref:P68287.02, plugging and screwing (N1330/332 - 1 nr)	nr	85.72	7.20		92.92
192	**Supply and fix folding shower seat with arms and back rest**					
	Folding shower seat with arms and back rest, Classic N&C ref:P68287.07, plugging and screwing (N1330/333 - 1 nr)	nr	198.91	10.08		208.99

© NSR 01 Aug 2008 - 31 Jul 2009   Y99

	Y30 : SANITARY CONVENIENCES		Mat. £	Lab. £	Plant £	Total £
Y3010						
194	**Supply and fix floor mounted shower seat with arms and back rest**					
	Floor mounted shower seat with arms and back rest, Ergoflex N&C ref:P68263.01, plugging and screwing (N1330/334 - 1 nr)	nr	337.45	15.84		353.29
196	**Supply and fix height adjustable shower seat with arms and back rest**					
	Height adjustable shower seat with arms and back rest, N&C ref:P68236.71, plugging and screwing (N1330/335 - 1 nr)	nr	1059.20	15.84		1075.04
Y3020	**CUBICLES**					
240	**Remove existing toilet or shower cubicle**					
	Removing proprietary WC partitions for re-use comprising one side panel, one fascia panel with door and capping piece (C2010/421 - 1 nr)	nr		15.10		15.10
242	**Adapt standard toilet cubicle for ambulant disabled user by adjusting door from inward to outward opening**					
	Taking off doors, adapting and rehanging on opposite side, inc fill/splice to existing hinges (C5170/845 - 1 nr)	nr	4.45	32.73		37.18
252	**Supply and fit shower comprising white four sided cubicle with half height doors and shower; to suit and including 900 x 900 x 30mm deep Phlexicare shower base; (not including wastefitting)**					
	Trays 900 x 900 mm, GRP, N&C Phlexicare ref:P68214.21 (Not inc waste fitting) (N1330/306 - 1 nr), 8.5KW Shower units, plugging and screwing (PC £120.00), electrical connection measured separately (N1330/344 - 1 nr), Shower cubicle, Athena white with half height doors to fit tray size 900 x 900mm; N&C Phlexicare ref:68213.02; plugging and screwing (N1330/350 - 1 nr)	nr	1358.25	99.82		1458.07
254	**Supply and fit shower comprising white four sided cubicle with half height doors and shower; to suit and including 1000 x 1000 x 30mm deep Phlexicare shower base; (not including wastefitting)**					
	Trays 1000 x 1000 mm, GRP, N&C Phlexicare ref:P68214.23 (Not inc waste fitting) (N1330/308 - 1 nr), 8.5KW Shower units, plugging and screwing (PC £120.00), electrical connection measured separately (N1330/344 - 1 nr), Shower cubicle, Athena white with half height doors to fit tray size 1000 x 1000mm; N&C Phlexicare ref:68213.04; plugging and screwing (N1330/351 - 1 nr)	nr	1443.10	99.82		1542.92
256	**Supply and fit shower comprising white four sided cubicle with half height doors and shower; to suit and including 1300 x 720 x 30mm deep Phlexicare shower base; (not including wastefitting)**					
	8.5KW Shower units, plugging and screwing (PC £120.00), electrical connection measured separately (N1330/344 - 1 nr), Shower cubicle, Athena white with half height doors to fit tray size 1300 x 720mm; N&C Phlexicare ref:68213.07; plugging and screwing (N1330/352 - 1 nr)	nr	1172.78	46.27		1219.05

© NSR 01 Aug 2008 - 31 Jul 2009   Y100

	Y30 : SANITARY CONVENIENCES		Mat. £	Lab. £	Plant £	Total £
Y3020						
258	Supply and fit shower comprising clear four sided cubicle with full height doors and shower; to suit and including 900 x 900 x 30mm deep Phlexicare shower base; (not including wastefitting)					
	Trays 900 x 900 mm, GRP, N&C Phlexicare ref:P68214.21 (Not inc waste fitting) (N1330/306 - 1 nr), 8.5KW Shower units, plugging and screwing (PC £120.00), electrical connection measured separately (N1330/344 - 1 nr), Shower cubicle, Athena clear with full height doors to fit tray size 900 x 900mm; N&C Phlexicare ref:68213.22; plugging and screwing (N1330/353 - 1 nr)	nr	1466.54	102.70		1569.24
260	Supply and fit shower comprising clear four sided cubicle with full height doors and shower; to suit and including 1000 x 1000 x 30mm deep Phlexicare shower base; (not including wastefitting)					
	8.5KW Shower units, plugging and screwing (PC £120.00), electrical connection measured separately (N1330/344 - 1 nr), Shower cubicle, Athena clear with full height doors to fit tray size1000 x 1000mm; N&C Phlexicare ref:68213.24; plugging and screwing (N1330/354 - 1 nr)	nr	1293.71	49.15		1342.86
262	Supply and fit shower comprising clear four sided cubicle with full height doors and shower; to suit and including 1300 x 720 x 30mm deep Phlexicare shower base; (not including wastefitting)					
	8.5KW Shower units, plugging and screwing (PC £120.00), electrical connection measured separately (N1330/344 - 1 nr), Shower cubicle, Athena clear with full height doors to fit tray size1300 x 720mm; N&C Phlexicare ref:68213.27; plugging and screwing (N1330/355 - 1 nr)	nr	1281.07	49.15		1330.22
264	Supply and fit shower comprising white corner cubicle with half height doors and shower; to suit and including 900 x 900 x 30mm deep Phlexicare shower base; (not including wastefitting)					
	8.5KW Shower units, plugging and screwing (PC £120.00), electrical connection measured separately (N1330/344 - 1 nr), Shower cubicle, Delphi white Corner cubicle, with half height doors to fit tray size 900 x 900mm; N&C Phlexicare ref:68210.02; plugging and screwing (N1330/360 - 1 nr)	nr	836.40	49.15		885.55
266	Supply and fit shower comprising white corner cubicle with half height doors and shower; to suit and including 1000 x 1000 x 30mm deep Phlexicare shower base; (not including wastefitting)					
	8.5KW Shower units, plugging and screwing (PC £120.00), electrical connection measured separately (N1330/344 - 1 nr), Shower cubicle, Delphi white Corner cubicle, with half height doors to fit tray size 1000 x 1000mm; N&C Phlexicare ref:68210.04; plugging and screwing (N1330/361 - 1 nr)	nr	877.79	49.15		926.94
268	Supply and fit shower comprising white corner cubicle with half height doors and shower; to suit and including 1300 x 720 x 30mm deep Phlexicare shower base; (not including wastefitting)					
	8.5KW Shower units, plugging and screwing (PC £120.00), electrical connection measured separately (N1330/344 - 1 nr), Shower cubicle, Delphi white Corner cubicle, with half height doors to fit tray size 1300 x 720mm; N&C Phlexicare ref:68210.07; plugging and screwing (N1330/362 - 1 nr)	nr	864.79	49.15		913.94

© NSR 01 Aug 2008 - 31 Jul 2009   Y101

	Y30 : SANITARY CONVENIENCES		Mat. £	Lab. £	Plant £	Total £
Y3020						
270	**Supply and fit shower comprising clear corner cubicle with full height doors and shower; to suit and including 900 x 900 x 30mm deep Phlexicare shower base; (not including wastefitting)**					
	8.5KW Shower units, plugging and screwing (PC £120.00), electrical connection measured separately (N1330/344 - 1 nr), Shower cubicle, Delphi clear Corner cubicle, with full height doors to fit tray size 900 x 900mm; N&C Phlexicare ref:68210.22; plugging and screwing (N1330/363 - 1 nr)	nr	988.86	52.03		1040.89
272	**Supply and fit shower comprising clear corner cubicle with full height doors and shower; to suit and including 1000 x 1000 x 30mm deep Phlexicare shower base; (not including wastefitting)**					
	8.5KW Shower units, plugging and screwing (PC £120.00), electrical connection measured separately (N1330/344 - 1 nr), Shower cubicle, Delphi clear Corner cubicle, with full height doors to fit tray size 1000 x 1000mm; N&C Phlexicare ref:68210.24; plugging and screwing (N1330/364 - 1 nr)	nr	1030.25	52.03		1082.28
274	**Supply and fit shower comprising clear corner cubicle with full height doors and shower; to suit and including 1300 x 720 x 30mm deep Phlexicare shower base; (not including wastefitting)**					
	8.5KW Shower units, plugging and screwing (PC £120.00), electrical connection measured separately (N1330/344 - 1 nr), Shower cubicle, Delphi clear Corner cubicle, with full height doors to fit tray size 1300 x 720mm; N&C Phlexicare ref:68210.27; plugging and screwing (N1330/365 - 1 nr)	nr	1017.61	52.03		1069.64
280	**Supply and fix 2000 x 1500mm demountable toilet cubicle with Olympus Classic Doc M Pack and all necessary fittings. Incl connection to existing services**					
	Olympus Classic Doc M Toilet Package N&C ref:P68300.48 (N1367/050 - 1 nr), Doc M Toilet Cubicle N&C ref:P68300.55 (N1367/074 - 1 nr), Sealing around sanitary appliances (P2240/100 - 1 m), Pipes 32 mm nominal size (KP104P), screwing to timber (R1120/150 - 1 m), Extra over for fittings with two ends (R1120/170 - 1 nr), Pipes 110 mm nominal size (100.4), plugging and screwing to masonry (R1130/300 - 2 m), Extra over for bends (101.4) (R1130/335 - 1 nr), Pipes 15 mm nominal size, plugging and screwing to masonry (S1220/105 - 4 m), Extra over for connections to appliances and equipment, straight tap connector (S1220/131 - 2 nr), Extra over for fittings with two ends (S1220/152 - 3 nr)	nr	1391.29	350.15		1741.44

© NSR 01 Aug 2008 - 31 Jul 2009

Y102

	Y30 : SANITARY CONVENIENCES		Mat. £	Lab. £	Plant £	Total £
Y3020						
282	**Supply and fix 2000 x 900 x 2400mm high timber stud and plasterboard toilet compartment in existing toilets, in 2 sides to corner; fully tiled walls, vinyl safety flooring with ambulant disabled toilet and all necessary fittings and connection to existing services. (Not incl doors, door furniture & frame)**					
	Wall or partition members 50 x 75 mm (G2010/122 - 30 m), Plates 50 x 75 mm (G2010/162 - 5 m), Plugging at 300 mm centres (G2060/665 - 6 m), Screwing at 300 mm centres (G2060/670 - 6 m), Walls over 300 mm wide to timber (M2044/150 - 12 m2), Walls ne 300 mm wide to timber (M2044/160 - 4 m), Walls over 300 mm wide to plaster (M4070/300 - 12 m2), Walls ne 300 mm wide to plaster (M4070/310 - 4 m), Floors over 300 mm wide level or to falls to screed (M5028/050 - 1.8 m2), Plaster general surfaces over 300 mm girth (M6015/140 - 6.76 m2), Olympus Ambulant Doc M Toilet Package N&C ref:P68300.49 (N1367/072 - 1 nr), Sealing around sanitary appliances (P2240/100 - 0.6 m), Pipes 110 mm nominal size (100.4), plugging and screwing to masonry (R1130/300 - 2 m), Extra over for bends (101.4) (R1130/335 - 1 nr), Pipes 15 mm nominal size, plugging and screwing to masonry (S1220/105 - 3 m), Extra over for connections to appliances and equipment, straight tap connector (S1220/131 - 2 nr), Extra over for fittings with two ends (S1220/152 - 3 nr)	nr	1085.12	1039.52	5.32	2129.96
284	**Supply and fix 2000 x 900 x 2400mm high masonry toilet compartment in existing toilets, in 2nr sides to corner; fully tiled walls, vinyl safety flooring with ambulant disabled toilet and all necessary fittings and connection to existing services. (Not incl doors, door furniture & frame)**					
	Walls (F1054/500 - 5 m2), Bonding to brickwork (F1054/650 - 4.8 m), Lintels ne 0.0125 m2 sectional area keyed on exposed faces (F3110/050 - 1.5 m), Walls over 300 mm wide to brickwork, blockwork or stonework (M2025/050 - 24 m2), Walls over 300 mm wide to plaster (M4070/300 - 12 m2), Walls ne 300 mm wide to plaster (M4070/310 - 4 m), Floors over 300 mm wide level or to falls to screed (M5028/050 - 1.8 m2), Plaster general surfaces over 300 mm girth (M6015/140 - 6.76 m2), Olympus Ambulant Doc M Toilet Package N&C ref:P68300.49 (N1367/072 - 1 nr), Sealing around sanitary appliances (P2240/100 - 0.6 m), Pipes 110 mm nominal size (100.4), plugging and screwing to masonry (R1130/300 - 2 m), Extra over for bends (101.4) (R1130/335 - 1 nr), Pipes 15 mm nominal size, plugging and screwing to masonry (S1220/105 - 3 m), Extra over for connections to appliances and equipment, straight tap connector (S1220/131 - 2 nr), Extra over for fittings with two ends (S1220/152 - 3 nr)	nr	1060.75	1102.58	13.30	2176.63

© NSR 01 Aug 2008 - 31 Jul 2009

Y103

	Y30 : SANITARY CONVENIENCES		Mat. £	Lab. £	Plant £	Total £
Y3020						
290	**Supply and fix 2200 x 2400 x 2400mm high timber stud and plasterboard toilet cubicle, in 2 sides to corner; fully tiled walls, vinyl safety flooring with Olympus classic Doc M pack and all necessary fittings and connection to existing services. (Not incl doors, door furniture & frame)**					
	Wall or partition members 50 x 75 mm (G2010/122 - 55 m), Plates 50 x 75 mm (G2010/162 - 8 m), Plugging at 300 mm centres (G2060/665 - 9 m), Screwing at 300 mm centres (G2060/670 - 9 m), Walls over 300 mm wide to timber (M2044/150 - 18.08 m2), Walls ne 300 mm wide to timber (M2044/160 - 4 m), Walls over 300 mm wide to plaster (M4070/300 - 20 m2), Walls ne 300 mm wide to plaster (M4070/310 - 4 m), Floors over 300 mm wide level or to falls to screed (M5028/050 - 5.3 m2), Plaster general surfaces over 300 mm girth (M6015/140 - 5.3 m2), Olympus Classic Doc M Toilet Package N&C ref:P68300.48 (N1367/050 - 1 nr), Sealing around sanitary appliances (P2240/100 - 0.6 m), Pipes 32 mm nominal size (KP104P), screwing to timber (R1120/150 - 1 m), Extra over for connections to bossed or tapped fittings, solvent joints to plastic (R1120/160 - 1 nr), Extra over for fittings with two ends (R1120/170 - 1 nr), Pipes 110 mm nominal size (100.4), plugging and screwing to masonry (R1130/300 - 2 m), Extra over for bends (101.4) (R1130/335 - 4 nr), Pipes 15 mm nominal size, plugging and screwing to masonry (S1220/105 - 4 m), Extra over for connections to appliances and equipment, straight tap connector (S1220/131 - 2 nr), Extra over for fittings with two ends (S1220/152 - 3 nr)	nr	1967.04	1728.00	7.39	3702.43
292	**Supply and fix 2200 x 2400 x 2400mm high masonry toilet compartment in existing toilets, in 2 sides to corner; fully tiled walls, vinyl safety flooring with Olympus classic Doc M pack and all necessary fittings and connection to existing services. (Not incl doors, door furniture & frame)**					
	Walls (F1054/500 - 9.04 m2), Bonding to brickwork (F1054/650 - 4.8 m), Lintels ne 0.0125 m2 sectional area keyed on exposed faces (F3110/050 - 1.5 m), Walls over 300 mm wide to brickwork, blockwork or stonework (M2025/050 - 18.08 m2), Walls over 300 mm wide to plaster (M4070/300 - 20 m2), Walls ne 300 mm wide to plaster (M4070/310 - 4 m), Floors over 300 mm wide level or to falls to screed (M5028/050 - 5.3 m2), Plaster general surfaces over 300 mm girth (M6015/140 - 5.3 m2), Olympus Classic Doc M Toilet Package N&C ref:P68300.48 (N1367/050 - 1 nr), Sealing around sanitary appliances (P2240/100 - 0.6 m), Pipes 32 mm nominal size (KP104P), screwing to timber (R1120/150 - 1 m), Extra over for connections to bossed or tapped fittings, solvent joints to plastic (R1120/160 - 1 nr), Extra over for fittings with two ends (R1120/170 - 1 nr), Pipes 110 mm nominal size (100.4), plugging and screwing to masonry (R1130/300 - 2 m), Extra over for bends (101.4) (R1130/335 - 4 nr), Pipes 15 mm nominal size, plugging and screwing to masonry (S1220/105 - 4 m), Extra over for connections to appliances and equipment, straight tap connector (S1220/131 - 2 nr), Extra over for fittings with two ends (S1220/152 - 3 nr)	nr	1850.32	1604.27	12.05	3466.64
Y3030	**SUNDRY BATHROOM FITTINGS**					
300	**Remove existing mirror and make good**					
	Removing mirrors, make good finishings (C2010/356 - 1 nr)	nr		2.88		2.88

© NSR 01 Aug 2008 - 31 Jul 2009   Y104

Y30 : SANITARY CONVENIENCES			Mat. £	Lab. £	Plant £	Total £
Y3030						
302	**Supply and fix stainless steel mirror 600 x 450**					
	Mirrors, stainless steel 600 x 450 N&C ref: P68341.01 (N1025/255 - 1 nr)	nr	54.43	4.32		58.75
304	**Supply and fix stainless steel mirror 900 x 400mm**					
	Mirrors, stainless steel 900 x 400 N&C ref: P68341.11 (N1025/260 - 1 nr)	nr	73.33	4.32		77.65
306	**Supply and fix acrylic mirror 600 x 450**					
	Mirrors, acrylic 600 x 450 N&C ref: P68341.21 (N1025/265 - 1 nr)	nr	53.44	4.32		57.76
308	**Supply and fix acrylic mirror 900 x 400**					
	Mirrors, acrylic 900 x 400 N&C ref: P68341.31 (N1025/270 - 1 nr)	nr	70.26	4.32		74.58
310	**Remove existing soap dish and make good**					
	Removing soap dishes (C2010/395 - 1 nr)	nr		1.73		1.73
312	**Supply and fix soap dish**					
	Soap dish plugged and screwed (N1377/600 - 1 nr)	nr	25.80	5.76		31.56
320	**Remove existing toilet roll holder and make good**					
	Removing toilet roll holders (C2010/397 - 1 nr)	nr		1.73		1.73
322	**Supply and fix toilet roll holder**					
	Toilet roll holder, plugged and screwed (N1377/605 - 1 nr)	nr	25.80	5.76		31.56
332	**Supply and fix towel rail**					
	Towel rail, plugged and screwed (N1377/615 - 1 nr)	nr	37.28	12.10		49.38
342	**Supply and fix toilet tissue dispenser**					
	Toilet tissue dispenser, N&C ref:S29440.01, plugged and screwed (N1377/620 - 1 nr)	nr	17.54	12.10		29.64
352	**Supply and fix paper towel dispenser**					
	Paper towel dispenser, N&C ref:P68343.01, plugged and screwed (N1377/622 - 1 nr)	nr	17.54	12.10		29.64
362	**Supply and fix soap dispenser**					
	Soapl dispenser, N&C ref:P68342.01, plugged and screwed (N1377/624 - 1 nr)	nr	29.12	12.10		41.22
368	**Supply and fix white plastic colostomy bag shelf**					
	White plastic colostomy bag shelf, plugged and screwed (N1377/630 - 1 nr)	nr	61.88	5.76		67.64

© NSR 01 Aug 2008 - 31 Jul 2009   Y105

	Y30 : SANITARY CONVENIENCES		Mat. £	Lab. £	Plant £	Total £
Y3030						
370	**Supply and fix white plastic shelf 525 x 230mm**					
	White plastic shelf, 525 x 230mm, plugged and screwed (N1377/632 - 1 nr)	nr	87.02	5.76		92.78
Y3040	**GRAB RAILS**					
400	**Supply and fix Single folding Grab rail**					
	Single folding rail (Ref P68516.01), fixing with plugs and screws (N1383/850 - 1 nr)	nr	72.68	5.76		78.44
402	**Supply and fix Twin folding Grab rail**					
	Twin folding rail (Ref P68517.51), fixing with plugs and screws (N1383/852 - 1 nr)	nr	81.48	5.76		87.24
404	**Supply and fix fixed support arm**					
	Pressalit fixed support arm (Ref P68351.51), fixing with rawlbolts (N1383/854 - 1 nr)	nr	96.08	8.64		104.72
406	**Supply and fix adjustable support arm**					
	Pressalit adjustable support arm (Ref P68351.11), fixing with rawlbolts (N1383/856 - 1 nr)	nr	157.08	8.64		165.72
408	**Supply and fix fixed support arm with leg**					
	Pressalit fixed support arm with leg (Ref P68351.01), fixing with rawlbolts (N1383/858 - 1 nr)	nr	98.58	8.64		107.22
410	**Supply and fix adjustable support arm with leg**					
	Pressalit adjustable support arm with leg (Ref P68351.01), fixing with rawlbolts (N1383/860 - 1 nr)	nr	139.58	8.64		148.22
412	**Supply and fix free standing support arm**					
	Pressalit free standing support arm (Ref P68351.61), fixing with rawlbolts (N1383/862 - 1 nr)	nr	240.58	8.64		249.22
414	**Supply and fix bath support rail**					
	Bath support rails (Ref: P68514.11), plugging and screwing (N1383/870 - 1 nr)	nr	58.42	8.64		67.06
416	**Supply and fix Grab rail any colour, 300mm**					
	Grab rails, nylon coated, any colour, 300 mm long, (Ref: P68600.04-10) plugging and screwing (N1383/872 - 1 nr)	nr	40.41	6.34		46.75
418	**Supply and fix Grab rail any colour, 400mm**					
	Grab rails, nylon coated, any colour, 400 mm long, (Ref: P68600.11-17) plugging and screwing (N1383/874 - 1 nr)	nr	42.77	6.34		49.11
420	**Supply and fix Grab rail any colour, 500mm**					
	Grab rails, nylon coated, any colour, 500 mm long, (Ref: P68600.18-24) plugging and screwing (N1383/876 - 1 nr)	nr	45.12	6.34		51.46

© NSR 01 Aug 2008 - 31 Jul 2009   Y106

Y30 : SANITARY CONVENIENCES			Mat. £	Lab. £	Plant £	Total £
Y3040						
422	**Supply and fix Grab rail any colour, 600mm**					
	Grab rails, nylon coated, any colour, 600 mm long, (Ref: P68600.25-31) plugging and screwing (N1383/878 - 1 nr)	nr	47.49	6.91		54.40
424	**Supply and fix Grab rail any colour, 700mm**					
	Grab rails, nylon coated, any colour, 700 mm long, (Ref: P68600.32-38) plugging and screwing (N1383/880 - 1 nr)	nr	50.32	6.91		57.23
426	**Supply and fix Grab rail any colour, 800mm**					
	Grab rails, nylon coated, any colour, 800 mm long, (Ref: P68600.39-45) plugging and screwing (N1383/882 - 1 nr)	nr	55.03	7.49		62.52
428	**Supply and fix Grab rail any colour, 900mm**					
	Grab rails, nylon coated, any colour, 900 mm long, (Ref: P68600.46-52) plugging and screwing (N1383/884 - 1 nr)	nr	60.23	7.49		67.72
430	**Supply and fix Grab rail any colour, Angled 45 deg 550mm**					
	Grab rails, nylon coated, any colour, 550 mm long, 45 degree angled (Ref: P68600.53-59) plugging and screwing (N1383/886 - 1 nr)	nr	74.86	7.49		82.35
432	**Supply and fix Grab rail any colour, Angled 90 deg 760 x 760mm**					
	Grab rails, nylon coated, any colour, 760 x 760 mm angled (Ref: P68601.16-22) plugging and screwing (N1383/888 - 1 nr)	nr	109.93	9.22		119.15
434	**Supply and fix Grab rail any colour, Angled 90 deg 660 x 1000mm**					
	Grab rails, nylon coated, any colour, 660 x 1000 mm angled (Ref: P68601.23-29) plugging and screwing (N1383/890 - 1 nr)	nr	139.67	9.22		148.89
436	**Supply and fix Grab rail any colour, vertical wall fixed 1400mm high**					
	Grab rails, nylon coated, any colour, 1400 mm high vertical wall fixed (Ref: P68604.67-73) plugging and screwing (N1383/892 - 1 nr)	nr	105.21	9.22		114.43
438	**Supply and fix Grab rail any colour, vertical wall fixed 1600mm high**					
	Grab rails, nylon coated, any colour, 1600 mm high vertical wall fixed (Ref: P68602.70-76) plugging and screwing (N1383/894 - 1 nr)	nr	107.57	9.22		116.79
440	**Supply and fix Grab rail any colour, vertical floor to wall fixed 1600mm high**					
	Grab rails, nylon coated, any colour, 1600 mm high vertical floor to wall (Ref: P68602.84-90) plugging and screwing (N1383/896 - 1 nr)	nr	111.82	9.22		121.04

© NSR 01 Aug 2008 - 31 Jul 2009   Y107

	Y30 : SANITARY CONVENIENCES		Mat. £	Lab. £	Plant £	Total £
Y3050	**KITCHEN FITTINGS**					
500	**Remove wall fixed cupboard**					
	Stripping off/removing/taking down fittings, setting aside for re-use (C6210/051 - 1 nr)	nr		8.62		8.62
502	**Reposition wall cupboard to correct height**					
	Taking down wall units, resecuring components and refixing (C6210/200 - 1 nr)	nr	1.95	27.69		29.64
504	**Remove cupboard hinges and replace with new to open 180 deg**					
	Taking off kitchen unit doors, renewing hinges and refitting (C6210/290 - 1 pr)	pr	3.68	2.88		6.56
506	**Remove cupboard handle and replace with new contrasting coloured handle**					
	Drawer/cupboard knobs or handles (C2015/855 - 1 nr), Cupboard handles (P2150/300 - 1 nr)	nr	4.50	2.88		7.38
508	**Remove base unit**					
	Stripping off/removing/taking down fittings, setting aside for re-use (C6210/051 - 1 nr)	nr		8.62		8.62
510	**Remove base unit plinth, adjust height of unit and replace plinth**					
	Taking out floor units, repairing and refixing (C6210/210 - 1 nr)	nr	1.95	25.17		27.12
520	**Replace standard stainless steel sink with shallow sink, chrome plated taps, waste, overflow and chain, adapt supply and disposal pipework, silicone mastic sealant**					
	Removing sinks, stainless steel, making good finishings (C2010/452 - 1 nr), Stripping out ne 25 mm plastics or copper pipework (C7550/500 - 1 m), Stripping out 25 to 50 mm plastics or copper pipework (C7550/502 - 0.5 m), Sinks, stainless steel with single drainers, chromium plated waste, plug and chain, pair of chromium plated pillar taps (PC £90.00), screwing with clips (N1345/420 - 1 nr), Sealing around sanitary appliances (P2240/100 - 2.5 m), Pipes 32 mm nominal size (KP104P), screwing to timber (R1120/150 - 0.5 m), Extra over for fittings with two ends (R1120/170 - 1 nr), Bottle traps 32 mm nominal size (611.125 P type outlet), joint to waste outlet (R1160/830 - 1 nr), Pipes 15 mm nominal size, screwing to timber (S1220/100 - 1 m), Extra over for connections to appliances and equipment, straight tap connector (S1220/131 - 1 nr), Fixed to surface, single core 4 mm2, screwing to timber (V9015/090 - 1 m), Labelled "SAFETY ELECTRICAL EARTH DO NOT REMOVE" (V9205/010 - 2 nr)	nr	123.86	199.80		323.66

	Y30 : SANITARY CONVENIENCES		Mat. £	Lab. £	Plant £	Total £
Y3050						
522	**Supply and fix standard stainless steel sink with shallow sink, chrome plated taps, waste, overflow and chain, new supply and disposal pipework, silicone mastic sealant**					
	Removing sinks, stainless steel, making good finishings (C2010/452 - 1 nr), Stripping out ne 25 mm plastics or copper pipework (C7550/500 - 1 m), Stripping out 25 to 50 mm plastics or copper pipework (C7550/502 - 0.5 m), Sinks, stainless steel with single drainers, chromium plated waste, plug and chain, pair of chromium plated pillar taps (PC £90.00), screwing with clips (N1345/420 - 1 nr), Sealing around sanitary appliances (P2240/100 - 2.5 m), Pipes 32 mm nominal size (KP104P), screwing to timber (R1120/150 - 0.5 m), Extra over for fittings with two ends (R1120/170 - 1 nr), Bottle traps 32 mm nominal size (611.125 P type outlet), joint to waste outlet (R1160/830 - 1 nr), Pipes 15 mm nominal size, screwing to timber (S1220/100 - 1 m), Extra over for connections to appliances and equipment, straight tap connector (S1220/131 - 1 nr), Fixed to surface, single core 4 mm2, screwing to timber (V9015/090 - 1 m), Labelled "SAFETY ELECTRICAL EARTH DO NOT REMOVE" (V9205/010 - 2 nr)	nr	123.86	199.80		323.66
530	**Supply and fix 300mm base unit with adjustable legs**					
	Base units 300 x 900 x 600 mm, plugging and screwing (PC £80) (N1115/050 - 1 nr)	nr	81.09	20.14		101.23
532	**Supply and fix 500mm base unit with adjustable legs**					
	Base units 500 x 900 x 600 mm, plugging and screwing (PC £100) (N1115/055 - 1 nr)	nr	101.09	22.15		123.24
534	**Supply and fix 600mm base unit with adjustable legs**					
	Base units 600 x 900 x 600 mm, plugging and screwing (PC £95) (N1115/060 - 1 nr)	nr	96.09	22.15		118.24
536	**Supply and fix 1000mm base unit with adjustable legs**					
	Base units 1000 x 900 x 600 mm, plugging and screwing (PC £130) (N1115/065 - 1 nr)	nr	131.09	25.17		156.26
538	**Supply and fix 1200mm base unit with adjustable legs**					
	Base units 1200 x 900 x 600 mm, plugging and screwing (PC £135) (N1115/070 - 1 nr)	nr	136.09	26.18		162.27
540	**Supply and fix 2100mm tall base unit with adjustable legs**					
	Base units 2100 x 900 x 600 mm, tall unit, plugging and screwing (PC £190) (N1115/075 - 1 nr)	nr	191.09	32.73		223.82
550	**Supply and fix 300mm wall unit**					
	Wall units 300 x 600 x 300 mm, plugging and screwing (PC £80) (N1115/100 - 1 nr)	nr	80.92	25.17		106.09
552	**Supply and fix 500mm wall unit**					
	Wall units 500 x 600 x 300 mm, plugging and screwing (PC £90) (N1115/110 - 1 nr)	nr	90.92	27.19		118.11

© NSR 01 Aug 2008 - 31 Jul 2009   Y109

	Y30 : SANITARY CONVENIENCES		Mat. £	Lab. £	Plant £	Total £
Y3050						
554	**Supply and fix 600mm wall unit**					
	Wall units 600 x 600 x 300 mm, plugging and screwing (PC £85) (N1115/120 - 1 nr)	nr	85.92	27.19		113.11
556	**Supply and fix 1000mm wall unit**					
	Wall units 1000 x 600 x 300 mm, plugging and screwing (PC £120) (N1115/130 - 1 nr)	nr	120.92	29.20		150.12
558	**Supply and fix 600mm wall corner unit**					
	Wall corner units 600 x 600 x 600 mm, plugging and screwing (PC £90) (N1115/150 - 1 nr)	nr	91.09	34.24		125.33
570	**Supply and fix worktop 600mm wide x 30mm thick**					
	Worktops 600 x 30 mm, screwing (PC £40/m) (N1123/235 - 1 m)	m	40.42	12.08		52.50
600	**Supply and fit Ropox wall mounted adjustable height worktop with manual adjustment 1000mm long**					
	Worktops 600 x 30 mm, screwing (PC £40/m) (N1123/235 - 1 m), Supply and fix worktop plinth for Ropox unit (N1125/157 - 2.2 m), Supply and fix 'Ropox' wall mounted adjustable height worktop assembly 1000mm manual (N1125/200 - 1 nr)	nr	948.14	27.06		975.20
602	**Supply and fit Ropox wall mounted adjustable height worktop with manual adjustment 1500mm long**					
	Worktops 600 x 30 mm, screwing (PC £40/m) (N1123/235 - 1.5 m), Supply and fix worktop plinth for Ropox unit (N1125/157 - 2.7 m), Supply and fix 'Ropox' wall mounted adjustable height worktop assembly 1500mm manual (N1125/202 - 1 nr)	nr	1100.61	37.42		1138.03
604	**Supply and fit Ropox wall mounted adjustable height worktop with electrical adjustment 1000mm long**					
	Worktops 600 x 30 mm, screwing (PC £40/m) (N1123/235 - 1 m), Supply and fix worktop plinth for Ropox unit (N1125/157 - 2.2 m), Supply and fix 'Ropox' wall mounted adjustable height worktop assembly 1000mm electrically operated (N1125/204 - 1 nr)	nr	1650.20	27.06		1677.26
606	**Supply and fit Ropox wall mounted adjustable height worktop with electrical adjustment 1500mm long**					
	Worktops 600 x 30 mm, screwing (PC £40/m) (N1123/235 - 1.5 m), Supply and fix worktop plinth for Ropox unit (N1125/157 - 2.7 m), Supply and fix 'Ropox' wall mounted adjustable height worktop assembly 1500mm electrically operated (N1125/206 - 1 nr)	nr	1828.01	37.42		1865.43
608	**Supply and fit Ropox floor mounted adjustable height worktop with manual adjustment 1000mm long**					
	Worktops 600 x 30 mm, screwing (PC £40/m) (N1123/235 - 1 m), Supply and fix worktop plinth for Ropox unit (N1125/157 - 2.2 m), Supply and fix 'Ropox' floor mounted adjustable height worktop assembly 1000mm manual (N1125/208 - 1 nr)	nr	932.30	21.30		953.60

© NSR 01 Aug 2008 - 31 Jul 2009   Y110

	Y30 : SANITARY CONVENIENCES		Mat. £	Lab. £	Plant £	Total £
Y3050						
610	**Supply and fit Ropox floor mounted adjustable height worktop with manual adjustment 1500mm long**					
	Worktops 600 x 30 mm, screwing (PC £40/m) (N1123/235 - 1.5 m), Supply and fix worktop plinth for Ropox unit (N1125/157 - 2.7 m), Supply and fix 'Ropox' floor mounted adjustable height worktop assembly 1500mm manual (N1125/210 - 1 nr)	nr	1076.85	29.36		1106.21
612	**Supply and fit Ropox floor mounted adjustable height worktop with electrical adjustment 1000mm long**					
	Worktops 600 x 30 mm, screwing (PC £40/m) (N1123/235 - 1 m), Supply and fix worktop plinth for Ropox unit (N1125/157 - 2.2 m), Supply and fix 'Ropox' floor mounted adjustable height worktop assembly 1000mm electrically operated (N1125/212 - 1 nr)	nr	1634.36	21.30		1655.66
614	**Supply and fit Ropox floor mounted adjustable height worktop with electrical adjustment 1500mm long**					
	Worktops 600 x 30 mm, screwing (PC £40/m) (N1123/235 - 1.5 m), Supply and fix worktop plinth for Ropox unit (N1125/157 - 2.7 m), Supply and fix 'Ropox' floor mounted adjustable height worktop assembly 1500mm electrically operated (N1125/214 - 1 nr)	nr	1804.25	37.42		1841.67
620	**Supply and fix 'Ropox' manual wall mounted adjuster frame to receive wall units, size; 600-900mm long**					
	Supply and fix 'Ropox' manual wall mounted adjuster frame 600 - 900mm long (N1125/220 - 1 nr)	nr	803.11	8.64		811.75
622	**Supply and fix 'Ropox' manual wall mounted adjuster frame to receive wall units, size; 1100-1400mm long**					
	Supply and fix 'Ropox' manual wall mounted adjuster frame 1100 - 1400mm long (N1125/222 - 1 nr)	nr	908.53	11.52		920.05
624	**Supply and fix 'Ropox' manual wall mounted adjuster frame to receive wall units, size; 1600-2000mm long**					
	Supply and fix 'Ropox' manual wall mounted adjuster frame 1600 - 2000mm long (N1125/224 - 1 nr)	nr	939.18	14.40		953.58
626	**Supply and fix 'Ropox' manual wall mounted adjuster frame to receive wall units, size; 2100-2400mm long**					
	Supply and fix 'Ropox' manual wall mounted adjuster frame 2100 - 2400mm long (N1125/226 - 1 nr)	nr	1170.95	17.28		1188.23
628	**Supply and fix 'Ropox' manual wall mounted adjuster frame to receive wall units, size; 2500-3000mm long**					
	Supply and fix 'Ropox' manual wall mounted adjuster frame 2500 - 3000mm long (N1125/228 - 1 nr)	nr	1274.19	20.16		1294.35
640	**Supply and fix 'Ropox' electric wall mounted adjuster frame to receive wall units, size; 600-900mm long (not incl electrical connection)**					
	Supply and fix 'Ropox' electric wall mounted adjuster frame 600 - 900mm long (N1125/240 - 1 nr)	nr	1194.38	8.64		1203.02

© NSR 01 Aug 2008 - 31 Jul 2009

Y111

	Y30 : SANITARY CONVENIENCES		Mat. £	Lab. £	Plant £	Total £
Y3050						
642	**Supply and fix 'Ropox' electric wall mounted adjuster frame to receive wall units, size; 1100-1400mm long (not incl electrical connection)**					
	Supply and fix 'Ropox' electric wall mounted adjuster frame 1100 - 1400mm long (N1125/242 - 1 nr)	nr	1509.96	11.52		1521.48
644	**Supply and fix 'Ropox' electric wall mounted adjuster frame to receive wall units, size; 1600-2000mm long (not incl electrical connection)**					
	Supply and fix 'Ropox' electric wall mounted adjuster frame 1600 - 2000mm long (N1125/244 - 1 nr)	nr	1556.09	14.40		1570.49
646	**Supply and fix 'Ropox' electric wall mounted adjuster frame to receive wall units, size; 2100-2400mm long (not incl electrical connection)**					
	Supply and fix 'Ropox' electric wall mounted adjuster frame 2100 - 2400mm long (N1125/246 - 1 nr)	nr	1801.76	17.28		1819.04
648	**Supply and fix 'Ropox' electric wall mounted adjuster frame to receive wall units, size; 2500-3000mm long (not incl electrical connection)**					
	Supply and fix 'Ropox' electric wall mounted adjuster frame 2500 - 3000mm long (N1125/248 - 1 nr)	nr	1919.67	20.16		1939.83

© NSR 01 Aug 2008 - 31 Jul 2009   Y112

			Mat. £	Lab. £	Plant £	Total £
	Y32 : GENERAL BUILDING WORKS					
Y3205	**TEMPORARY WORKS & SCAFFOLDING**					
002	**Erect and dismantle temporary roof to protect the work area, allowing for 1 week hire period**					
	Temporary roofs (Plastic tarpaulin covered) (A4410/090 - 1 m2), Temporary roofs (Plastic tarpaulin covered) (A4465/090 - 1 m2)	m2		12.96	0.90	13.86
004	**Erect and dismantle temporary dust screen, allowing for 1 week hire period**					
	Temporary dust screens (Plastic tarpaulin) (A4410/092 - 1 m2), Temporary dust screens (Plastic tarpaulin) (A4465/092 - 1 m2)	m2		3.46	0.90	4.36
006	**Contruct temporary internal screen comprising softwood studding fixed to existing structure, covered both sides, maintain and adapt as necessary, remove and make good on completion of work**					
	Internal dustproof screens covered both sides with 1000 G polythene building sheet (C1079/950 - 1 m2)	m2	15.31	27.71		43.02
008	**Extra over for access door complete with ironmongery and sealed all round**					
	Extra for access doors complete with ironmongery and sealed all round (C1079/952 - 1 nr)	nr	59.05	48.94		107.99
010	**Contruct temporary external screen comprising softwood studding fixed to existing structure, covered one side, maintain and adapt as necessary, remove and make good on completion of work**					
	External weatherproof screens covered one side with 12.7 mm Sterling board (C1079/980 - 1 m2)	m2	20.52	36.28		56.80
012	**Extra over for access door complete with ironmongery**					
	Extra for access doors complete with ironmongery (C1079/982 - 1 nr)	nr	69.46	48.94		118.40
020	**Erect and dismantle 2.0 x 2.0m tower scaffold to provide working platform 1.5 - 3.00m from ground level; allowing for 1 week hire period**					
	1.50 - 3.00 m From ground level (A4425/100 - 1 nr), 1.50 - 3.00 m From ground level (A4480/100 - 1 nr)	nr		21.60	113.50	135.10
022	**Erect and dismantle 2.0 x 2.0m tower scaffold to provide working platform 3.00 - 4.0m from ground level; allowing for 1 week hire period**					
	3.00 - 4.00 m From ground level (A4425/105 - 1 nr), 3.00 - 4.00 m From ground level (A4480/105 - 1 nr)	nr		25.92	133.00	158.92

© NSR 01 Aug 2008 - 31 Jul 2009   Y113

	Y32 : GENERAL BUILDING WORKS		Mat. £	Lab. £	Plant £	Total £
Y3205						
024	**Erect and dismantle 2.0 x 2.0m tower scaffold to provide working platform 4.00 - 5.0m from ground level; allowing for 1 week hire period**					
	4.00 - 5.00 m From ground level (A4425/110 - 1 nr), 4.00 - 5.00 m From ground level (A4480/110 - 1 nr)	nr		29.38	152.50	181.88
026	**Erect and dismantle 2.0 x 2.0m tower scaffold to provide working platform 5.00 - 6.0m from ground level; allowing for 1 week hire period**					
	5.00 - 6.00 m From ground level (A4425/115 - 1 nr), 5.00 - 6.00 m From ground level (A4480/115 - 1 nr)	nr		34.56	172.00	206.56
Y3210	**DEMOLITION**					
010	**Demolish masonry walls 102 thick**					
	Brick walls 102 mm thick (C1070/800 - 1 m2)	m2		12.93		12.93
012	**Demolish masonry walls 215 thick**					
	Brick walls 215 mm thick (C1070/805 - 1 m2)	m2		25.86		25.86
014	**Demolish masonry walls 275 thick with cavity**					
	Brick walls 275 mm thick including cavity (C1070/810 - 1 m2)	m2		19.39		19.39
016	**Demolish blockwork walls 100 thick**					
	Block walls 100 mm thick (C1070/820 - 1 m2)	m2		7.76		7.76
018	**Demolish paramount partitions**					
	Paramount partitions (C1070/930 - 1 m2)	m2		4.31		4.31
020	**Demolish stud partitions**					
	Stud partitions including finishings (C1070/935 - 1 m2)	m2		5.39		5.39
Y3220	**BRICKWORK/BLOCKWORK WALLS**					
100	**Construct masonry wall 102mm thick common brickwork**					
	Walls 102 mm thick (F1010/010 - 1 m2)	m2	20.51	14.25	0.52	35.28
102	**Construct masonry wall 215mm thick common brickwork**					
	Walls 215 mm thick (F1010/012 - 1 m2)	m2	43.08	29.29	1.07	73.44
104	**Construct isolated piers in masonry wall 102mm thick common brickwork**					
	Isolated piers 102 mm thick (F1010/020 - 1 m2)	m2	20.51	30.08	1.10	51.69
106	**Construct isolated piers in masonry wall 215mm thick common brickwork**					
	Isolated piers 215 mm thick (F1010/022 - 1 m2)	m2	43.08	43.53	1.60	88.21

© NSR 01 Aug 2008 - 31 Jul 2009   Y114

	Y32 : GENERAL BUILDING WORKS		Mat. £	Lab. £	Plant £	Total £
Y3220						
108	**Closing cavities in masonry wall 50 - 75mm wide; common brickwork**					
	Closing cavities 60 mm wide, brickwork 102 mm thick, vertical (F1010/062 - 1 m)	m	1.53	3.17	0.12	4.82
120	**Construct masonry wall 102mm thick engineering brickwork**					
	Walls 102 mm thick (F1027/010 - 1 m2)	m2	47.15	15.83	0.58	63.56
122	**Construct masonry wall 215mm thick engineering brickwork**					
	Walls 215 mm thick (F1027/012 - 1 m2)	m2	96.36	32.45	1.19	130.00
124	**Construct isolated piers in masonry wall 102mm thick engineering brickwork**					
	Isolated piers 102 mm thick (F1027/020 - 1 m2)	m2	47.15	34.03	1.25	82.43
126	**Construct isolated piers in masonry wall 215mm thick engineering brickwork**					
	Isolated piers 215 mm thick (F1027/022 - 1 m2)	m2	96.36	49.07	1.80	147.23
128	**Closing cavities in masonry wall 50 - 75mm wide; engineering brickwork**					
	Closing cavities 60 mm wide, brickwork 102 mm thick, vertical (F1027/062 - 1 m)	m	3.34	3.17	0.12	6.63
130	**Construct masonry wall 102mm thick facing brickwork**					
	Walls 102 mm thick, facework one side (F1034/100 - 1 m2)	m2	26.21	20.58	0.75	47.54
132	**Construct masonry wall 215mm thick facing brickwork**					
	Walls 215 mm thick, facework one side (F1034/104 - 1 m2)	m2	54.36	36.41	1.33	92.10
134	**Construct isolated piers in masonry wall 102mm thick facing brickwork, facework one side**					
	Isolated piers 102 mm thick, facework one side (F1034/120 - 1 m2)	m2	26.21	35.62	1.31	63.14
136	**Construct isolated piers in masonry wall 215mm thick facing brickwork, facework one side**					
	Isolated piers 215 mm thick, facework one side (F1034/124 - 1 m2)	m2	54.36	50.66	1.86	106.88
138	**Construct isolated piers in masonry wall 215mm thick facing brickwork, facework both sides**					
	Isolated piers 215 mm thick, facework both sides (F1034/126 - 1 m2)	m2	54.36	54.61	2.00	110.97
140	**Closing cavities in masonry wall 50 - 75mm wide; facing brickwork**					
	Closing cavities 60 mm wide, 102 mm thick, facework, vertical (F1034/187 - 1 m)	m	1.91	3.96	0.15	6.02

© NSR 01 Aug 2008 - 31 Jul 2009   Y115

Y32 : GENERAL BUILDING WORKS

			Mat. £	Lab. £	Plant £	Total £
Y3220						
150	**Construct masonry wall 75mm thick blockwork; facework one side**					
	Walls, facework one side (F1056/700 - 1 m2)	m2	13.99	11.08	0.41	25.48
152	**Construct masonry wall 75mm thick blockwork; facework both sides**					
	Walls, facework both sides (F1056/702 - 1 m2)	m2	13.99	12.66	0.46	27.11
154	**extra over for bonding to brickwork**					
	Bonding to brickwork, facework both sides (F1056/861 - 1 m)	m	0.59	6.33	0.23	7.15
160	**Construct masonry wall 100mm thick blockwork; facework one side**					
	Walls, facework one side (F1060/700 - 1 m2)	m2	19.73	11.08	0.41	31.22
162	**Construct masonry wall 100mm thick blockwork; facework both sides**					
	Walls, facework both sides (F1060/702 - 1 m2)	m2	19.73	12.66	0.46	32.85
164	**extra over for bonding to brickwork**					
	Bonding to brickwork, facework both sides (F1060/861 - 1 m)	m	0.76	7.12	0.26	8.14
170	**Construct masonry wall 150mm thick blockwork; facework one side**					
	Walls, facework one side (F1064/700 - 1 m2)	m2	29.29	12.66	0.46	42.41
172	**Construct masonry wall 150mm thick blockwork; facework both sides**					
	Walls, facework both sides (F1064/702 - 1 m2)	m2	29.29	14.25	0.52	44.06
174	**extra over for bonding to brickwork**					
	Bonding to brickwork, facework both sides (F1064/861 - 1 m)	m	1.04	8.71	0.32	10.07
180	**Forming cavities in hollow walls 50 wide**					
	Forming cavities in hollow walls 50 mm wide, galvanised steel butterfly wall ties to BS 1243 fig 1, spec 3.3, 4 nr/m2 (F3010/050 - 1 m2)	m2	0.84	1.58		2.42
182	**Forming cavities in hollow walls 50 wide; 25 thick insulation**					
	Forming cavities in hollow walls 50 mm wide, galvanised steel butterfly wall ties to BS 1243 fig 1, spec 3.3, 4 nr/m2, with cavity insulation of 25mm thick expanded polystyrene slabs fixed with clips to wall ties (F3010/055 - 1 m2)	m2	14.84	3.17		18.01

© NSR 01 Aug 2008 - 31 Jul 2009 Y116

	Y32 : GENERAL BUILDING WORKS		Mat. £	Lab. £	Plant £	Total £
Y3220						
186	**Forming cavities in hollow walls 75 wide**					
	Forming cavities in hollow walls 75 mm wide, galvanised steel butterfly wall ties to BS 1243 fig 1, spec 3.3, 4 nr/m2 (F3010/065 - 1 m2)	m2	0.72	1.58		2.30
188	**Forming cavities in hollow walls 75 wide; 50 thick insulation**					
	Forming cavities in hollow walls 75 mm wide, galvanised steel butterfly wall ties to BS 1243 fig 1, spec 3.3, 4 nr/m2, with cavity insulation of 50 mm thick expanded polystyrene slabs fixed with clips to wall ties (F3010/068 - 1 m2)	m2	26.88	3.17		30.05
Y3230	**TIMBER STUDDED WALLS**					
200	**Construct timber stud partition 2.1 to 2.4m high, taped and filled joints**					
	Wall or partition members 50 x 75 mm (G2010/122 - 9 m), Linings to walls 2.1 to 2.4 m high (K1010/010 - 2 m)	m	53.23	45.95		99.18
202	**Construct timber stud partition 2.4 to 2.7m high, taped and filled joints**					
	Wall or partition members 50 x 75 mm (G2010/122 - 10 m), Linings to walls 2.4 to 2.7 m high (K1010/020 - 2 m)	m	59.08	52.28		111.36
210	**Construct timber stud partition 2.1 to 2.4m high, insulated, taped and filled joints**					
	Wall or partition members 50 x 75 mm (G2010/122 - 9 m), Linings to walls 2.1 to 2.4 m high (K1010/010 - 2 m), Between members 80 mm thick, horizontal (P1010/016 - 2.4 m2)	m	59.35	54.40		113.75
212	**Construct timber stud partition 2.4 to 2.7m high, insulated, taped and filled joints**					
	Wall or partition members 50 x 75 mm (G2010/122 - 10 m), Linings to walls 2.4 to 2.7 m high (K1010/020 - 2 m), Between members 80 mm thick, horizontal (P1010/016 - 2.7 m2)	m	65.97	61.78		127.75
220	**Extra over for abutments**					
	Abutments (K1010/050 - 1 m)	m	0.18	1.58		1.76
222	**Extra over for angles**					
	Internal angles (K1010/060 - 1 m)	m	0.12	1.58		1.70

© NSR 01 Aug 2008 - 31 Jul 2009   Y117

Y32 : GENERAL BUILDING WORKS	Mat. £	Lab. £	Plant £	Total £

© NSR 01 Aug 2008 - 31 Jul 2009 — Y118

			Mat. £	Lab. £	Plant £	Total £
	Y33 : MECHANICAL AND ELECTRICAL					
Y3310	**SWITCHES/SOCKETS**					
050	**Replace 1 gang 1way light switch**					
	Removing plate switches/socket outlet plates for replacement (C8610/447 - 1 nr), Flush plate switches (K4870 WHI), 5A, 1 gang, 1 way, screwing to mounting box (V9090/740 - 1 nr)	nr	1.80	5.72		7.52
052	**Replace 1 gang 2 way light switch**					
	Removing plate switches/socket outlet plates for replacement (C8610/447 - 1 nr), Flush plate switches (K4871 WHI), 5A, 1 gang, 2 way, screwing to mounting box (V9090/741 - 1 nr)	nr	2.86	6.56		9.42
054	**Replace 2 gang 2 way light switch**					
	Removing plate switches/socket outlet plates for replacement (C8610/447 - 1 nr), Flush plate switches (K4872 WHI), 5A, 2 gang, 2 way, screwing to mounting box (V9090/742 - 1 nr)	nr	3.75	7.41		11.16
058	**Replace ceiling switch**					
	Stripping out pull cord ceiling switches, isolating, making good finishings (C8610/200 - 1 nr), Ceiling switches (3192 WHI), 5A, 1 way, screwing to timber (V9090/795 - 1 nr)	nr	6.03	9.00		15.03
060	**Supply and fit ceiling pull switch to supplement wall switch**					
	Lighting intermediate switch, 1.5 mm2 single core cables (V9020/204 - 1 nr), Ceiling switches (3192 WHI), 5A, 1 way, screwing to timber (V9090/795 - 1 nr)	nr	75.27	31.97		107.24
062	**Reposition one way light switch, extending wiring to new position (not incl redecoration)**					
	Stripping out surface light switches, isolating, making good finishings (C8610/160 - 1 nr), Cutting or forming chases for services 1 nr 19 mm dia in brickwork, making good (P3110/202 - 1 m), Cutting or forming holes, mortices, sinkings and chases, concealed service, socket outlet point, making good (P3230/802 - 1 nr), 1 - 4 nr Cables (V8700/010 - 1 nr), Extra for 1.5 mm2 single core cables (where cable length varies from 10 m allowance) (V9020/230 - 3 m), Extra for 25 x 8 mm galvanised mild steel channel (V9020/250 - 1 m), Blanking plates (K3827 WHI), 1 gang, screwing to mounting box (V9090/720 - 1 nr)	nr	15.62	53.32	1.86	70.80
064	**Reposition two way light switch, extending wiring to new position (not incl redecoration)**					
	Stripping out surface light switches, isolating, making good finishings (C8610/160 - 1 nr), Cutting or forming chases for services 1 nr 19 mm dia in brickwork, making good (P3110/202 - 1 m), Cutting or forming holes, mortices, sinkings and chases, concealed service, socket outlet point, making good (P3230/802 - 1 nr), 1 - 4 nr Cables (V8700/010 - 1 nr), Extra for 1.5 mm2 single core cables (where cable length varies from 10 m allowance) (V9020/230 - 3 m), Extra for 25 x 8 mm galvanised mild steel channel (V9020/250 - 1 m), Blanking plates (K3827 WHI), 1 gang, screwing to mounting box (V9090/720 - 1 nr)	nr	15.62	53.32	1.86	70.80

© NSR 01 Aug 2008 - 31 Jul 2009   Y119

Y33 : MECHANICAL AND ELECTRICAL			Mat. £	Lab. £	Plant £	Total £
Y3310						
066	**Reposition one way light switch, renewing embedded wirng circuit n.e.10m. (Not incl redecoration)**					
	Stripping out surface light switches, isolating, making good finishings (C8610/160 - 1 nr), Cutting or forming chases for services 1 nr 19 mm dia in brickwork, making good (P3110/202 - 4 m), Cutting or forming holes, mortices, sinkings and chases, concealed service, socket outlet point, making good (P3230/802 - 1 nr), 1 - 4 nr Cables (V8700/010 - 1 nr), Lighting outlets, 1.5 mm2 single core cables (V9020/196 - 1 nr), Extra for 25 x 8 mm galvanised mild steel channel (V9020/250 - 4 m)	nr	45.55	88.70	7.44	141.69
068	**Reposition two way light switch, renewing embedded wirng circuit n.e.10m. (Not incl redecoration)**					
	Stripping out surface light switches, isolating, making good finishings (C8610/160 - 1 nr), Cutting or forming chases for services 1 nr 19 mm dia in brickwork, making good (P3110/202 - 4 m), Cutting or forming holes, mortices, sinkings and chases, concealed service, socket outlet point, making good (P3230/802 - 1 nr), 1 - 4 nr Cables (V8700/010 - 1 nr), Lighting outlets, 1.5 mm2 single core cables (V9020/196 - 1 nr), Extra for 25 x 8 mm galvanised mild steel channel (V9020/250 - 4 m)	nr	45.55	88.70	7.44	141.69
100	**Replace single socket with switched socket**					
	Removing plate switches/socket outlet plates for replacement (C8610/447 - 1 nr), Surface switched socket outlets (K2757 WHI), 1 gang, screwing to mounting box (V9090/797 - 1 nr)	nr	3.71	8.25		11.96
102	**Replace double socket with switched socket**					
	Removing plate switches/socket outlet plates for replacement (C8610/447 - 1 nr), Surface switched socket outlets (K2747 WHI), 2 gang, screwing to mounting box (V9090/798 - 1 nr)	nr	6.83	9.52		16.35
104	**Reposition single socket, extending wiring to new position (not incl redecoration)**					
	Stripping out flush socket outlets, isolating, making good finishings (C8610/250 - 1 nr), Cutting or forming chases for services 1 nr 19 mm dia in brickwork, making good (P3110/202 - 1 m), Cutting or forming holes, mortices, sinkings and chases, concealed service, socket outlet point, making good (P3230/802 - 1 nr), 1 - 4 nr Cables (V8700/010 - 1 nr), Extra for 2.5 mm2 twin core and earth cables (where cable length varies from 10 m allowance) (V9020/235 - 1 m), Extra for 25 x 8 mm galvanised mild steel channel (V9020/250 - 1 m), Blanking plates (K3827 WHI), 1 gang, screwing to mounting box (V9090/720 - 1 nr)	nr	7.61	52.58	1.86	62.05

© NSR 01 Aug 2008 - 31 Jul 2009   Y120

	Y33 : MECHANICAL AND ELECTRICAL		Mat. £	Lab. £	Plant £	Total £
Y3310						
106	**Reposition double socket, extending wiring to new position (not incl redecoration)**					
	Stripping out flush socket outlets, isolating, making good finishings (C8610/250 - 1 nr), Cutting or forming chases for services 1 nr 19 mm dia in brickwork, making good (P3110/202 - 1 m), Cutting or forming holes, mortices, sinkings and chases, concealed service, socket outlet point, making good (P3230/802 - 1 nr), 1 - 4 nr Cables (V8700/010 - 1 nr), Extra for 2.5 mm2 twin core and earth cables (where cable length varies from 10 m allowance) (V9020/235 - 1 m), Extra for 25 x 8 mm galvanised mild steel channel (V9020/250 - 1 m), Blanking plates (K3827 WHI), 1 gang, screwing to mounting box (V9090/720 - 1 nr)	nr	7.61	52.58	1.86	62.05
108	**Reposition single socket, renewing embedded wirng circuit n.e.10m. (Not incl redecoration)**					
	Stripping out flush socket outlets, isolating, making good finishings (C8610/250 - 1 nr), Cutting or forming chases for services 1 nr 19 mm dia in brickwork, making good (P3110/202 - 4 m), Cutting or forming holes, mortices, sinkings and chases, concealed service, socket outlet point, making good (P3230/802 - 1 nr), 1 - 4 nr Cables (V8700/010 - 1 nr), Heating and power socket, switch socket and the like, single switch, 2.5 mm2 single core cables (V9020/208 - 1 nr), Extra for 25 x 8 mm galvanised mild steel channel (V9020/250 - 4 m)	nr	73.54	94.62	7.44	175.60
110	**Reposition double socket, renewing embedded wirng circuit n.e.10m. (Not incl redecoration)**					
	Stripping out flush socket outlets, isolating, making good finishings (C8610/250 - 1 nr), Cutting or forming chases for services 1 nr 19 mm dia in brickwork, making good (P3110/202 - 4 m), Cutting or forming holes, mortices, sinkings and chases, concealed service, socket outlet point, making good (P3230/802 - 1 nr), 1 - 4 nr Cables (V8700/010 - 1 nr), Heating and power socket, switch socket and the like, single switch, 2.5 mm2 single core cables (V9020/208 - 1 nr), Extra for 25 x 8 mm galvanised mild steel channel (V9020/250 - 4 m)	nr	73.54	94.62	7.44	175.60
120	**Replace cooker control unit**					
	Stripping out cooker control units, isolating, making good finishings (C8610/270 - 1 nr), Cooker control units (K5040 WHI), 45 amp, plugging and screwing to masonry (V9090/840 - 1 nr)	nr	30.48	15.55		46.03
122	**Reposition cooker control unit, extending wiring to new position (not incl redecoration)**					
	Stripping out cooker control units, isolating, making good finishings (C8610/270 - 1 nr), Cutting or forming chases for services 1 nr 19 mm dia in brickwork, making good (P3110/202 - 1 m), Cutting or forming holes, mortices, sinkings and chases, concealed service, socket outlet point, making good (P3230/802 - 1 nr), 1 - 4 nr Cables (V8700/010 - 1 nr), Extra for 2.5 mm2 twin core and earth cables (where cable length varies from 10 m allowance) (V9020/235 - 1 m), Extra for 25 x 8 mm galvanised mild steel channel (V9020/250 - 1 m), Blanking plates (K3827 WHI), 1 gang, screwing to mounting box (V9090/720 - 1 nr), Connection units for cookers (K5045 WHI), steel box, plugging and screwing to masonry (V9090/845 - 1 nr)	nr	15.41	56.49	1.86	73.76

© NSR 01 Aug 2008 - 31 Jul 2009   Y121

	Y33 : MECHANICAL AND ELECTRICAL		Mat. £	Lab. £	Plant £	Total £
Y3310						
126	**Replace heating control unit**					
	Stripping out central heating control switches, isolating, making good finishings (C8610/210 - 1 nr), To domestic central heating system (V3060/600 - 1 nr)	nr	78.55	17.14		95.69
128	**Reposition heating control unit, extending wiring to new position (not incl redecoration)**					
	Stripping out central heating control switches, isolating, making good finishings (C8610/210 - 1 nr), Cutting or forming chases for services 1 nr 19 mm dia in brickwork, making good (P3110/202 - 1 m), Cutting or forming holes, mortices, sinkings and chases, concealed service, socket outlet point, making good (P3230/802 - 1 nr), Boiler controller (V3050/096 - 1 nr), Extra for 2.5 mm2 twin core and earth cables (where cable length varies from 10 m allowance) (V9020/235 - 1 m), Extra for 25 x 8 mm galvanised mild steel channel (V9020/250 - 1 m)	nr	5.59	47.09	1.86	54.54
130	**Replace heating thermostat**					
	Room thermostat (V3772/070 - 1 nr)	nr	12.93	11.42		24.35
132	**Reposition heating thermostat, extending wiring to new position (not incl redecoration)**					
	Stripping out central heating control switches, isolating, making good finishings (C8610/210 - 1 nr), Cutting or forming chases for services 1 nr 19 mm dia in brickwork, making good (P3110/202 - 1 m), Cutting or forming holes, mortices, sinkings and chases, concealed service, socket outlet point, making good (P3230/802 - 1 nr), Room thermostats (V3760/070 - 1 nr), Extra for 2.5 mm2 twin core and earth cables (where cable length varies from 10 m allowance) (V9020/235 - 1 m), Extra for 25 x 8 mm galvanised mild steel channel (V9020/250 - 1 m)	nr	5.59	45.39	1.86	52.84
134	**Replace telephone socket**					
	Removing telephone sockets for replacement (C8610/445 - 1 nr), Flush telephone socket (MK KO422 WHI), steel box, plugging and screwing to masonry (W1050/200 - 1 nr)	nr	13.13	16.73		29.86
136	**Reposition telephone socket, extending wiring to new position (not incl redecoration)**					
	Removing telephone sockets for replacement (C8610/445 - 1 nr), Extra for 16 x 16 mm pvc mini-trunking (V9020/255 - 1 m), Flush telephone socket (MK KO422 WHI), steel box, plugging and screwing to masonry (W1050/200 - 1 nr), 4 pair telephone cable (W4010/050 - 1 m)	nr	14.25	21.29		35.54
138	**Replace TV arial socket, flush fitted**					
	Stripping out flush socket outlets, isolating, making good finishings (C8610/250 - 1 nr), Flush aerial sockets (MK 3520 WHI), steel box, plugging and screwing to masonry (W2050/205 - 1 nr)	nr	11.14	19.26		30.40

© NSR 01 Aug 2008 - 31 Jul 2009

Y122

Y33 : MECHANICAL AND ELECTRICAL		Mat. £	Lab. £	Plant £	Total £
Y3310					
140	**Reposition TV arial socket, flush fitted, extending wiring to new position (not incl redecoration)**				
	Stripping out flush socket outlets, isolating, making good finishings (C8610/250 - 1 nr), Cutting or forming chases for services 1 nr 19 mm dia in brickwork, making good (P3110/202 - 1 m), Cutting or forming holes, mortices, sinkings and chases, concealed service, socket outlet point, making good (P3230/802 - 1 nr), Conduits 16 mm diameter in chases, plugging and screwing to masonry (W1010/050 - 1 m), Fixed to surface, screwing to timber (W2010/050 - 1 m), Fix only flush aerial sockets, plugging and screwing to masonry (W2050/207 - 1 nr) nr	3.60	60.23	1.86	65.69
142	**Replace TV arial socket, surface fitted**				
	Stripping out surface socket outlets, isolating, making good finishings (C8610/240 - 1 nr), Surface aerial sockets (MK 3520 WHI), moulded box, plugging and screwing to masonry (W2050/200 - 1 nr) nr	9.96	9.10		19.06
144	**Reposition TV arial socket, surface fitted, extending wiring to new position (not incl redecoration)**				
	Stripping out surface socket outlets, isolating, making good finishings (C8610/240 - 1 nr), Extra for 16 x 16 mm pvc mini-trunking (V9020/255 - 1 m), Fixed to surface, screwing to timber (W2010/050 - 1 m), Fix only surface aerial sockets, plugging and screwing to masonry (W2050/202 - 1 nr) nr	1.91	16.52		18.43
150	**Remove rewireable fused consumer unit and replace with new 4 way consumer unit**				
	SP, 6-32 amp (V1010/060 - 4 nr), Four way, SPN/DP distribution board or consumer unit, 100A (V1070/090 - 1 nr) nr	113.47	108.28		221.75
152	**Remove rewireable fused consumer unit and replace with new 6 way consumer unit**				
	SP, 6-32 amp (V1010/060 - 6 nr), Six way, SPN/DP distribution board or consumer unit 100A (V1070/101 - 1 nr) nr	140.19	125.20		265.39
154	**Remove rewireable fused consumer unit and replace with new 8 way consumer unit**				
	SP, 6-32 amp (V1010/060 - 8 nr), Eight way, SPN/DP distribution board or consumer unit, 100A (V1070/105 - 1 nr) nr	164.18	142.12		306.30
160	**Reposition rewireable fused consumer unit up to 6 way**				
	Stripping out consumer units, isolating, making good finishings (C8610/140 - 1 nr), Four - six way, SPN/DP (V1060/212 - 1 nr) nr	0.33	62.61		62.94
162	**Reposition rewireable fused consumer unit 8 - 12 way**				
	Stripping out consumer units, isolating, making good finishings (C8610/140 - 1 nr), Eight - twelve way, SPN/DP (V1060/214 - 1 nr) nr	0.33	68.11		68.44

	Y33 : MECHANICAL AND ELECTRICAL		Mat. £	Lab. £	Plant £	Total £
Y3320	**LIGHTING**					
200	**Provide additional battenholder, electrical connection n.e. 5m**					
	Lighting outlets, 1.5 mm2 single core cables (V9020/196 - 0.5 nr), Straight pattern (1173 WHI) (V9045/500 - 1 nr)	nr	25.59	22.54		48.13
202	**Provide additional cord and lampholder, electrical connection n.e. 5m**					
	Lighting outlets, 1.5 mm2 single core cables (V9020/196 - 0.5 nr), Flexible cords ne 1 m drop, lampholder (1170 WHI) (V9050/545 - 1 nr)	nr	24.22	21.69		45.91
204	**Replace cord, ceiling rose and lampholder**					
	Stripping out ceiling roses and pendants, isolating, making good finishings (C8610/330 - 1 nr), Flexible cords ne 1 m drop, ceiling rose (1161 WHI), lampholder (1170 WHI), screwing to timber (V9050/550 - 1 nr)	nr	5.56	15.02		20.58
206	**Provide additional single fluorescent luminaire with daylight lamp, 1200mm, electrical connection n.e. 5m**					
	Lighting outlets, 1.5 mm2 single core cables (V9020/196 - 0.5 nr), General purpose type (PPC/40), 1200 mm nominal length, single, screwing to timber (V9060/650 - 1 nr), General purpose 1200 mm nominal length, daylight (V9075/705 - 1 nr)	nr	69.35	55.78		125.13
208	**Provide additional single fluorescent luminaire with white lamp, 1200mm, electrical connection n.e. 5m**					
	Lighting outlets, 1.5 mm2 single core cables (V9020/196 - 0.5 nr), General purpose type (PPC/40), 1200 mm nominal length, single, screwing to timber (V9060/650 - 1 nr), General purpose 1200 mm nominal length, White (V9075/708 - 1 nr)	nr	63.27	55.78		119.05
210	**Provide additional double fluorescent luminaire with daylight lamps, 1200mm, electrical connection n.e. 5m**					
	Lighting outlets, 1.5 mm2 single core cables (V9020/196 - 0.5 nr), General purpose type (PPC/240), 1200 mm nominal length, twin, screwing to timber (V9060/655 - 1 nr), General purpose 1200 mm nominal length, daylight (V9075/705 - 2 nr)	nr	116.82	57.47		174.29
212	**Provide additional double fluorescent luminaire with white lamps, 1200mm, electrical connection n.e. 5m**					
	Lighting outlets, 1.5 mm2 single core cables (V9020/196 - 0.5 nr), General purpose type (PPC/240), 1200 mm nominal length, twin, screwing to timber (V9060/655 - 1 nr), General purpose 1200 mm nominal length, White (V9075/708 - 2 nr)	nr	104.66	57.47		162.13
214	**Provide additional single fluorescent luminaire with daylight lamp, 1800mm, electrical connection n.e. 5m**					
	Lighting outlets, 1.5 mm2 single core cables (V9020/196 - 0.5 nr), General purpose type (PPC/675), 1800 mm nominal length, single, screwing to timber (V9060/660 - 1 nr), General purpose 1800 mm nominal length, daylight (V9075/710 - 1 nr)	nr	88.56	55.78		144.34

© NSR 01 Aug 2008 - 31 Jul 2009   Y124

	Y33 : MECHANICAL AND ELECTRICAL		Mat. £	Lab. £	Plant £	Total £
Y3320						
216	**Provide additional single fluorescent luminaire with white lamp, 1800mm, electrical connection n.e. 5m**					
	Lighting outlets, 1.5 mm2 single core cables (V9020/196 - 0.5 nr), General purpose type (PPC/675), 1800 mm nominal length, single, screwing to timber (V9060/660 - 1 nr), General purpose 1800 mm nominal length, white (V9075/712 - 1 nr)	nr	83.07	55.78		138.85
218	**Provide additional double fluorescent luminaire with daylight lamps, 1800mm, electrical connection n.e. 5m**					
	Lighting outlets, 1.5 mm2 single core cables (V9020/196 - 0.5 nr), General purpose type (PPC/2675), 1800 mm nominal length, twin, screwing to timber (V9060/665 - 1 nr), General purpose 1800 mm nominal length, daylight (V9075/710 - 2 nr)	nr	134.32	57.47		191.79
220	**Provide additional double fluorescent luminaire with white lamp, 1800mm, electrical connection n.e. 5m**					
	Lighting outlets, 1.5 mm2 single core cables (V9020/196 - 0.5 nr), General purpose type (PPC/2675), 1800 mm nominal length, twin, screwing to timber (V9060/665 - 1 nr), General purpose 1800 mm nominal length, white (V9075/712 - 2 nr)	nr	123.34	57.47		180.81
230	**Provide new diffuser to fluorescent light, 1200 - 1800mm single**					
	Taking down single diffusers 1200 to 1800 mm, and renewing (C8610/395 - 1 nr)	nr	6.50	3.81		10.31
232	**Provide new diffuser to fluorescent light, 1200 - 1800mm double**					
	Taking down double diffusers 1200 to 1800 mm, and renewing (C8610/400 - 1 nr)	nr	9.10	5.08		14.18
Y3330	**DOOR ENTRY SYSTEMS**					
300	**Remove existing surface fixed door entry pad and reposition.**					
	Stripping out door entry handset, isolating, making good finishings (C8610/225 - 1 nr), 6 pair telephone cable (W4010/100 - 2 m), Fix only door entry handset, connection to wiring (W4020/150 - 1 nr)	nr	2.16	12.49		14.65
304	**Remove existing flush fixed door entry pad and reposition, in masonry wall**					
	Cutting out decayed, defective and cracked brickwork and renewing single brick with common brick in cement mortar (1:6), pointing (C4005/125 - 1 nr), Stripping out door entry handset, isolating, making good finishings (C8610/225 - 1 nr), Cutting or forming chases for services 1 nr 19 mm dia in brickwork, making good (P3110/202 - 1 m), Cutting or forming holes, mortices, sinkings and chases, concealed service, equipment and control gear point, making good (P3230/806 - 1 nr), 6 pair telephone cable (W4010/100 - 2 m), Fix only door entry handset, connection to wiring (W4020/150 - 1 nr)	nr	3.39	57.08	1.86	62.33
310	**Provide and fit 4 pair telephone cable for access control system**					
	4 pair telephone cable (W4010/050 - 1 m)	m	0.27	0.85		1.12

© NSR 01 Aug 2008 - 31 Jul 2009   Y125

	Y33 : MECHANICAL AND ELECTRICAL		Mat. £	Lab. £	Plant £	Total £
Y3330						
312	**Provide and fit 6 pair telephone cable for access control system**					
	6 pair telephone cable (W4010/100 - 1 m)	m	0.43	0.85		1.28
320	**Supply and fix Phlexicare door entry system ref: P6881020**					
	Supply and fit door entry system (N&C Ref: P6881020) (W4020/170 - 1 nr)	nr	370.25	44.46		414.71
322	**Extra over for CCTV, monitor and remote control ref: P6881021**					
	extra over for CCTV and monitor (N&C Ref: P6881021) (W4020/172 - 1 nr), extra over for remote control (N&C Ref: P6881023) (W4020/176 - 1 nr)	nr	607.57	44.46		652.03
324	**Extra over for pinhole camera and remote control ref: P6881022**					
	extra over for pinhole camera (N&C Ref: P6881022) (W4020/174 - 1 nr), extra over for remote control (N&C Ref: P6881023) (W4020/176 - 1 nr)	nr	586.48	33.34		619.82
330	**Supply and fix Phlexicare wireless door entry system ref: P6881026**					
	Supply and fit wireless door entry system (N&C Ref: P6881026) (W4020/180 - 1 nr)	nr	338.67	22.23		360.90
Y3340	**ALARM AND ALERT SYSTEMS**					
010	**Supply and fit mains powered audio/visual alarm system with in-cubicle reset and 2nr 50mm red bangles, incl electrical supply**					
	Heating and power socket, switch socket and the like, single switch, 2.5 mm2 single core cables (V9020/208 - 1 nr), Phlexicare Audio/visual alarm system, plug in mains power unit and ceiling pull switch. N&C ref:P64410.01 (W4120/050 - 1 nr)	nr	167.30	56.32		223.62
015	**Supply and fit wireless audio/visual alarm system**					
	Phlexicare Wireless Audio/visual alarm system, comprising master pull cord and master lamp unit. N&C ref:P64410.10/11 (W4130/050 - 1 nr)	nr	288.96	14.82		303.78
017	**Extra over for slave lamp unit**					
	Extra over for Slave lamp unit N&C ref:P64410.12 (W4130/055 - 1 nr)	nr	189.00	6.34		195.34
020	**Supply and fit deaf alarm with power pack, strobe and vibrating disc**					
	Deaf alarm, power pack, strobe and vibrating disc (W6000/014 - 1 nr)	nr	130.00	46.53		176.53
030	**Supply and fit RNID smoke alarm system with vibration and flashing light alert**					
	RNID smoke alarm system type A128 (W4120/060 - 1 nr)	nr	85.00	21.15		106.15

© NSR 01 Aug 2008 - 31 Jul 2009   Y126

	Y33 : MECHANICAL AND ELECTRICAL		Mat. £	Lab. £	Plant £	Total £
Y3340						
035	**Supply and fit RNID smoke alarm system with vibration and flashing light alert and ionisation smoke detector**					
	RNID smoke alarm system with ionisation detector type A223 (W4120/065 - 1 nr)	nr	96.00	25.38		121.38
040	**Supply and fit RNID smoke alarm system with vibration and flashing light alert and optical smoke detector**					
	RNID smoke alarm system with optical detector type A224 (W4120/070 - 1 nr)	nr	105.00	25.38		130.38
045	**Extra over for additional strobe light to existing smoke alarm system**					
	Additional strobe light unit to existing smoke alarm system (W4140/070 - 1 nr)	nr	26.00	6.34		32.34
050	**Extra over for additional heat detector to existing smoke alarm system**					
	Additional heat detector to existing smoke alarm system (W4140/075 - 1 nr)	nr	18.30	6.34		24.64
055	**Extra over for additional optical smoke detector to existing smoke alarm system**					
	Additional optical smoke detector to existing smoke alarm system (W4140/080 - 1 nr)	nr	17.38	6.34		23.72
060	**Extra over for additional ionisation smoke detector to existing smoke alarm system**					
	Additional ionisation smoke detector to existing smoke alarm system (W4140/085 - 1 nr)	nr	8.23	6.34		14.57
065	**Extra over for additional vibrating pad to existing smoke alarm system**					
	Additional vibrating pad to existing smoke alarm system (W4140/090 - 1 nr)	nr	12.00	6.34		18.34
100	**Supply and fit additional sounder to existing fire alarm system**					
	Additional sounder to existing fire alarm system (W4140/050 - 1 nr), 1.0 mm2, Two core (W5040/387 - 3 m)	nr	34.53	34.28		68.81
105	**Supply and fit additional beacon to existing fire alarm system**					
	Additional beacon to existing fire alarm system (W4140/055 - 1 nr), 1.0 mm2, Two core (W5040/387 - 3 m)	nr	42.53	34.28		76.81
110	**Supply and fit additional Xenon beacon to existing fire alarm system**					
	Xenon beacon to existing fire alarm system (W4140/060 - 1 nr), 1.0 mm2, Two core (W5040/387 - 3 m)	nr	42.53	34.28		76.81

© NSR 01 Aug 2008 - 31 Jul 2009   Y127

	Y33 : MECHANICAL AND ELECTRICAL		Mat. £	Lab. £	Plant £	Total £
Y3340						
115	**Supply and fit additional combined sounder and beacon to existing fire alarm system**					
	Combined sounder and beacon to existing fire alarm system (W4140/065 - 1 nr), 1.0 mm2, Two core (W5040/387 - 3 m)	nr	61.08	34.28		95.36
150	**Supply and fix flashing door bell alert**					
	Supply and fix flashing door chime (W4140/102 - 1 nr)	nr	24.95	4.23		29.18
155	**Supply and fix combined door bell and telephone alert**					
	Supply and fix combined doorbell and telephone alert (W4140/100 - 1 nr)	nr	33.95	4.23		38.18
160	**Supply and fix telephone alert with strobe light**					
	Supply and fix telephone alert with strobe light (W4140/104 - 1 nr)	nr	16.40	4.23		20.63
165	**Supply and fix combined telephone and door emtry alert**					
	Supply and fix telephone/door entry alert (W4140/106 - 1 nr)	nr	24.15	4.23		28.38
Y3350	**SUNDRY ELECTRICAL EQUIPMENT**					
100	**Reposition electric hand drier, incl extending/renewing wiring n.e 3m**					
	Stripping out electric hand driers (C8610/340 - 1 nr), 2 - 2.4kW, 240 volt, automatic start (V3706/027 - 1 nr), Extra for 2.5 mm2 twin core and earth cables (where cable length varies from 10 m allowance) (V9020/235 - 3 m)	nr	12.33	62.73		75.06
106	**Remove existing electric hand drier, supply and fit new hand drier with push button start, incl extending/renewing wiring n.e 3m**					
	Stripping out electric hand driers (C8610/340 - 1 nr), 1.5 kW, 240 volt, push button start. ABS cover (V3718/020 - 1 nr), Extra for 2.5 mm2 twin core and earth cables (where cable length varies from 10 m allowance) (V9020/235 - 3 m)	nr	88.28	62.73		151.01
108	**Remove existing electric hand drier, supply and fit new hand drier with automatic start, incl extending/renewing wiring n.e 3m**					
	Stripping out electric hand driers (C8610/340 - 1 nr), 2.0 kW, 240 volt, automatic start ABS polycarbonate cover (V3718/027 - 1 nr), Extra for 2.5 mm2 twin core and earth cables (where cable length varies from 10 m allowance) (V9020/235 - 3 m)	nr	115.83	62.73		178.56
Y3360	**MECHANICAL WORKS**					
400	**Remove radiator, set aside for reuse, reposition n.e. 1.0m from original position, inc new pipework and fittings, 15mm dia copper pipework**					
	15 mm Diameter pipework, copper (S2710/300 - 1 nr)	nr	10.36	121.56		131.92

© NSR 01 Aug 2008 - 31 Jul 2009   Y128

	Y33 : MECHANICAL AND ELECTRICAL		Mat. £	Lab. £	Plant £	Total £
Y3360						
402	Remove radiator, set aside for reuse, reposition n.e. 1.0m from original position, inc new pipework and fittings, 22mm dia copper pipework					
	22 mm Diameter pipework, copper (S2710/302 - 1 nr)	nr	19.74	131.40		151.14
404	Remove radiator, set aside for reuse, reposition n.e. 1.0m from original position, inc new pipework and fittings, 28mm dia copper pipework					
	28 mm Diameter pipework, copper (S2710/304 - 1 nr)	nr	27.71	140.08		167.79
406	Remove radiator, set aside for reuse, reposition n.e. 1.0m from original position, inc new pipework and fittings, 28mm dia copper pipework					
	28 mm Diameter pipework, copper (S2710/304 - 1 nr)	nr	27.71	140.08		167.79
408	Remove radiator, set aside for reuse, reposition n.e. 1.0m from original position, inc new pipework and fittings, 15mm dia steel pipework					
	15 mm Diameter pipework, mild steel (S2710/306 - 1 nr)	nr	54.35	129.66		184.01
410	Remove radiator, set aside for reuse, reposition n.e. 1.0m from original position, inc new pipework and fittings, 20mm dia steel pipework					
	20 mm Diameter pipework, mild steel (S2710/308 - 1 nr)	nr	65.89	166.13		232.02
412	Remove radiator, set aside for reuse, reposition n.e. 1.0m from original position, inc new pipework and fittings, 25mm dia steel pipework					
	25 mm Diameter pipework, mild steel (S2710/310 - 1 nr)	nr	80.47	182.33		262.80
430	Supply and fix Radiator guard 600 x 600 x 150; as Pendock Profiles Prima range					
	Prima radiator guard, 600 x 600 x 150 with 2nr cut outs (P2072/100 - 1 nr)	nr	96.23	6.69		102.92
432	Supply and fix Radiator guard 800 x 600 x 150; as Pendock Profiles Prima range					
	Prima radiator guard, 800 x 600 x 150 with 2nr cut outs (P2072/102 - 1 nr)	nr	100.04	6.69		106.73
434	Supply and fix Radiator guard 1000 x 600 x 150; as Pendock Profiles Prima range					
	Prima radiator guard, 1000 x 600 x 150 with 2nr cut outs (P2072/104 - 1 nr)	nr	107.67	8.37		116.04
436	Supply and fix Radiator guard 1200 x 600 x 150; as Pendock Profiles Prima range					
	Prima radiator guard, 1200 x 600 x 150 with 2nr cut outs (P2072/106 - 1 nr)	nr	111.50	8.37		119.87

© NSR 01 Aug 2008 - 31 Jul 2009   Y129

	Y33 : MECHANICAL AND ELECTRICAL		Mat. £	Lab. £	Plant £	Total £
Y3360						
438	**Supply and fix Radiator guard 1500 x 600 x 150; as Pendock Profiles Prima range**					
	Prima radiator guard, 1500 x 600 x 150 with 2nr cut outs (P2072/108 - 1 nr)	nr	122.99	9.04		132.03
450	**Supply and fix Radiator guard 600 x 600 x 200; as Pendock Profiles Prima range**					
	Prima radiator guard, 600 x 600 x 200 with 2nr cut outs (P2072/120 - 1 nr)	nr	106.80	6.69		113.49
452	**Supply and fix Radiator guard 800 x 600 x 200; as Pendock Profiles Prima range**					
	Prima radiator guard, 800 x 600 x 200 with 2nr cut outs (P2072/122 - 1 nr)	nr	110.84	6.69		117.53
454	**Supply and fix Radiator guard 1000 x 600 x 200; as Pendock Profiles Prima range**					
	Prima radiator guard, 1000 x 600 x 200 with 2nr cut outs (P2072/124 - 1 nr)	nr	118.99	8.37		127.36
456	**Supply and fix Radiator guard 1200 x 600 x 200; as Pendock Profiles Prima range**					
	Prima radiator guard, 1200 x 600 x 200 with 2nr cut outs (P2072/126 - 1 nr)	nr	123.04	8.37		131.41
458	**Supply and fix Radiator guard 1500 x 600 x 200; as Pendock Profiles Prima range**					
	Prima radiator guard, 1500 x 600 x 200 with 2nr cut outs (P2072/128 - 1 nr)	nr	135.21	9.04		144.25
470	**Supply and fix boxing to pipework up to 50mm; as Pendock profiles MX range; 75/150**					
	Pendock MX pipe boxing; 75/150; fixed to battens (P2072/140 - 1 m)	m	19.62	5.76		25.38
472	**Extra over for stop ends**					
	Pendock MX pipe boxing; 75/150; stop ends (P2072/142 - 1 nr)	nr	5.09	1.44		6.53
474	**Extra over for external corners**					
	Pendock MX pipe boxing; 75/150; external corner (P2072/144 - 1 nr)	nr	34.58	2.02		36.60
476	**Extra over for internal corners**					
	Pendock MX pipe boxing; 75/150; internal corner (P2072/146 - 1 nr)	nr	25.56	2.02		27.58
480	**Supply and fix boxing to pipework 50 - 100mm; as Pendock profiles MX range; 150/150**					
	Pendock MX pipe boxing; 150/150; fixed to battens (P2072/150 - 1 m)	m	25.54	5.76		31.30

© NSR 01 Aug 2008 - 31 Jul 2009   Y130

	Y33 : MECHANICAL AND ELECTRICAL		Mat. £	Lab. £	Plant £	Total £
Y3360						
482	**Extra over for stop ends**					
	Pendock MX pipe boxing; 150/150; stop ends (P2072/152 - 1 nr)	nr	5.62	1.44		7.06
484	**Extra over for external corners**					
	Pendock MX pipe boxing 150/150; external corner (P2072/154 - 1 nr)	nr	34.58	2.02		36.60
486	**Extra over for internal corners**					
	Pendock MX pipe boxing; 150/150; internal corner (P2072/156 - 1 nr)	nr	25.56	2.02		27.58
Y3370	**M & E ANCILLARIES**					
300	**Lift and relay floorboards**					
	Stripping off/removing/taking down tongued and grooved jointed floor boards, any thickness, denail joists in preparation for replacement, setting aside for re-use areas ne 300 mm wide (C5140/555 - 1 m), Floors 25 mm thick ne 300 mm wide to timber (C5150/710 - 1 m)	m	0.11	10.58		10.69
302	**Cutting chases for wiring and make good**					
	Cutting or forming chases for services 1 nr 25 mm dia in brickwork, making good (P3110/252 - 1 m)	m	0.37	5.17	1.96	7.50
304	**Install galvanised steel conduit**					
	Extra for 20 mm diameter heavy gauge steel conduits (V9020/260 - 1 m)	m	3.97	8.15		12.12
306	**Install galvanised steel channel**					
	Extra for 25 x 8 mm galvanised mild steel channel (V9020/250 - 1 m)	m	0.60	1.48		2.08
308	**Install PVC mini trunking**					
	Extra for 16 x 16 mm pvc mini-trunking (V9020/255 - 1 m)	m	0.85	3.71		4.56

© NSR 01 Aug 2008 - 31 Jul 2009   Y131

Y33 : MECHANICAL AND ELECTRICAL	Mat. £	Lab. £	Plant £	Total £

© NSR 01 Aug 2008 - 31 Jul 2009 — Y132

			Mat. £	Lab. £	Plant £	Total £
	Y34 : LIFTS AND LIFTING DEVICES					
Y3410	**Passenger Lifts; Configure Ltd**					
010	**Inclined curved platform lift, internal**					
	Supply and fit inclined curved platform lift; internal (X1010/010 - 1 nr)	nr	1350.00			1350.00
012	**Inclined curved platform lift, external**					
	Supply and fit inclined curved platform lift; external (X1010/015 - 1 nr)	nr	11800.00			11800.00
014	**Inclined straight platform lift, internal**					
	Supply and fit inclined straight platform lift; internal (X1010/020 - 1 nr)	nr	6800.00			6800.00
016	**Inclined straight platform lift, external**					
	Supply and fit inclined straight platform lift; external (X1010/025 - 1 nr)	nr	8400.00			8400.00
018	**Short rise vertical platform lift**					
	Supply and fit short rise vertical platform lift; (X1010/030 - 1 nr)	nr	6600.00			6600.00
020	**2 stop vertical platform lift**					
	Supply and fit 2 stop vertical platform lift (X1010/035 - 1 nr)	nr	22000.00			22000.00
022	**walk through step lift**					
	Supply and fit walk through step lift (X1010/040 - 1 nr)	nr	5750.00			5750.00

© NSR 01 Aug 2008 - 31 Jul 2009   Y133

Y34 : LIFTS AND LIFTING DEVICES	Mat. £	Lab. £	Plant £	Total £